Clinical Procedures for Medical Technology Specialists

Clinical Procedures for Medical Technology Specialists

Laurence J. Street

CRC Press is an imprint of the
Taylor & Francis Group, an **informa** business

CRC Press
Taylor & Francis Group
6000 Broken Sound Parkway NW, Suite 300
Boca Raton, FL 33487-2742

Printed in the United States of America on acid-free paper
10 9 8 7 6 5 4 3 2 1

International Standard Book Number: 978-1-4200-8199-2 (Hardback)

Library of Congress Cataloging-in-Publication Data

Street, Laurence J.
Clinical procedures for medical technology specialists / Laurence J. Street.
p. ; cm.
Includes bibliographical references and index.
ISBN 978-1-4200-8199-2 (hardcover : alk. paper)
1. Clinical medicine. 2. Biomedical technicians. I. Title.
[DNLM: 1. Inpatients. 2. Patient Care Team. 3. Hospital Departments--organization & administration. 4. Patient-Centered Care--methods. WX 162.5 S915c 2011]

RC48.S77 2011
616--dc22 2010017448

Visit the Taylor & Francis Web site at
http://www.taylorandfrancis.com

and the CRC Press Web site at
http://www.crcpress.com

To my family: "I do it all for you."

and

In memory of Turko, a wonderful and amazing companion, protector, and family member.

Contents

Preface: Medical Procedures for Allied Health Professionals

The medical system in the twenty-first century grows ever more complex on all fronts. Technology develops rapidly, with patient diagnostic and treatment equipment becoming more capable, faster, more flexible, and more interconnected. Research constantly brings new or refined techniques of diagnosis and treatment, and modes of delivery are refined to try to keep pace. The level of acuity in most hospitals continues to increase so that patients are often sent home who would have been in general wards in the past, patients in general wards are those who would have been in intensive care, and those now intensive care may not have survived at all just a few decades ago.

As the front lines of medicine advance, the supporting team grows larger and more comprehensive, and it becomes critical that all of the team members have a grasp of the functioning of the system as a whole. This means not only seeing the "big picture," but also having a working knowledge of the various subsections of the system. A knowledgeable team can interact more effectively, and understanding helps avoid confusion and potential errors. At one time, hospital staff consisted solely of doctors and nurses. Support staff were added to do cleaning and facility maintenance, laboratory and x-ray technologists came into the picture as technologies advanced, biomedical engineering technologists and respiratory therapists provided support for specialized equipment, and physiotherapists were available for treatment and rehabilitation. Specialization grew, with individual lab techs working only on blood chemistry or pathology or urinalysis. X-ray techs became radiology techs, then diagnostic ultrasound, CT, MRI, or nuclear medicine specialists. A cardiology tech might focus only on stress testing or ambulatory ECG monitoring.

Just as nurses can use patient monitoring equipment better if they understand the basic principles of operation of the various components, biomedical engineering technologists can more readily help with setup and troubleshooting of the equipment if they have some knowledge of why the patient is being monitored and what the parameters signify. A respiratory technologist can discuss surgical cases with an anaesthesiologist more readily if they understand why the surgery is being performed and what some of the potential complications might be. Imaging technologists can fine-tune exposures and patient placement if they know something of the disease processes being investigated, and can thus work with radiologists more effectively. Dieticians can help with patient meal plans if they have a working knowledge of previous and future treatment plans and current patient conditions.

This text provides an overview of the most common medical procedures performed in modern clinical settings. The material is arranged as an encyclopedia, with entries in alphabetical order using a common name.

Specialized terms are avoided where possible in order to make the material accessible to a wide range of readers.

A consistent format is used with procedure descriptions. Common names of the procedure are given, followed by an outline of the disease or injury for which the procedure is performed. Anatomical structures and pathological and physiological processes that are involved are listed, but not in great depth because this information is available from many other sources. Staffing requirements are outlined, as well as equipment and supply needs. Any preprocedure requirements are listed, and then the procedure itself is described, followed by a discussion of expected outcomes or results and some of the potential complications.

While a number of other texts describe medical procedures from a nursing or physician perspective, these sources tend to have more detail than is required for paramedical professionals. Excess detail tends to obscure any message.

Other material, important to the primary caregivers but not necessarily to support staff, is included in these alternate sources. Such considerations as informed patient consent and education, medical billing, and communicating good and bad news to patients and families are vital for nurses and doctors but much less so for laboratory or medical records technologists.

Appendices include the above-mentioned glossary, as well as some general anatomy, a table of normal values, a bibliography, and a listing of Internet resources.

The author and publisher strive to produce texts that are current, accessible, relevant, and useful to the intended audience. If you have any suggestions for future editions of this text, please contact the author at: Chilliwack General Hospital, 10333 Royalwood Blvd., Rosedale, BC, V0X 1X1, Canada; ljstreet@shaw.ca.

Laurence J. Street

Acknowledgments

Thank you to Michael Slaughter, Executive Editor, CRC Press/Taylor & Francis. Thank you—again—for your patience, Michael! Thank you also to the whole team at CRC/Taylor & Francis for turning my files into a book.

And thank you to my family, especially my wife, Sheri, for support, encouragement, and patience through this whole process.

Author

Laurence J. Street earned a diploma in biomedical engineering technology at the British Columbia Institute of Technology in Vancouver, BC, Canada in 1979, and has since worked continuously in a variety of hospitals, from the very large Shaughnessey/BC Children's/Vancouver Women's complex, to a regional hospital in the British Columbia interior, to (1991 to present) the Chilliwack General Hospital, a midsized community hospital outside Vancouver, and part of the Fraser Health Authority. In the latter two cases, he was solely responsible for setting up and implementing the biomedical engineering departments.

Mr. Street received a B.Sc. in zoology from the University of British Columbia (1974), as well as a BC teaching certificate, also from UBC (1975). He taught junior high school science at Trafalgar Jr. Secondary in Nelson, BC (1975–1976), and a number of electronics and math courses to electronics and computer technology students at East Kootenay Community College in Cranbrook, BC, including several courses that he developed himself (1983–1991).

Mr. Street's hospital work involves the repair and maintenance of all patient care electronic devices in the hospital, as well as close involvement in planning for future technological directions and the evaluation and acquisition of equipment. He also provides in-service education to medical staff regarding the safe and effective use of patient care devices. Because this is a well-equipped acute care and teaching hospital and is also part of a regional medical community, he works with an extremely wide variety of both older and very modern equipment. He has taken a large number of factory training courses on the various devices that he services.

1

Introduction

A hospital is a complex system. A building, a community, a residence, a treatment facility, a business and a school—all of these aspects are interconnected, and each is critical to the overall function of the system.

This book is intended to provide information about some of the clinical procedures that take place within a hospital, but to put those procedures in context, a look at the overall hospital system is in order.

The Hospital Team

All hospitals, from the smallest to the largest, share a common organizational structure, because they all have the same objectives: to care for patients.

Hospital Departments

Medicine

The medical department of a hospital consists of physicians and physicians-in-training.

General practitioners (GPs) and specialists can admit and discharge patients, and are in charge of their care, determining what medications are required and what tests and diagnostic procedures may be needed. They perform physical examinations, review vital signs and patient conditions, interpret test results, and decide on the best course of treatment for each patient. Various specialists may be involved in the care of a patient, depending on the complexity of their case.

Some GPs perform surgery, though specialist surgeons handle the majority of surgical cases.

Medical specialties include dermatology (skin and related structures), cardiology (heart and circulatory system), emergency medicine, family medicine, geriatrics (older patients), laboratory medicine, neurology (brain and nervous system), obstetrics and gynecology, oncology (cancer), otorhinolaryngology (ear, nose and throat), pathology, pediatrics, psychiatry, radiology, and urology (kidneys, bladder, and related organs).

Most physicians maintain a practice outside the hospital, but some are directly employed by the hospital, usually radiologists and pathologists.

Medical residents are MDs training for a specialty, and sometimes actually reside in the hospital, though this is less common than it once was. Most hospitals provide a quiet room with a bed where resident doctors can catch a bit of sleep between shifts.

Interns are medical school graduates who are working in a hospital for a relatively short time, usually one year, in order to gain clinical experience. An MD, often one of the residents, always supervises them.

There is generally a Chief of Medical Staff for the facility, and depending on the size of the hospitals, various medical departments may each have a department head.

Nursing

Most of the nursing staff in hospitals is made up of registered nurses (RNs) or licensed practical nurses (PNs, LPNs, or various other terms), as well as nursing aides (also sometimes called orderlies, patient care aides, and other terms). Together, the nursing staff members are the people who provide most of the direct care to patients. RNs have the highest level of responsibility, while aides have the lowest and LPNs in between.

Nurses often specialize in areas such as operating rooms (ORs), maternity, psychiatry, geriatrics, oncology, or pediatrics. There is an increasing trend toward longer training periods for nurses, particularly RNs, with many programs now consisting of four years of study instead of the traditional two. Nurses are usually expected or even required to participate in continuing education programs. The various medical departments of the hospital have head nurses or nurse clinicians, and many have a staff member who is focused on organizing and implementing education for fellow nurses.

Administration

Managing the overall operation of a hospital is the function of Administration. Depending on the definition of the term, administration may include, in addition to direct administrators, the financial department including accounts payable and receivable, human resources, purchasing, and stores. Stores may include shipping and receiving as well.

Administration is responsible for organizing and coordinating the staff and facilities of the hospital, ensuring that patients receive the best possible care while also operating the system within a reasonable budget. They also represent the hospital to the public and to various funding bodies and other outside agencies such as insurance companies and professional organizations.

Clinical Support

The clinical support portion of a hospital includes a wide variety of large and small departments; these include:

- Cardiology, responsible for taking and organizing ECG records. These may include 12-lead ECGs, stress tests, ambulatory ECG monitoring, plethysmography, and others. Cardiology is sometimes a subdepartment of the clinical laboratory.
- Respiratory, responsible for maintaining and sometimes operating patient ventilators, anesthetic machines, oxygen concentrators, suction units, and gas regulating devices, among others. RTs (respiratory therapists or technologists) may also run a respiratory rehabilitation program for people with acute or chronic respiratory problems, and they may be in charge of systems such as pulmonary function analyzers. They may be specialists for intubating patients, or for drawing blood for gas analyses.

- Diagnostic imaging (DI) or radiology was once simply the X-ray department, but its role has expanded along with the technology used. Basic X-rays are still used frequently, but more complex systems are available such as CT (computerized tomography), MRI (magnetic resonance imaging), PET (positron emission tomography) scans, and nuclear medicine scans. Ultrasound machines can produce good images of soft tissues with less potential cell damage than X-rays. DI technologists, who are often specialized for the various imaging modalities, operate imaging equipment. Radiologists interpret results and prepare reports on the findings, as well as supervising some of the imaging procedures or performing some of the more complex procedures.
- The clinical or medical laboratory analyzes blood and tissue samples from patients and provides reports on their findings. Laboratory technologists, often specialized, perform lab work, and a physician pathologist supervises the lab. The lab has various subdepartments.
  - Pathology is concerned with disease effects. Histopathology studies diseased tissues, while cytopathology focuses (often literally, using microscopes) on disease-related cells.
  - Microbiology studies various disease-causing organisms, with subsets of bacteriology, virology, parasitology, and mycology (fungi). Tissue and blood samples may be examined directly for such organisms, or they may be cultured in special incubators for further testing.
  - Hematology is involved with blood, testing for various parameters such as counts of the various blood cells and coagulation measurements. Hematology usually also looks after the collection, storage, and distribution of blood and blood products.
  - Biochemistry handles samples from patients, testing for specific chemicals, using very high-tech computerized and automated analyzers.
  - The hospital morgue may be a part of the lab, administratively though usually not physically. Deceased patients are taken to the morgue for autopsy, if required, and storage until family can make funeral arrangements. Morgues may maintain samples of organs for research or legal purposes.
  - Some manufacturers offer point-of-care testing devices that can analyze blood samples for a wide variety of values right at the bedside, integrating the results into the patient record being generated by physiological monitoring. This provides immediate results, which can be very important in critical cases, and represents a merging of direct patient care and laboratory functions.
  - Lab work may involve long time periods for testing, or a very short turnaround time may be required, for example if blood gases from a patient need to be analyzed in order to determine an urgent course of treatment. Tissue samples may need to be examined and reported on during the course of surgery, so that the surgeon can determine if further intervention is required.

- Dietary departments coordinate with nurses and physicians to ensure that specific dietary needs of individual patients are being met. This requires an in-depth knowledge of nutrition and how various foods affect the health of patients with different diseases or conditions.

- Rehabilitation services include physical (or physio) therapy and occupational therapy. This work is aimed at maximizing the recovery of motion and function for patients after surgery or a course of disease. A variety of methods are used, including supervised exercises, heat or cold therapy, range of motion work, ultrasound and other technologies, and massage and manipulation.
- Many hospitals have a spiritual care department, providing direct support to patients, family members, and staff, or coordinating such support from outside sources, or simply providing quiet spaces within the hospital for those wishing to pray or contemplate.
- Infection control staff may be part of the clinical laboratory, but certainly work closely with the lab. Hospital patients are especially vulnerable to infections, and staff are often exposed to potentially harmful pathogens, so it is critical that sufficient precautions be taken to minimize these risks. Infection control staff work to ensure that both nursing and support staff are thoroughly trained in infection control procedures, that infection control protocols are adequate and that they are followed, and that hospital facilities are clean and meet current standards. They coordinate with nursing, housekeeping, laboratory, and central processing.
- Biomedical Engineering is responsible for the support of the patient care electronic equipment in the hospital. Some Biomed departments also look after such things as electric patient beds, gas and suction regulators, and patient lifts, as well as clinical laboratory equipment. Departments are often divided into clinical and imaging sections, with clinical looking after such devices as physiological monitors, defibrillators, infant incubators, and IV pumps, while imaging takes care of the X-ray, CT, MRI, and ultrasound equipment. Biomed may be involved in the evaluation of potential new equipment, and also ongoing training of nursing staff in the most effective use of equipment. Equipment repair, preventive maintenance, and performance testing are functions of Biomedical Engineering.
- Prosthetics departments are usually found only in larger or specialized hospitals, where patients need to be fitted with custom artificial arms, legs, or other body parts. These parts may by simple and for cosmetic purposes only, or they may be more functional. Some facilities are working with mechanical limbs that are controlled by the patient's remaining muscle activity or even brain waves.
- The medical records department provides critical legal documentation of the course of treatment of patients in the hospital. It includes their demographics, medical history, details of their medical condition, charts of vital signs and medications administered, tests and treatments performed, and results of these tests and treatments. Each country requires specific formats, procedures, and materials to be used in medical records, as well as storage conditions and the length of time that the records must be kept.
- Most hospitals have a library of medical texts and journals, and they have access to the collections of larger hospital or university libraries. The Internet provides a vast amount of professional resources for hospital staff to utilize. Library staff collect and organize material and assist other staff members in finding and using material.
- Patients and family members are often placed in difficult social or emotional circumstances due to the illness or injury involved, and social workers in the hospital help them deal with these issues, as well as acting as liaison to outside agencies and facilities.

Facility Support

Any large building or collection of buildings, including a hospital, requires a team to keep everything running and to meet the physical needs of the occupants.

- Plant services, or physical plant, or building maintenance, provides for the structure and mechanical operation of the hospital. Heating, ventilation and air conditioning, plumbing, electrical wiring and lighting, alarm systems, hospital beds, and patient lifts are all maintained by plant staff. They may also look after emergency power generators and medical gas systems. Trades people such as plumbers, electricians, carpenters, and painters may be part of the plant services staff, to do repairs and some new installations. Larger construction projects are usually contracted to outside companies, working in coordination with plant services.
- Security is a vital part of hospital operations, both for staff and for patients and visitors. Security staff provide surveillance and protection, and help ensure that the building and its occupants are safe and that theft and vandalism are prevented.
- Housekeeping keeps the facility clean, taking care of garbage removal, room and floor cleaning, washroom maintenance, and cleaning of special areas such as operating rooms. Housekeeping may include laundry services, but this may also be contracted to an outside company.
- Food Services provides nutritionally balanced meals to hospital patients, working with dieticians and medical staff to ensure that any special needs are met. Food services staff may also operate employee and visitor cafeterias and snack bars, though these are often contracted out.
- Information Management/Information Technology/Information Services/Data Processing/"Computers" are all various terms for the department that provides, operates, and maintains the computer systems of the hospital. These systems are used for both nonpatient care functions such as financials, education, personnel, and e-mail, as well as patient care functions including logging patient data from monitors and lab tests, ADT (admission, discharge, and transfer), and patient charting. Information Management is responsible for the selection and installation of software packages, and interfacing the main hospital computer system with various specialized medical computer systems, as well as the Internet.
- Volunteers are an important part of any hospital, with services ranging from simply visiting with patients, to operating carts with snacks and magazines, to running gift shops and organizing fund-raising activities.

Outside Agencies

Hospitals interact with a variety of organizations and agencies in the course of operation.

- Government regulatory bodies, such as the Food and Drug Administration in the United States, the Health Products and Foods Branch of Health Canada, and the Medicines and Healthcare Products Regulatory Agency in the United Kingdom, set standards which medical devices and medications must meet before they can be used for treating patients. These bodies carry legal authority, and it is incumbent upon the device or medication manufacturers to prove that their products meet requirements. The regulatory bodies also perform their own tests to confirm compliance.

- A number of nongovernmental organizations exist to perform tests on medical devices. They may have their own standards for equipment, and many levels of government and health care organizations such as individual hospitals or health maintenance organizations use these standards in developing specifications for equipment purchases. Some of these testing and safety organizations include Underwriters Laboratories, the Canadian Standards Association, ECRI (formerly the Emergency Care Research Institute), the American National Standards Institute, the Association for the Advancement of Medical Instrumentation, ASTM International (formerly the American Society for Testing and Materials), and the International Electrotechnical Commission.
- Operating a hospital requires a huge range of supplies and equipment, and all the vendors of these things must be dealt with in an economical, efficacious, and ethical manner.
- Most hospitals rely to a greater or lesser degree on funding from various charities, fund-raising groups, or the donations from individuals or organizations. Some of these may be branches of the hospital, or they may be independent. In any case, ethical guidelines must be in place to deal with the groups properly.
- Various professional organizations such as the American, Canadian, and British Medical Associations have some influence over hospital staff, and their concerns must be taken into consideration.
- Many hospital employees are members of unions, and as such, their unions must be dealt with in regard to negotiating contracts and resolving grievances.
- Government health agencies, health maintenance organizations, other medical insurance companies, the armed forces, police departments, and workers compensation boards all may contribute toward the costs of hospital patient care, and as such, they have a greater or lesser degree of influence on how those costs are managed.

Surgery and Surgical Teams

Some surgery can be performed by a single person, but anything more complex than a lesion removal usually requires a team of people. With more complex procedures, the team grows.

The Team

A surgeon is the core of any surgical team, and there may be one or more other surgeons involved, though one person will be the lead (Figure 1). Some situations, such a live donor kidney transplant, mean that two teams are working in close proximity and coordination with each other. The functions of the surgeon(s) is described in more detail in the encyclopedic section of the book. Surgeons might perform general, relatively simple procedures, or they may specialize in specific areas such as orthopedic, cardiovascular, neurological, cosmetic, ophthalmic, or thoracic surgery.

An anesthesiologist will usually be present to take care of anaesthetizing and monitoring the patient (Figure 2). The anesthesiologist also often interviews the patient immediately before surgery to confirm any drug allergies, to try to allay any possible apprehensions, and to ensure that the patient is aware of the surgery that is about to be

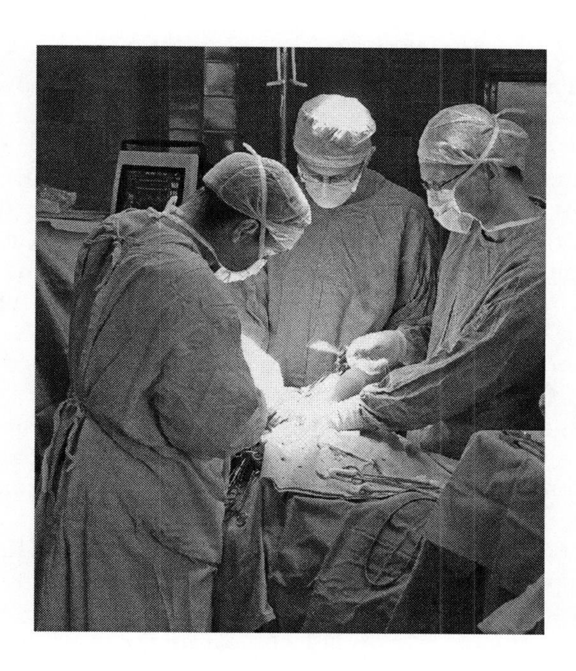

FIGURE 1
Surgeons at work.

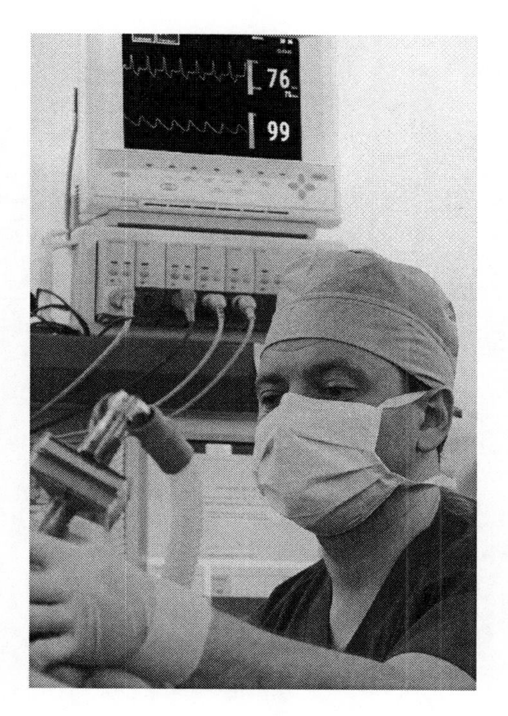

FIGURE 2
Anesthesiologist holding mask on patient.

performed. They may administer the initial sedative and later intubate the patient, select, administer and regulate gaseous agents, determine and monitor the level of consciousness and vital signs of the patient, bring the patient out of anesthesia at the end of the procedure, and then check on the patient's recovery.

Operating room nurses assist surgeons, providing instruments and supplies as needed, removing items from the surgical field as required, ensuring that fluids are removed as effectively as possible, and generally keeping things working smoothly. They may also do such mundane things as drying the surgeon's forehead or changing CDs for background music, but generally their functions are critical for the team. OR nurses may be directly assisting, or they may "float," moving about the OR as required. Nurses often set up the OR before surgery, preparing any supplies and equipment and arranging everything in the OR to suit the particular procedure and surgeon.

Respiratory therapists/technologists may assist the anesthesiologist in intubating patients and in running the anesthetic machines. They may perform functional tests of the machine before procedures and check that anesthetic-related supplies and equipment are ready.

During open heart or heart transplant surgery, a specialized technologist or nurse will likely be present to operate the heart–lung machine (Figure 3).

Biomedical engineering technologists may be in the operating room or on close standby to provide advice and support regarding technical functions of the high-tech equipment being used.

Laboratory technologists may be stationed near the operating room to help in examining tissues and organs removed during surgery, so that determinations can be made regarding the need for further intervention.

Imaging technologists may be called into the OR to obtain X-ray or ultrasound images as part of the surgical process.

FIGURE 3
A perfusionist operating a heart–lung machine.

In a teaching hospital, students of any of the surgical team disciplines may be present in the OR, as well as observing through viewing windows or video links.

Equipment vendor representatives often come into the OR to assist in the use of a new device, providing advice and recommendations to hospital staff to help them get the most out of the device.

Housekeeping staff members are an integral part of the team, because they have the responsibility of cleaning up after each surgery (often a messy job that requires special training in biohazardous material handling) and ensuring that the room is clean and antiseptic before the next procedure.

An OR manager takes care of scheduling, staffing, and equipment and supply ordering and control.

General Types of Surgery

Surgery can be divided into various categories depending on the overall goals of the procedures:

- Removal of diseased or damaged tissues or organs.
- Replacement of body parts, including joints, eye lenses, and heart valves, with artificial substitutes.
- Transplantation of organs and other body parts.
- Diagnostic or exploratory surgery. Such procedures may be combined with removal surgery depending on findings.
- Orthopedic surgery, in which bones and cartilage are adjusted, removed, repaired, or repositioned.
- Caesarean births.
- Cosmetic surgery.

Minimally Invasive Surgery

Many surgical procedures can be done with minimally invasive or endoscopic techniques, in which small incisions allow the insertion of one or more tubes into an area of the patient's body (Figure 4). The surgeon can visualize the specific site on which they are working, and thin instruments can be inserted within the tubes to provide cutting, cauterization, suctioning, and suturing. A gas such as carbon dioxide is used to inflate the area surrounding the surgical site, and illumination is provided by fiber optics either integrated into the access tubes or built into the exploratory or operating instruments. Accessory tubes provide irrigation and suction.

Post-Surgery

After the surgery is completed and the patient has been brought back to consciousness (or has stabilized following conscious surgery), the patient must be monitored carefully for some time before being returned to a general ward in the hospital or, in the case of day care surgery, before being allowed to go home. Vital signs are monitored and checked frequently, pain medication is administered as required, and the surgical site is observed to make sure there is no excessive bleeding or other problems. Some catheters and IV lines

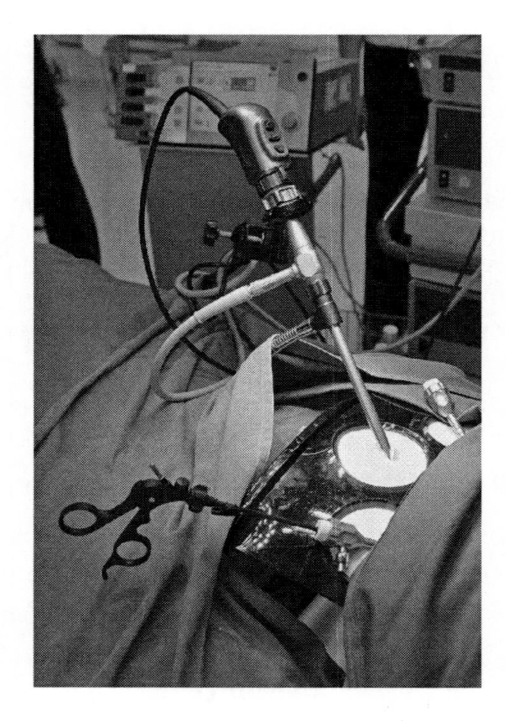

FIGURE 4
Instruments being used in a laparoscopic surgical procedure. The largest device has a camera head on top; the one in the lower left is used for grasping or cutting tissue.

may be removed as part of the postoperative care, and patients may be given ice chips or small sips of water depending on their condition. The length of postoperative stay depends on the condition of the individual patient and the degree of trauma produced during surgery. Postop areas are in close proximity to the ORs, so that surgical staff and facilities are readily available if they are needed.

Anesthesia

Surgery has been performed for thousands of years, as evidenced by ancient writings as well as physical signs such as prehistoric skulls with incised holes, some of which were healed over. But for most of the history of surgery, it was performed as a last resort, and the patients endured intense agony during the process, often to such a degree that they died of shock. Alcohol and opium were used to lessen the pain, but patients still remained conscious during surgery. It was not until the advent of ether that physicians and dentists were able to induce somewhat-controlled unconsciousness in their patients so that surgery became more bearable—for both patient and surgeon! Ether was the first general anesthetic.

Anesthesia means, literally, "without sensation" and in medical practice can refer to several different processes.

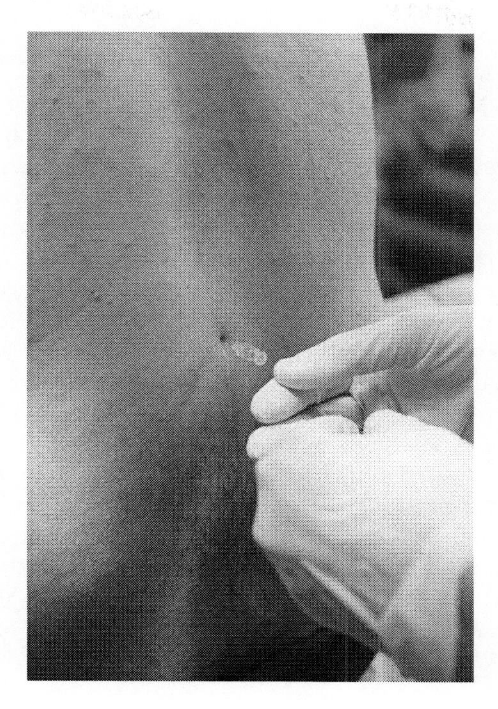

FIGURE 5
Needle inserted into the spinal canal for anesthesia.

Types of Anesthesia

Various prescription and nonprescription medications are able to reduce pain when taken internally, either orally or by intravenous injection. These medications usually are nonspecific; that is, they relieve pain in whatever area of the body it is being experienced. Ideally the agents work without impairing other mental or physical functions, but more powerful drugs do have some such side effects. Examples of these agents include acetylsalicylic acid (Aspirin®), acetaminophen (Tylenol®), ibuprofen (Advil®), and opium derivatives such as heroin, codeine, oxycodone, methadone, and morphine.

Local anesthetics are, as the name implies, agents that have a very localized effect. They may be administered topically or by injection. Ideally they affect only the area of the body near the administration site, and have only a temporary effect. Local anesthetic agents range from simple ice packs, to aloe vera juice, to cocaine and derivatives such as Novocain and benzocaine. Dentists use injected local anesthetics to help make their work more comfortable for patients, and an injection into the spinal canal can block some of the pain associated with labor and delivery or other abdominal surgery (Figure 5).

General anesthetics are those that produce unconsciousness. Hypnotic (allowing the patient to "ignore" pain) or amnesiac (causing the patient to not remember pain) medications may be included in this definition. Amnesiac drugs have the benefit of allowing the patient to be conscious during procedures so that they can respond to medical staff and cooperate with the procedures; for the patient, after the procedure is complete, not remembering pain or discomfort is effectively like not having that pain or discomfort at all. These drugs also tend to have fewer side affects than agents that produce unconsciousness. Examples of amnesiac drugs include midazolam, Propofol, and scopolamine.

Sedatives reduce the patient's level of consciousness to levels that are sufficient for a particular procedure. The level of sedation may be classified as deep, moderate, or minimal. In deep sedation, patients can respond to clear, sharp instructions, but are generally unresponsive. They may require breathing assistance. With minimal sedation, patients can respond normally, but are in a relaxed state, and do not require breathing assistance. The original sedative was alcohol, but its influence is widely variable, and it produces a number of undesirable side effects. Current sedatives include barbiturates, benzodiazepines, Ketamine, and Propofol. Some of these drugs may serve as both a sedative and an amnesiac.

One of the first general anesthetic agents that produced unconsciousness was ether, which was wondrous at the time because it allowed patients to undergo surgical procedures without being in agony. However, ether had undesirable features such as being difficult to control, having short-acting effects, and being extremely flammable. Newer substances were developed that reduced or eliminated these features, and the most common general anesthetic agents in use now are isoflurane, desflurane, and sevoflurane.

Related to anesthetic drugs, though not actually producing analgesia, are muscle relaxant substances. These drugs reduce or eliminate the contractions of skeletal muscles in order to make surgery easier and to prevent unwanted movement, especially in response to electrosurgery impulses. Intubation of a patient is much easier when their muscles are relaxed. Some of these drugs were derived from the curare used by South American Indians for their poisoned arrows or blowgun darts. Examples of muscle relaxant agents are Succinylcholine, Mivacurium, Vecuronium, and Gallamine.

Anesthesia in Surgery

Most major surgery involves the use of multiple agents. A sedative is administered, usually intravenously, in advance of the surgery, to relax the patient. Once the patient is in position on the operating table, sedation is increased until the patient can be intubated, and the general anesthetic agent is administered via the tracheal tube. After unconsciousness is achieved, muscle relaxants are given, and when all levels are sufficient, surgery can commence.

The condition of the patient must be monitored closely during surgery, both to ensure vital signs remain stable and to maintain an adequate level of unconsciousness, neither too high nor too low. Vital signs can change due to the general condition of the patient and to the trauma of the surgery itself. Level of consciousness is important because, if too high, the patient can become aware of the surgery being performed, and if too low, cardiovascular function can be compromised or even halted.

Side note: In the United States, a physician who administers anesthetics is referred to as an anesthesiologist, while a specialist nurse who does so is called an anesthetist. In Canada, the United Kingdom, and Australia, the terms anesthetist and anesthesiologist are both used to refer to a physician who is trained in anesthetic administration. For the purposes of this book, the term anesthesiologist will be used to mean any trained person who is administering anesthesia.

Anesthetic Machines

In the course of any major surgery, anesthesia must be monitored and maintained, and along with this the patient must be monitored and maintained as well. Anesthetic machines, handled by an anesthesiologist, provide these functions (Figure 6).

The primary function of an anesthetic machine is to regulate the flow of gases to the patient, including medical air, oxygen, and anesthetic agents. Various mechanical and/or

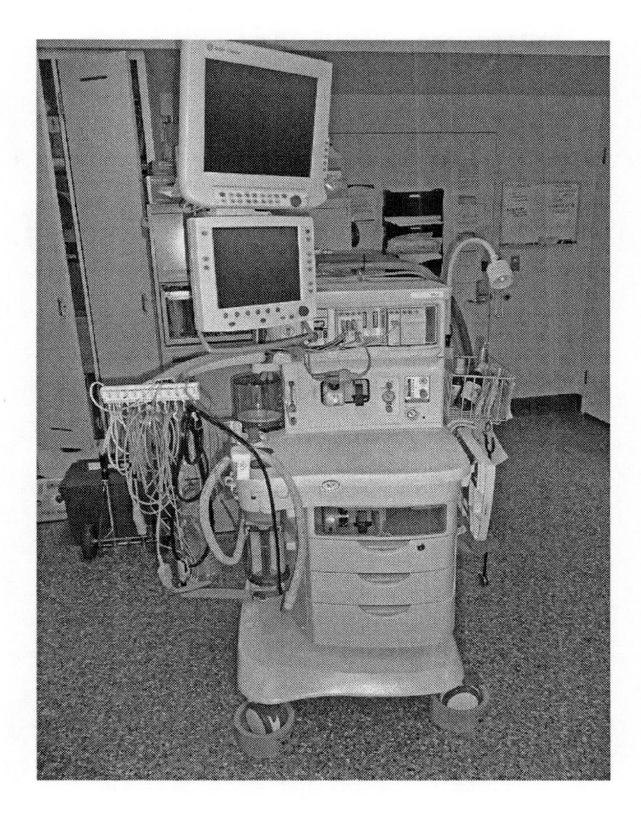

FIGURE 6
An anesthetic machine.

electronic components provide this regulation as well as giving readouts of the current values. Current systems record readings for later examination if necessary.

Anesthetic machines are usually operated electrically and may have gas pressures provided by hospital distribution systems, by attached pressurized tanks, or by internal compressors, or a combination of these. Some units have internal batteries that can continue normal functions in case of a line power failure, though usually only for a limited time.

Anesthetic agents such as isoflurane, desflurane, and sevoflurane are liquids at room temperature and must be vaporized in a controlled manner before they can be mixed with air and/or oxygen. Special vaporizer units attached to the anesthetic machine both contain the agent and vaporize it as necessary. Because each agent has different physical properties, the vaporizers are somewhat different for each.

Since patients undergoing surgery are often unable to breathe on their own, the machine provides a means of mechanical ventilation (Figure 8). The parameters of ventilation are controllable to various degrees depending on the model, but include rate, flow, and volume as a minimum. These parameters and measurements are displayed and also often logged.

In order to reduce the amount of anesthetic agent used, the exhaled breath of the patient is recycled, which means that carbon dioxide must be removed before the mixture can be returned to the patient. This is accomplished by passing the gas through a canister containing soda lime, which absorbs CO_2. The canister is clear, and the soda lime changes color as it becomes saturated with CO_2, so that the operators can change it when necessary (Figure 7).

FIGURE 7
CO_2 absorber.

Since anesthetic agents can be harmful, there must be a system in place to remove any excess from the area.

Various physiological parameters must be monitored during surgery, and this may be done with a stand-alone external system, or with one integrated into the anesthetic machine. In either case, ECG, respiration, and blood pressure can be monitored, plus blood

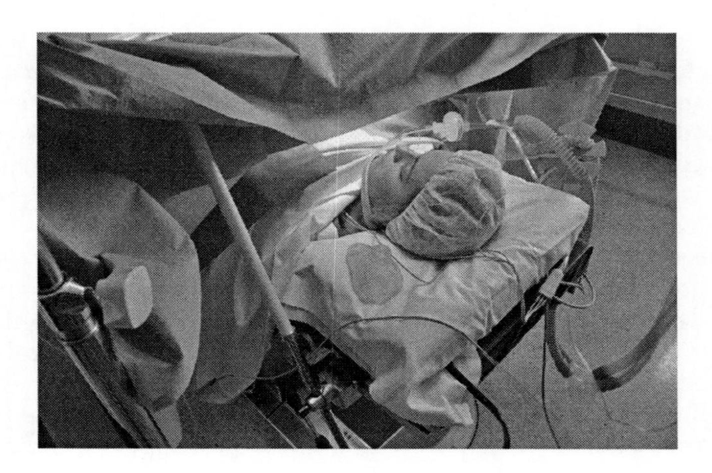

FIGURE 8
Patient undergoing surgery, with the anesthetic machine breathing circuit connected through the patient's mouth into the trachea.

oxygen saturation (SpO2) and expired CO_2 levels. The system records measurements and has settable alarms in case parameters go beyond specific limits.

Level of consciousness is an important factor during surgery, and various means have been used to try to determine this. Nerve/muscle stimulators apply an electrical signal to a part of the patient's body such as the hand or foot, and the degree of muscle contraction elicited is a measure of depth of anesthesia, though not necessarily level of consciousness. Most manufacturers now offer a means of giving a quantitative value for the depth of consciousness, usually by measuring and analyzing specific EEG waveforms. Some controversy exists as to the efficacy of these methods; however, they do provide more information than was previously available.

Selected Hospital Medical Units

Every hospital provides a different range of services, some of which are common to almost all facilities, and others that may only be offered in specialized hospitals. Generally, larger hospitals offer a wider range of services, although a small hospital in a relatively isolated area may provide a more complete range than a similar-sized suburban hospital, where specialty services are readily available in nearby city facilities.

This section will look at the range of clinical services offered by hospitals in general, and then examine a few specific areas in detail.

General Hospital Clinical Units

This is by no means a comprehensive list of clinical units and services, but serves as an example of what a relatively full-featured general hospital might offer. Units marked with an asterisk (*) are examined more closely in subsequent sections.

Addiction Treatment—provides emergency treatment of patients in addiction crises, support and treatment during the withdrawal process, and counseling and ongoing treatment to help patients remain free of substance abuse.

Cardiology—a diagnostic department that does testing of patients with confirmed or suspected heart disease. Many of these tests are described in the Clinical Procedures section of this book.

Diagnostic Imaging—provides X-ray, CT, MRI, ultrasound and other services to give clinicians a noninvasive look inside the patient.

*Emergency Services—care for patients suffering from severe trauma or critical illnesses, though the definitions of severe and critical may be open to interpretation.

Endocrinology and Diabetes Clinic—testing, diagnosis, treatment, and ongoing counseling of patients with endocrine disorders such as Addison's Disease, growth hormone abnormalities, and diabetes.

ENT (Ear, Nose & Throat)—concerned with diseases and traumas involving the self-described systems.

Family Medicine—a holistic approach to working with patients in the context of their family and community.

Gastroenterology—concerned with the digestive system: esophagus, stomach, small and large intestines, and associated organs.

Geriatrics—provides care for elderly patients.

*Intensive Care Units—ICUs treat patients with severe illnesses or traumas, often after being stabilized in the ER, but also after major surgery or following a medical crisis. ICUs may be specialized depending on their specific type of patients.

Laboratory—provides testing and analysis of tissue and fluid samples from patients.

Long Term Care—care of patients with conditions that require hospitalization for extended periods of time, such as those recovering from devastating illnesses or traumas, or those with severe chronic diseases or conditions.

*Maternity and Neonatal Care—prenatal and perinatal care for expectant mothers and their infants.

Neurosciences—concerned with diseases and conditions involving the brain and nervous system.

Oncology—counseling, cancer diagnosis, and treatment, including radiation therapy and chemotherapy.

*Operating Rooms—surgical suites.

Ophthalmology—conditions of the eye. A large portion of many ophthalmology units may be involved with performing cataract removal and lens replacement surgery.

Orthopedics—diseases, injuries, and conditions of bones and connective tissues, especially joints.

Outpatients—a unit that provides care for patients who need specific medical treatments but do not require overnight stays. This may include minor surgeries, administration of critical drugs, or procedures such as gastroscopies or colonoscopies.

Pediatrics—dealing with children and their specific, unique needs and conditions. Various diseases are almost exclusively found in children, as well.

Palliative Care—providing terminally ill patients and their families with counseling and support, and working to ensure patients are as comfortable and pain-free as possible, while respecting their wishes or those of their families should they not be able to respond themselves.

Physical Medicine—also known as rehabilitation or physiotherapy and occupational therapy, helps maximize patient function and comfort while reducing recovery time, through physical manipulation and exercise, and application of various therapies.

Psychiatry—dealing with patients with mental illnesses, ensuring the optimal treatment and counseling while protecting the safety of patients and others.

Pulmonary Medicine and Respiratory Care—diagnosis and rehabilitation of patients with suspected or known pulmonary or respiratory problems. Also may be involved with support and application of patient ventilators and anesthetic machines.

*Renal Unit—providing support and treatment for patients with temporary or permanent kidney function impairment or failure. Treatment may include hemodialysis or peritoneal dialysis.

Women's Health—may also be known as gynecology, though units often now deal with health issues for women in a more holistic manner. Women's Health units may include maternity.

Urology—dealing with medical issues involving the urinary system, including the kidneys, bladder, and prostate.

Emergency Rooms

- Function—ER is the first point of contact for many patients entering the hospital, whether they come on their own, accompanied by family or friends, or via ambulance. Patients are triaged as soon as possible upon entry, a triage nurse separating them into groups. Traditionally, triage was a battlefield process that meant division of patients into three groups: those who need immediate medical intervention; those who can wait for treatment; and those who either do not require treatment or are beyond help. Current hospital "triage" usually breaks down into more than three groups, for example: those with severe trauma or catastrophic illness, who may die in a short period of time if not treated; those with significant trauma or disease, who require treatment as soon as possible; those who are in need of significant treatment but are currently stable; those who need some form of more minor medical treatment but can wait; and those who do not require any treatment and can be discharged. Paramedics may perform triage on the scene or in the ambulance prior to arrival to the ER and will communicate their observations to ER staff, along with all other pertinent medical information.
- Layout—ERs consist of various fairly common areas, because their function is basically universal, differing only in scope. The following areas will be found in almost any ER.
    - Ambulatory entry where patients come into the ER under their own power or with a friend or family member. This entry is usually near the triage nurse, the admitting clerk, security, and the waiting room.
    - Ambulance entry. In a larger ER, there may be multiple ambulance entries. They usually have large, automatic doors and are near the treatment bays that are equipped to handle the most severe cases. There may be an area where multiple gurneys can be stationed along with the paramedic attendants. Paramedics bringing patients to the ER may assist in initial treatments, or there may be a hospital policy preventing them from doing so. The paramedics wait with the patient until the patient can be passed over to ER staff, at which time they convey any pertinent information regarding the patient's history, condition, and medications, as well as details of any treatment they have administered. Ambulances may be in radio contact with the ER, either directly or through their dispatch office, to allow for as much preparation as possible for the arrival of the patient in the ER.
    - Patients not requiring immediate attention as well as anyone accompanying the patient are accommodated in a waiting area. There are usually washroom facilities nearby, and vending machines may be provided.
    - The triage nurse is stationed near the ambulatory entrance in order to immediately assess patients when they arrive. If possible, the triage nurse will take a brief medical history of the patient.
    - Admissions gathers information from patients or someone accompanying them, including their name, their contact information, insurance details if available, family physician name and contact details, and any other information required

before the patient can be formally admitted to the hospital. This information is entered into the hospital computer system so that it is available to other areas of the hospital such as the OR, lab, or medical wards as required.

- Patients entering Emergency wards may be irrational due to metal illness or substance abuse, or they may be involved in criminal activities that can spill over into the hospital. Security staff is immediately available to help control such situations, and will have special training in dealing with ER cases, including negotiation, physical restraint, and communicable disease precautions.
- The ER will have a number of treatment bays, some of which may be specialized for specific types of cases. These may include severe trauma, cardiac arrest, burns, communicable diseases, or broken bones. The equipment, supplies, facilities, and physical arrangement of the specialized bays have all been optimized for those specific types of cases. Some but not all treatment bays will have physiological monitors to measure and record such parameters as ECG, respiration, blood pressure, blood oxygen saturation (SpO2), and temperature. Other less critical bays will have basic equipment for measuring vital signs, and most bays will have equipment for close examination of ears, noses, and throats. Some of the bays may be designed for psychiatric patients, and others may have precautions in place for patients with communicable diseases. An area is usually set aside for cast application and removal, though these functions may be performed by a separate cast clinic. Facilities for emergency surgery will be available, again usually in specific bays.
- The ER nursing station is the core of the department, where staff members can observe the whole area for which they are responsible. Desks are provided for patient charting and examination of records and test results, and computer terminals allow communication in both directions with other areas of the hospital, including the lab for results of tests. The nursing station may also have special terminals for examining images from the medical imaging department such as X-rays, and CT, MRI, or ultrasound scans. Physiological monitors from each bay so equipped may be connected to a central station in the nursing station where specific waveforms and numeric data for each monitored patient can be examined, monitored, and recorded. Alarms from the monitors can be checked and adjusted as necessary at the central station as well.
- Patients will often need to be moved to other parts of the hospital, and it makes the ER function better if these other areas are readily accessible. Medical imaging is often located immediately next to the ER, so that patients can have X-ray or other examinations done quickly. The clinical lab is usually nearby as well, so that samples can be obtained and delivered easily, and test results made available in a timely manner. Since ER patients often end up being transferred to other areas of the hospital such as the operating rooms or ICU, these areas are usually in relatively close proximity as well.
- Easily accessible storage areas hold supplies and equipment. These must be checked and replenished as necessary, either as the department consumes them or when their expiry date has passed.

- Staffing—The primary staff of an ER consists of physicians and nurses. MDs may specialize in emergency medicine, and residents and interns are assigned to the

ER on a rotation. Radiologists and various specialists as well as family physicians may be in the ER contributing to patient care at various times. Nurses are also often specialized in emergency medicine and usually stay assigned to the ER for long periods of time. This is also the case for nursing assistants. Nursing students may rotate through the ER, but usually only for observation. Other staff in the ER may include patient transport personnel; imaging, ECG, and laboratory technologists; housekeeping staff; admitting and ward clerks; and other support staff such as biomedical engineering technologists, plant services personnel, and information systems techs.

- Equipment—As mentioned above, many of the treatment bays in the ER will have physiological monitors mounted in the bay. An ER will likely have one or more ventilators available for patients with compromised respiratory functions. IV pumps are necessary for medication and fluid delivery, and defibrillators must be readily available for treatment of patients with severe cardiac arrhythmias or cardiac arrest. Fluid warmers are used to warm IV solutions or blood to near body temperature to help avoid hypothermia in patients who are in critical condition and require relatively large amounts of intravenous fluid administration. The function of these devices is described in more detail in Appendix B.
- Supplies—A wide range of medications, instruments, and supplies must be right at hand and well organized, and blood products for transfusion must be accessible at all times. X-ray equipment may also be stationed within the ER for use as needed. Medical gases including medical air, oxygen, and nitrous oxide will be available from outlets in each treatment bay, and wall suction and/or portable suction units will also be easy to access.

Intensive Care

- Function—Intensive care units provide direct critical care to patients. Staffing is usually more than one to one; that is, there are more staff members providing care than there are patients. Patients may be admitted to ICU from Emergency after they have been basically stabilized, from the OR following major surgery with the possibility of significant complications, or from general medical or surgical wards when a patient's condition becomes critical.
- Layout—An ICU consists of a number of patient rooms, each with its own physiological monitor and special lighting and gas supplies. There is enough room around the bed for several team members should they be required for treatment procedures. Some rooms may have isolation features such as negative air pressure to keep pathogens from escaping from the room and individual hand cleaning stations and mask, gown, and glove supplies. A nursing station provides an area for a central monitoring station, charting facilities, and areas for staff consultation, plus hospital computer system terminals and diagnostic imaging viewing stations (Figure 9). A large ICU may be divided into two or more sections, each with its own central station. An equipment and supplies storage room will be readily accessible, and since family and friends of the patient are often very worried and want to be nearby as much as possible, most ICUs have their own visitor waiting area.
- Staff—The primary staffing in any ICU consists of nurses and physicians. Nurses are usually specialized in ICU work, and physicians are generally specialists in an area related to the focus of the particular ICU, though GPs may come to the unit to

FIGURE 9
A dual-display ICU central monitoring station.

check on their patients and consult with other staff. Residents will rotate through ICU, but interns and other students will likely only be present for short observation periods. Secondary staffing in an ICU may consist of: imaging technologists, because patients are often too critical to be able to be moved to the imaging department; respiratory technologists to assist with the application and ongoing support of ventilators; dieticians to help assess and meet the special nutritional needs of ICU patients; infection control specialists, because many ICU patients have communicable diseases and otherwise are especially susceptible to infections; biomedical engineering technologists, to troubleshoot any problems with the high-tech equipment in the ICU, and to support the application and most effective use of the equipment; and housekeeping personnel to ensure that the area is kept clean.

- Equipment—ICUs have very high concentrations of medical electronics equipment. Each bedside will have a physiological monitor that is capable of measuring, displaying, and recording such parameters as ECG, respiration, temperature, blood oxygen saturation (SpO2), blood pressure either noninvasively using an automatic arm cuff system or invasively via a catheter inserted into various points in the patient's circulatory system, exhaled CO_2 levels, cardiac output, level of consciousness, and perhaps blood chemistry using a point-of-care analyzer built into or interfaced with the main monitor. Information gathered by the monitor is displayed on its video screen in either graphical or numeric format, or both (Figure 10).

  The bedside monitors will be connected to a central station where nurses can watch the waveforms and data from each patient, set and respond to various alarms from the bedside monitors, and record pertinent waveforms or data to form part of the patient chart. Some systems allow information from a second bedside monitor to be displayed on one bedside monitor if a staff member needs

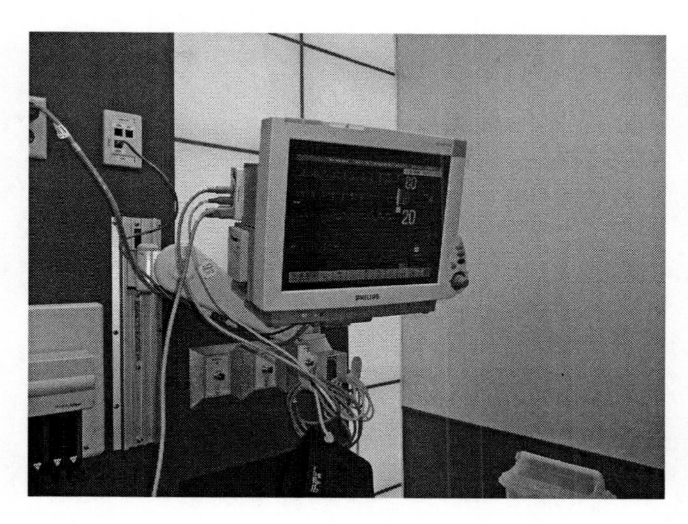

FIGURE 10
An ICU bedside physiological monitor.

to check on the other patient without leaving the side of the first. Systems may also be connected to the hospital computer system and to the Internet, allowing remote viewing of patient monitoring data. ICUs will also have defibrillators, ventilators, intravenous and PCA (patient-controlled analgesia) pumps, and a variety of other minor equipment. Portable X-ray and ultrasound machines may be dedicated to the unit. If the particular ICU has patients that need to be monitored but who are able to walk some, they may have a telemetry system providing ECG and perhaps SpO2 and noninvasive blood pressure measurements remotely via radio signals from a transmitter pack worn by the patient. Computers form a vital link to the ICU, for sending and receiving lab data, pharmacy orders, service requests, patient charts and admission information, and routine hospital communications. The function of the patient care electronic devices is described in more detail in Appendix B.

- Special types—Smaller hospitals often have a just a single ICU that deals with critical patients up to the level of acuity that staff and facilities can handle. Larger hospitals will likely have several ICUs specializing in specific types of patients. These specialty ICUs often have their own acronyms and might include cardiac (CICU), cardiac step-down (CSICU), telemetry (TelICU), medical (MICU), neonatal (NICU), pediatric (PICU), burn (BICU), neurological including spinal cord injury (NeuroICU or SICU), psychiatric (PICU), geriatric (GICU), and respiratory (RICU).

Maternity

- Function—Maternity provides care for women during pregnancy, delivery, and immediate postnatal period. Counseling and education is available regarding prenatal nutrition and health, birthing alternatives, birthing preparation, infant care, lactation, fertility, and family planning. Complications of pregnancy due to maternal or fetal medical issues are dealt with, and testing such as fetal monitoring and nonstress procedures are available. Many hospitals have designed or renovated their maternity units to be more comfortable and homelike for patients.

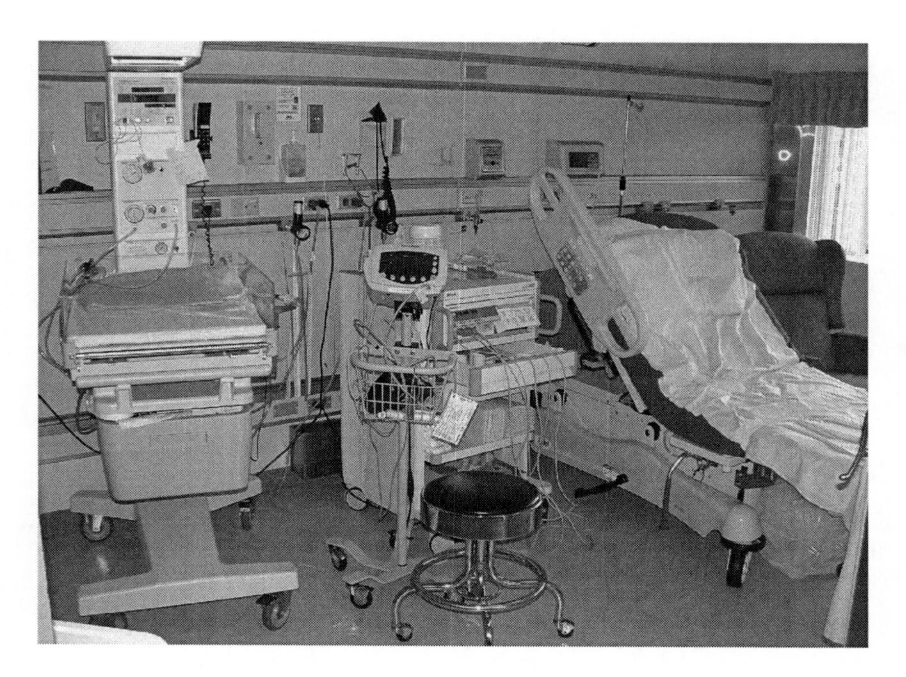

FIGURE 11
A typical labor–delivery room. Equipment includes the tall infant resuscitation unit on the left, the vital signs monitor on the stand, a fetal monitor on the cart, and the specialized labor–delivery bed with patient- or staff-adjustable positions and a removable section at the foot to allow staff access to assist in delivery.

- Layout—Patient rooms in Maternity are for expectant mothers who may require predelivery medical care due to complications of pregnancy and for new mothers for recovery following birth. Some mothers in good health may leave the hospital the same day as they gave birth or within a day or two, while others may have medical issues that necessitate longer stays. Hospitals often have facilities that allow the newborn to stay in the same room as the mother until discharge, while others have separate nurseries for all babies. Infants requiring special care will stay in the nursery, either in a bassinette or an infant incubator. Special facilities such as birthing beds will be available in the department, sometimes in special labor–delivery rooms and sometimes in the regular patient rooms (Figure 11 and Figure 12).

  An operating room for Caesarean section births may be part of the Maternity unit, with the same facilities as a regular OR (Figure 13).

  A nursery will be in close proximity to the patient rooms, with the traditional viewing windows (Figure 14).

  Rooms may be available for procedures such as nonstress testing and other examinations. A nursing station will have facilities for charting and other such routine nursing procedures, and may have a central station for either fetal monitor data or neonatal physiological monitoring information, or both. A small kitchen is usually present for fresh food preparation for pre- and postdelivery mothers, fathers, and birth coaches, as well as a lounge for those involved in the birthing process. The lounge may double as a visitor waiting room, or the two may be separate. If

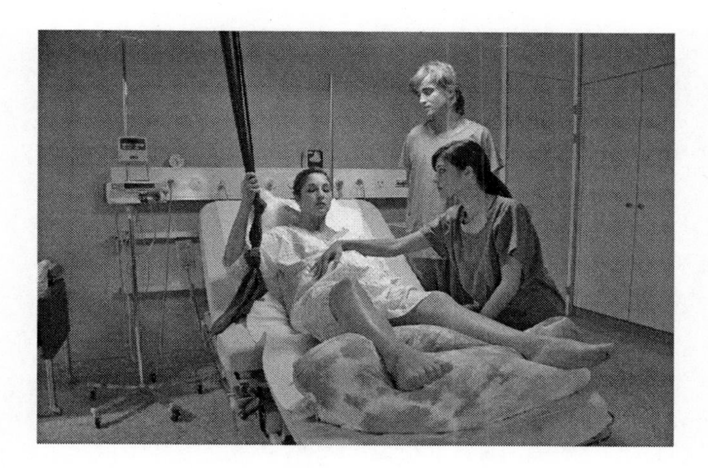

FIGURE 12
A patient in labor.

the hospital has a neonatal intensive care unit, it is usually in close proximity to Maternity, if not an integral part of the area. Some maternity units have special facilities, such as pools for water births.

- Staff—Maternity is primarily staffed by nurses. Physicians may or may not be present for vaginal deliveries, though any complications usually mean a family physician, resident, or gynecologist will be involved. Surgeons perform C-sections. Professional labor coaches or midwives are often available, as well as social

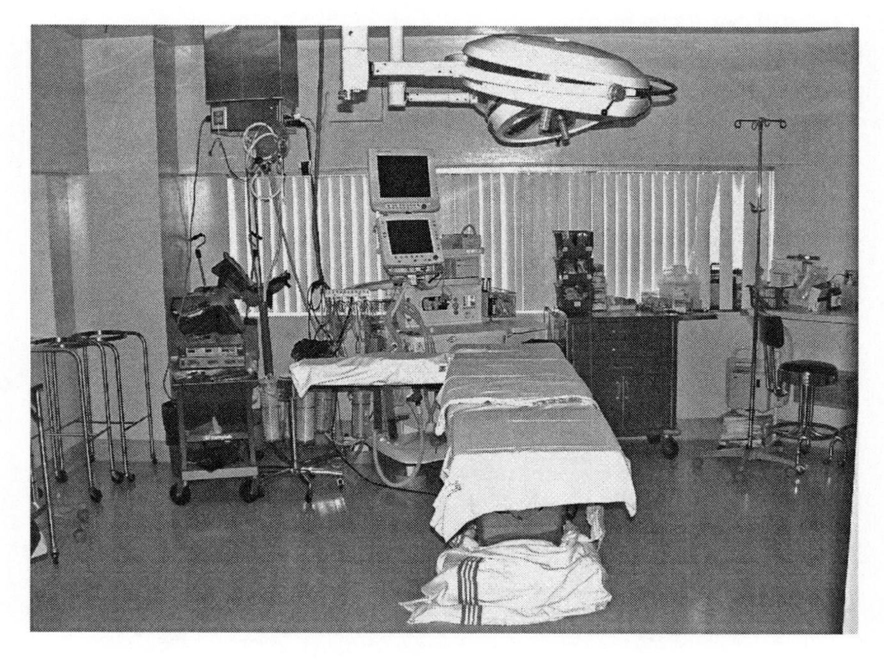

FIGURE 13
An operating room specialized for Caesarian sections. It includes an electrosurgery machine on the left, an anesthetic machine at the back, a surgical table, and overhead surgical lights.

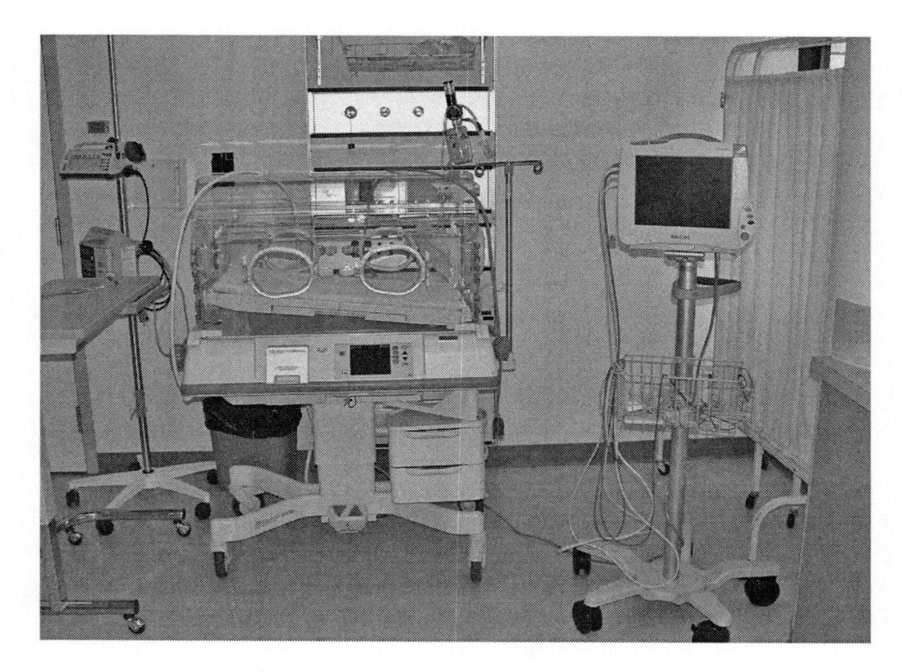

FIGURE 14
Part of a nursery with an infant incubator and mobile physiological monitor.

workers, dieticians, and lactation advisors. Housekeeping staff is vital for maintaining order and cleanliness, especially following birth.

- Equipment—Fetal monitors allow the measurement of fetal heart rate and the relative strength of labor contractions (Figure 15). They may also include pulse oximetry and noninvasive blood pressure measurement.

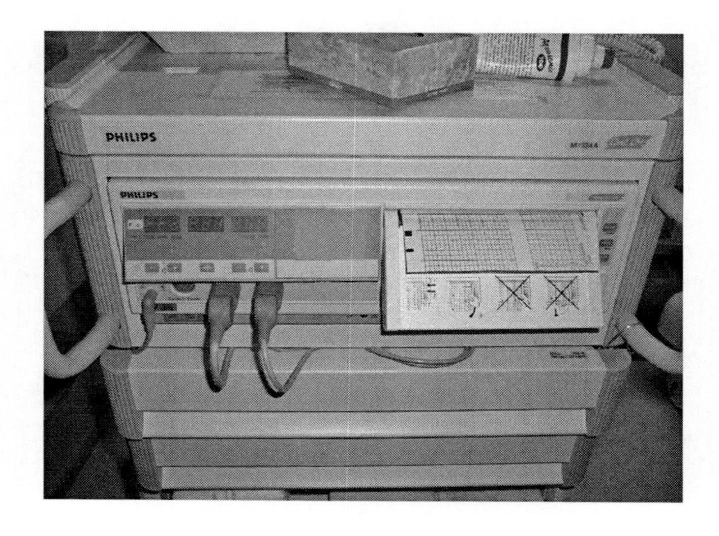

FIGURE 15
A fetal monitor.

Some units connect a number of fetal monitors to a central station, and they may also be equipped with telemetry so that mothers in labor can move around more freely. Handheld ultrasonic fetal heart detectors allow for quick checks of fetal status. Special devices, sometimes called infant resuscitators, with radiant heating elements, lights, oxygen and suction, APGAR timers, and facilities for X-rays are available for neonates who are having some degree of difficulty following birth. Infant incubators in the nursery area have Plexiglas walls that allow easy observation of the child inside while maintaining controlled levels of temperature, humidity, and oxygen (Figure 16).

The incubators have access ports to allow for changing diapers or performing other procedures without allowing a significant amount of outside air into the unit. Some newborns have an excess of bilirubin in their systems, and UV therapy lamps, often referred to as bililights or phototherapy units, help break down the bilirubin in the infant's skin. Bililights may consist of overhead or side-mounted lamp fixtures, or they may utilize fiber optic technology to deliver UV light to the inside of a blanket placed around the baby. Breast pumps are available for mothers who cannot directly breastfeed their babies but wish to provide them with their own breast milk. Special birthing beds are equipped with electrical controls to adjust bed position in many configurations, as well as built-in stirrups to aid in birthing, and a removable center foot section to allow access by the delivering physician or midwife. Maternity wards have the usual complement of basic equipment such as IV pumps, PCA pumps, thermometers, pulse oximeters, and blood pressure monitors. Security is often a special concern in maternity units, especially the nursery, so security systems are often in place. These may consist of limited access doors, alarms, video camera, and tags worn by babies that allow

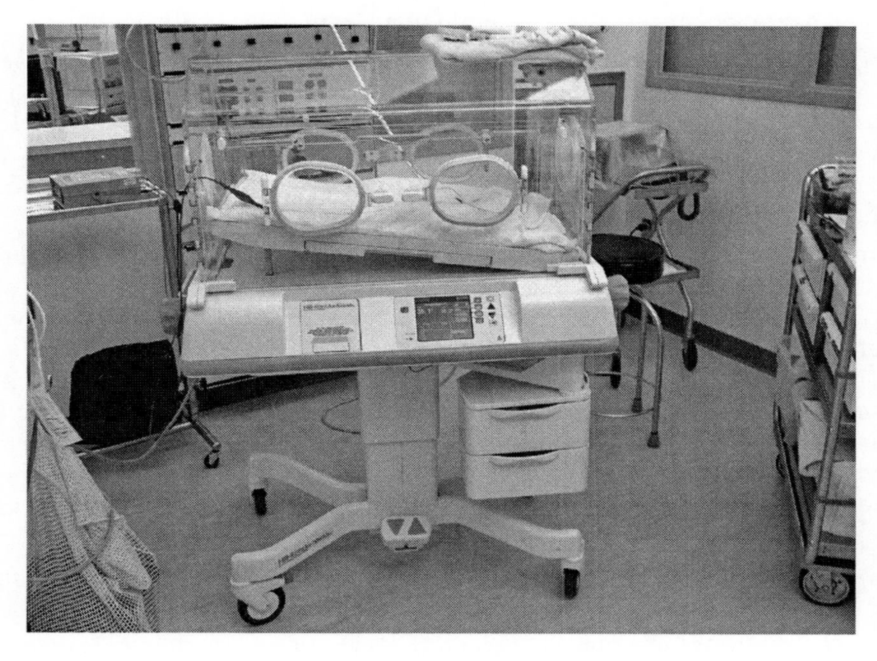

FIGURE 16
An infant incubator.

remote tracking. Such tags may trigger automatic alarms if they pass beyond certain boundaries. The function of patient care electronic devices is described in more detail in Appendix B.

Operating Rooms

- Function—Operating rooms are the venues for scheduled and emergency surgical procedures of all kinds. Some larger hospitals may have more than one operating room area, for different areas of specialty within the hospital.
- Layout, overall—The surgical suite of a hospital consists of a number of individual operating rooms, often arranged in a more-or-less circular pattern, with a wide hallway surrounding the rooms. Each room will have many characteristics and components in common, but often at least some rooms are specially outfitted for specific types of surgery such as urology, endoscopic procedures, trauma surgery, cardiac surgery, organ transplantation surgery, or neurosurgery. A holding area at the entrance to the ORs provides space for patients when the individual room is not quite ready for them, and to allow surgeons, anesthesiologist, or other OR staff to speak with the patient before surgery if necessary. This area may also serve as a waiting room for family. The OR office coordinates use of the rooms and performs room and staff scheduling, provides reminder calls for patients, and ensures that OR supplies and equipment are always ready. A sterile supply area is often in the center of the various operating rooms, to allow easy access to and from each room. A postanesthesia recovery area will be adjacent to the OR suite, and basic lab and diagnostic imaging facilities will often be included in the area to allow for rapid testing and imaging. Biomedical engineering may have an area in or near the OR suite to provide quick response should there be problems with equipment.
- Layout, individual general OR—Operating rooms usually have two access doors: a large double door opening to the outside hallway for entry and exit of patient stretchers and a smaller door. Rooms must be spacious enough to allow easy access to all sides of the operating table, to permit potentially five or more staff members to circulate within the room as required, and to allow for all the various devices and supplies that will be used in the course of surgery (Figure 17).

  Equipment may be mounted on wheeled carts that can be moved near the table as required, but it is becoming more and more common for ORs to have special booms suspended from the ceiling, from which most of the critical equipment can be mounted on various arms and platforms. These booms also contain gas supplies, electrical outlets, computer network connections, video interfaces, and more. Anesthetic machines are usually too heavy and bulky to be mounted on the booms and so are usually freestanding. Diagnostic imaging viewing equipment, either light-boxes or high-resolution video monitors, are usually stationed on one wall of the room, and a hospital computer terminal may be present, again usually against a wall, out of the way of the main surgical field.
- Equipment—Operating rooms have the widest variety of equipment and supplies of any area of the hospital.
  - The anesthetic machine is the central device, providing delivery and monitoring of anesthetic gases and precisely controlled mechanical ventilation of the patient (Figure 18).

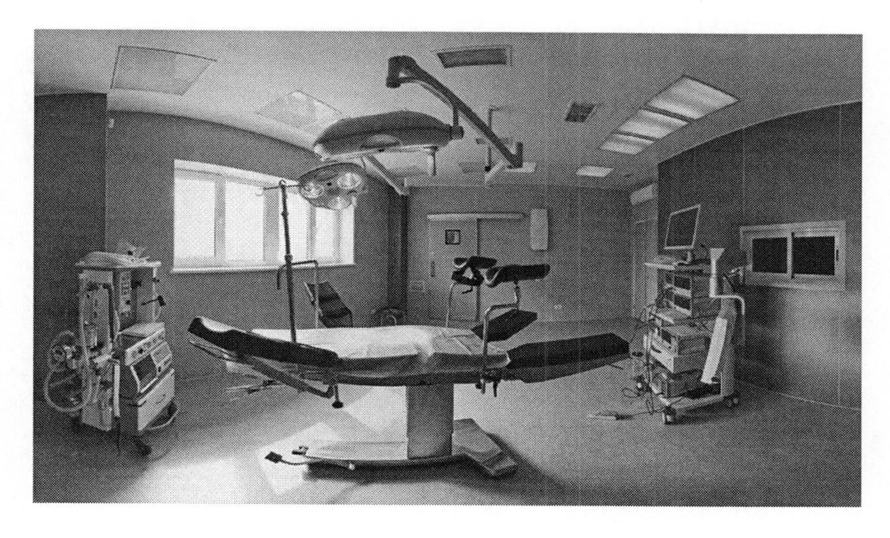

FIGURE 17
A typical operating room with anesthetic machine on the left, OR table and surgical lights in the center, and an endoscopic equipment cart on the right.

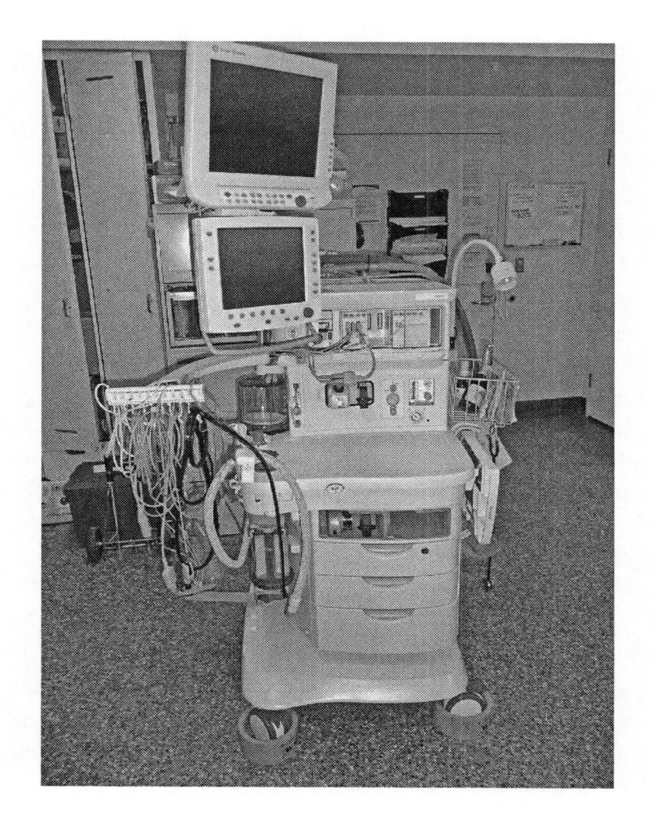

FIGURE 18
Anesthetic machine.

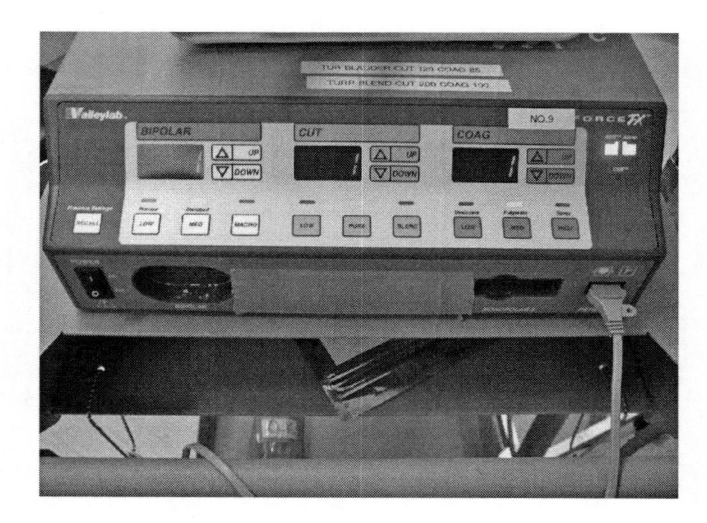

FIGURE 19
Electrosurgery machine.

- Physiological monitors that measure and display a wide variety of vital signs from the patient may be stand-alone or integrated into the anesthetic machine.
- Much of the actual process of cutting is performed using electrosurgery machines, which also provide cauterization and fulguration (Figure 19).
- Surgical lasers are used in a somewhat similar manner to electrosurgery machines.
- Because both of these types of devices can produce potentially harmful smoke particles in the course of their use, special smoke evacuator units are used in conjunction with them.
- The OR table gives a solid platform for procedures and can be moved up, down, and sideways, as well as being able to tilt lengthwise (pitch) and laterally (roll). OR tables are controlled by foot pedals.
- Surgical robots are used for some operations, especially delicate procedures on very small structures. Systems can allow surgeons to control movements of the robotic manipulators on a much smaller scale that they would be able to perform manually. Manipulators can be much smaller than human hands, which means that they can enter areas of the body with much smaller incisions than would be otherwise necessary. Surgical robots used in conjunction with video systems and remote access computers have the potential of allowing surgery to be performed on a patient who is in a location far away from the human surgeon.
- Anesthesiologists use nerve and/or muscle stimulators to aid in evaluating the level of consciousness of the patient.
- Automatic tourniquets are used in surgical procedures on limbs, in order to reduce blood flow to the area.
- Endoscopy stacks consist of light sources, insufflators, video cameras and recorders, possibly driver units for drills and other power accessories, and one

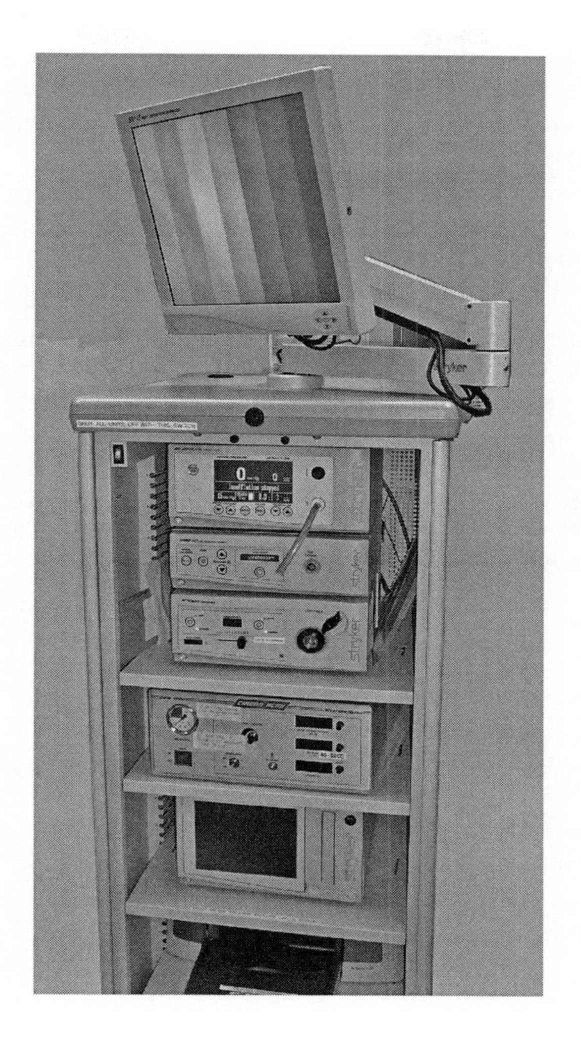

FIGURE 20
Endoscopy cart with (from top) video display, insufflator, camera console, light source, a specialized insufflator for gynecological surgery, and a video storage unit.

or more video displays (Figure 20). They may be mounted in a standalone cart, or on the OR boom.

- An operating microscope is used for microsurgery, including such procedures as cataract removal and lens replacement. These microscopes are quite large and contain their own light source. The viewing head can move in all directions; this plus magnification and focus are controlled by foot-pedals. A video system may be integrated into the microscope in order to display and record images.
- A variety of surgical tools will be available, including various scalpels, hemostats, forceps, suturing and stapling equipment, saws, drills, hammers, wrenches, and screwdrivers. The range of tools varies with the procedure being performed.

- Gas supplies are on the wall of the room, or more commonly on the OR boom, and include oxygen, medical air, nitrous oxide, and vacuum.
- Certain hardware may be required for surgery, for example the components of an artificial hip or knee, screws, nails, pins, and plates.
- Drugs and IV solutions, as well as a range of other surgical supplies such as sponges and dressings, will be on a tray or readily available from the central sterile supply room.
- Defibrillators will be immediately accessible in case of emergency.
- Patient body temperature may be subject to rapid fluctuations during surgery, so warming/cooling units are available, often with both functions integrated into one device.
- Many hospitals have a system called PACS (picture archiving and communication system) that allows central storage and remote display of any diagnostic images that have been produced for patients in the hospital. A large, high-resolution PACS video display will be available in the OR, though this function may be integrated into a general-purpose video system.
- A hospital information computer terminal will be in the room or nearby, to allow access to patient demographic information, medication history, and test results. As with PACS, the hospital computer system may be integrated into a general-purpose unit.
- Particularly in teaching hospitals, ORs may be equipped with high-resolution observation and recording video equipment, allowing students or remote specialists to observe procedures.
- Last but not least, a stereo is an integral part of most ORs, because many surgeons prefer to work with music in the background.
- The function of some of the patient care electronic devices is described in more detail in Appendix B.

- Lighting—A variety of equipment meets the specific lighting requirements of surgery. Large, bright overhead lights can be moved around and focused on particular areas, and designers attempt to minimize heat production while providing bright light with consistent color characteristics (Figure 21).

  Surgeons may utilize headlights to illuminate the specific area on which they are working. Fiber optic light sources are used in a variety of situations because they provide good light quality with minimal heat and can be directed through small, flexible channels to difficult-to-access locations. ORs are usually equipped with handheld laryngoscope and ophthalmoscopes, which have built-in illumination.
- Displays—A variety of video displays may be found in operating rooms, though there is a trend toward integrating the displays of various components. Displays may include video from endoscopes, operating microscopes, PACS, patient record and lab result information, physiological monitoring, and anesthesia monitoring and control. Video feeds may be provided to multiple large wall-mounted screens within the room, or to screens in various teaching and observation areas within the hospital.
- Staff—The surgical team is described in detail earlier in this section.

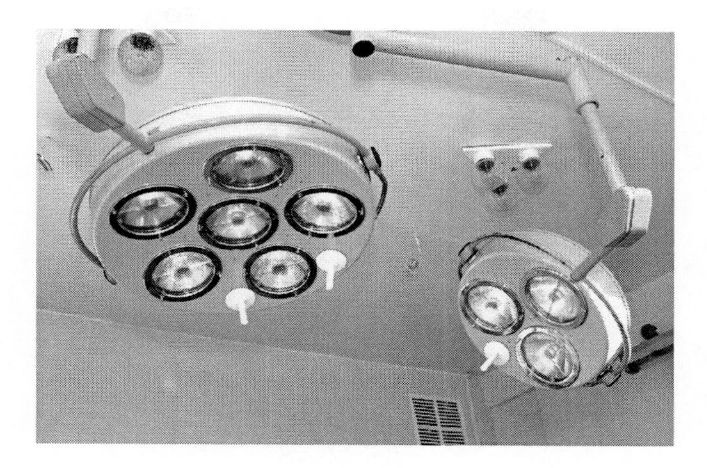

FIGURE 21
Surgical lights.

- Specialized OR types—Suites may be specially equipped and set up for specific types of surgery, including ophthalmologic procedures, orthopedics, Caesarean sections, dental surgery, or urology.

Renal Units

- Function—Patients with kidney disease come to renal units for counseling, dietary advice, and dialysis. Two basic forms of dialysis treatment are provided: peritoneal dialysis (PD), for patients with less severe kidney function impairment, and hemodialysis, for patients with severe impairment to complete kidney failure. Both are performed on an outpatient basis, though peritoneal dialysis can often be performed by the patient or a family member away from the hospital. Peritoneal dialysis involves the infusion of a dialysate solution into the abdominal (or peritoneal) cavity of the patient. In continuous ambulatory PD, the solution remains in place for several hours as the patient is free to go about daily living activities. At a later time, the solution, which has absorbed impurities, is then drained, the cavity is flushed, and a new batch of solution is infused. With automated PD, the solution is infused and exchanged automatically a number of times while the patient is sleeping at night. Hemodialysis requires that a port be in place to connect equipment to the circulatory system of the patient. Blood is taken from the port and passed through a hemodialysis machine, which removes impurities, and is then returned to the patient. This process takes several hours and must be repeated several times per week. Because of the size and complexity of hemodialysis machines, the treatment is mainly performed in clinics, though some patients may be able to afford to purchase and maintain a unit in their home (Figure 22).
- Layout—Since patients are required to be in the renal unit for extended periods with little movement, the units are usually designed to be otherwise pleasant, with comfortable beds, relaxing décor, and music or television available. A nursing station allows staff to monitor patients and perform record keeping and other duties, while significant areas must be available for equipment storage and maintenance.

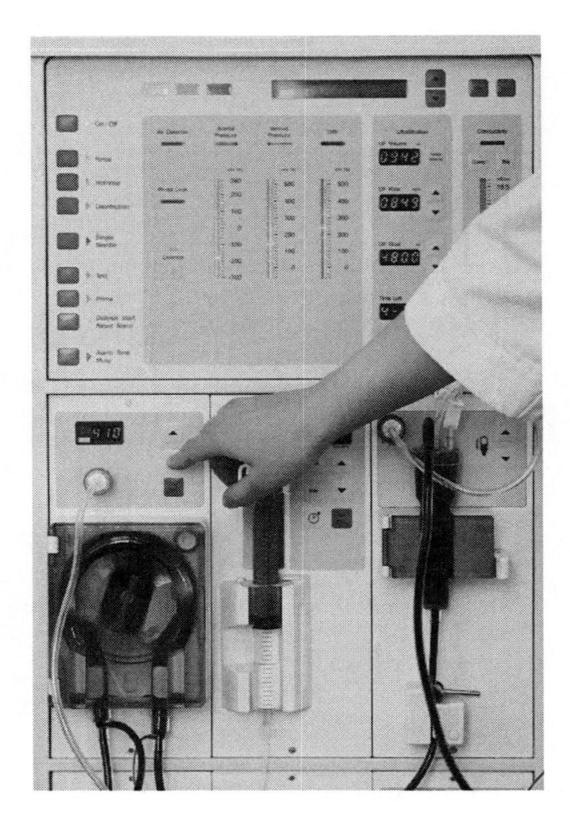

FIGURE 22
Hemodialysis machine.

- Staff—The renal unit is staffed by nurses and renal technologists, with biomedical engineering technologists present to perform equipment repair and maintenance.
- Equipment—The hemodialysis and peritoneal dialysis machines make up the majority of devices in the unit, with some vital signs monitors to keep track of patient conditions. The function of the dialysis machines is described in more detail in Appendix B.

2

Clinical Procedures

Index of Procedures

Introduction

The procedures described in this section are arranged alphabetically; all are included in the index for easy reference.

A standard format has been used, with information on alternate names for the procedure, the general purpose of the procedures, indications as to why the procedures are performed, and an outline of the relevant anatomy, pathology, and physiology. Following sections describe the team members normally involved, the equipment and supplies needed, the preparation required, a description of the procedure itself, the expected outcome and necessary follow-up, and any possible complications. Some procedures will have no information for some of the headings.

Preparation for general anesthesia surgery follows standard steps, including:

- Last-minute discussion with the patient or family regarding the procedure.
- Establishment of an IV line.
- Administration of a sedative.
- Placement of the patient on the OR table.
- Marking areas specific to the procedure.
- Applying antiseptic to the whole area.
- Arranging draping.
- Applying any required monitoring electrodes or sensors.
- Administering an increased dose of sedative to induce unconsciousness.
- Inserting an endotracheal tube for connection to the anesthetic machine.
- Administering a muscle-relaxing agent.
- Connecting the anesthetic machine and beginning gas anesthesia and artificial ventilation.

Some of these steps may follow a somewhat different order, and some may occur simultaneously, performed by different team members.

Surgical Teams

Surgical teams needed for a particular procedure are varied depending on the complexity of the case, policies of the institution, and the preferences of the lead surgeon. Teaching hospitals will often have a number of students observing or participating in procedures. For the purposes of this text, however, the team for a given procedure will be indicated as one of five standard teams, representing an approximate typical makeup of the team required for that procedure.

1. Individual.
2. Basic—consisting of one surgeon and an assistant, either another MD or (usually) a nurse.
3. Normal—one surgeon and an MD or nurse assistant, an anesthesiologist, and a float nurse.

4. Major—two surgeons, an anesthesiologist, one or more MD and/or nurse assistants, one or more float nurses, and specialized technologists as required.
5. Complex—three or more surgeons, several assistants, several float nurses, several specialized technologists. Some complex procedures may involve two teams, for example living donor transplantations or neonatal surgery following a caesarean section.

Surgical Equipment and Supplies

As with surgical teams, a wide range of equipment and supplies are used in different surgeries. The following will be considered present for all surgeries listed, even if all are not used in every procedure.

- Standard equipment for surgery consists of the operating table, OR lighting, an anesthetic machine, a physiological monitor, an intravenous pump, and an electrosurgery machine.
- Standard supplies may include items such as scalpels, hemostats, forceps, scissors, sponges, swabs, sutures, needles, needle holders, staples, IV solutions, antiseptics, medications, syringes, and drapes. These will often be available in packs containing all the supplies, and packs may be made specifically for different procedures (Figure 23).

Complications

Complications listed will be only those unique to the procedure being discussed. Such things as infections, blood clots, and suture failure are common to almost all surgical procedures and will not be listed separately for each procedure. If there are no complications likely other than these common ones, n/a will be indicated.

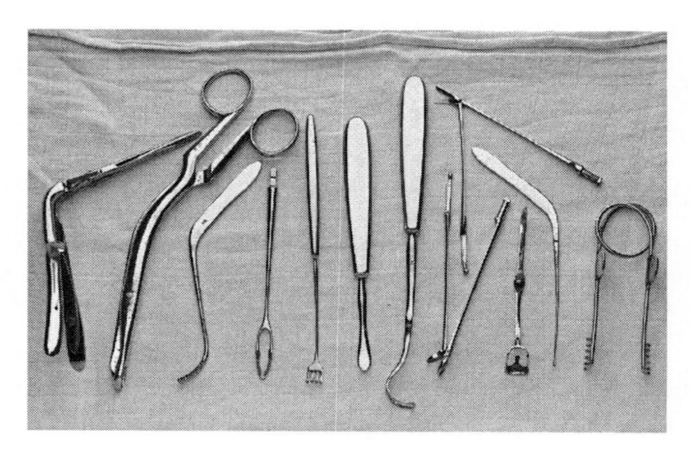

FIGURE 23
Surgical instruments.

Abdominal Wall Defect Repair

Alternate names—Omphalocele repair/closure or gastroschisis repair/closure.

Purpose

To close a congenital defect in the muscles of the abdominal wall.

Indications

The presence of this defect, which may be detected by prenatal ultrasound or by direct observation at birth.

Anatomy

The abdomen normally has a sheath-like layer of muscle that helps contain the abdominal organs.

Pathology

During fetal development, the abdominal muscles may not close completely, thus allowing internal organs, particularly intestines, to protrude from the abdomen. This has different names depending on location. A defect allowing protrusion into the umbilical cord itself is called omphalocele, while one with protrusion to the side of the umbilicus is called gastroschisis. Gastroschisis is more critical, requiring surgery immediately after birth. Omphalocele repair is usually performed as soon as is practical, but not on an emergency basis.

Physiology

n/a

Staffing

Normal surgical team.

Equipment and Supplies

Standard surgical.

Preparation

Standard surgical, plus dressings are placed over the exposed organs.

Procedure

Standard surgical procedures are followed, and a gastric tube is inserted. The abdominal opening is enlarged so that the intestines can be checked for damage and to allow room to work. Any damaged sections are removed and the healthy ends reattached. The protruding portions are moved into the abdominal cavity, and the site is closed.

Expected Outcome and Follow-Up

Normal function is restored. These conditions may be part of a general pattern of birth defects, many of which can be life-threatening and must be monitored carefully.

Complications

The above-mentioned accompanying birth defects can cause various problems separate from the repair surgery; these are more common with omphalocele, and a high percentage of infants with this condition do not survive to age one. Intestinal compromise from the defect and/or the necessary removal may result in digestive problems.

Ambulatory ECG Monitoring

Alternate names—Holter recording or monitoring.

Purpose

To obtain an electrocardiogram (ECG) recording over a relatively long period of time, usually 24 hours but sometimes 36 or 48.

Indications

Known or suspected cardiac problems.

Anatomy

The heart is a muscular organ consisting of left and right atria, left and right ventricles, four valves, and associated blood vessels (Figure 24). These blood vessels carry the blood pumped by the heart out to various body parts and also carry blood to the heart itself. The atria are thinner walled than the ventricles. Nerves and other pathways in the heart carry electrical signals to control pumping contractions.

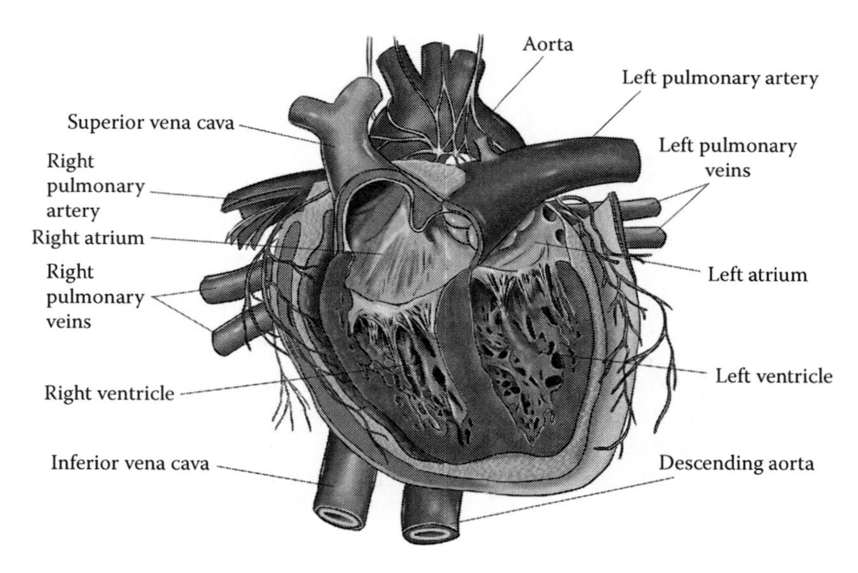

FIGURE 24
A color version of this figure follows page **176.** Cutaway view of heart and associated vessels.

Pathology

Heart pumping effectiveness can be compromised by various problems within the heart. The muscles of the atria and especially the ventricles can be damaged by disease or lack of blood supply (cardiac infarction). The cardiac valves can be damaged by disease or age, thus allowing backflow of blood. Finally, the conductive pathways that help produce and coordinate contractions may be damaged, so that contractions are either weaker or less coordinated, or both.

Physiology

The heart, being mostly muscle that is active all the time, requires an excellent blood supply to provide oxygen and nutrients and to remove wastes.

Signals from a center in the heart produce electrical signals that are carried throughout the heart to produce contractions. Various components provide delays to the signal so that contractions are coordinated and allow effective pumping.

Staffing

Cardiology technologist to instruct the patient, set up and remove the monitor, and download the recordings; a cardiologist to interpret the results.

Equipment and Supplies

Ambulatory ECG monitor/recorder, cables and electrodes, carrying pouch, recording downloading equipment, analysis and reporting software (Figure 25).

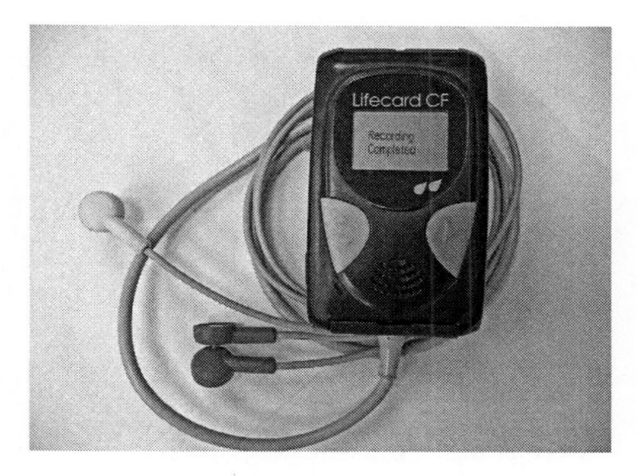

FIGURE 25
A digital Holter monitor with leads, showing ECG waveform on an LCD display.

Preparation

The electrode sites are shaved and cleaned carefully to ensure optimal conductivity. The patient is instructed in the use of the recorder, which usually has a button the patient can press if he experiences something unusual that he feels may be related to his heart. The patient is asked to keep a log of his activities.

Procedure

Electrodes are placed on the skin in specific locations, and the signal is tested to ensure quality. The patient goes about his normal activities for the prescribed period of time, after which the recorder is removed and recording stopped. The recording is downloaded to a display and analysis system, and reports are generated, either by the system itself or by the attending cardiologist, or both.

Expected Outcome and Follow-Up

A useful recording of ECG activity is obtained. Depending on the results, a course of treatment for cardiac disease may be initiated.

Complications

Some patients may be sensitive to the adhesive used in the electrodes, and skin irritation may occur.

Amniocentesis

Alternate names—n/a.

Purpose

To obtain a sample of amniotic fluid for genetic analysis of the fetus.

Indications

Family history of genetic defects, diagnostic imaging results suggesting genetic defects in the fetus.

Anatomy

The fetus develops within the mother's uterus, a muscle-walled organ that is pear shaped before conception but expands as the fetus grows (Figure 26). The fetus is contained by a thin-walled structure called the amniotic sac within the uterus; this sac is filled with fluid that helps protect the developing fetus. The fetus is connected to the mother via its umbilical cord, which carries blood in both directions between the fetus and the placenta. The placenta is a highly vascularized organ that is fetal tissue and has extensive connections to the wall of the uterus.

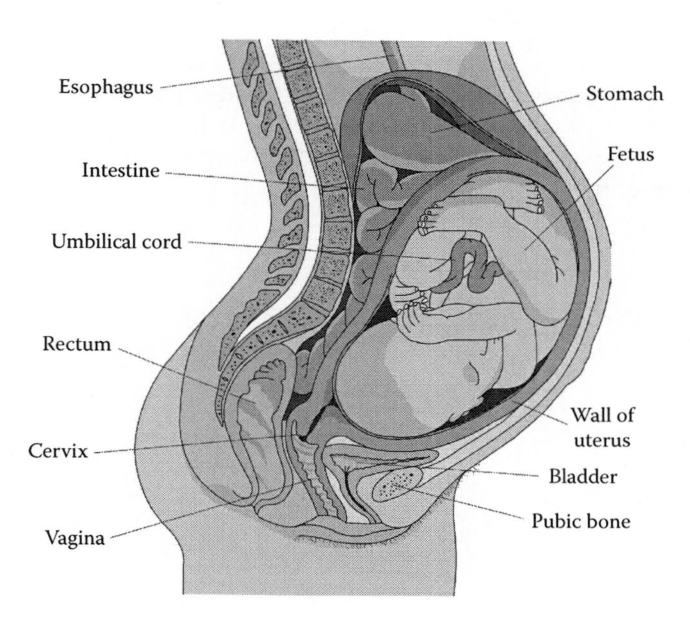

FIGURE 26
Fetus developing in uterus.

Pathology

Various genetic defects can cause a wide range of problems and syndromes including Down syndrome, Tay-Sachs disease, sickle cell disease, hemophilia, muscular dystrophy, and cystic fibrosis.

Physiology

Fetal cells are shed into the amniotic sac.

Staffing

Physician, assistant, diagnostic imaging technologist.

Equipment and Supplies

Amniocentesis needle and syringe, ultrasound machine.

Preparation

The area of needle insertion is cleaned and topical anesthetic applied.

Procedure

The needle is inserted into a space in the amniotic sac, guided by ultrasound images. A sample is withdrawn, and the needle is removed. The sample of fluid is sent to the lab for genetic analysis.

Expected Outcome and Follow-Up

Obtaining a sample adequate for testing. If the lab results indicate a serious genetic disorder in the fetus, the parents must decide whether or not to continue the pregnancy; if pregnancy is continued, they will have the opportunity to prepare themselves for the problems that may be encountered.

Complications

Miscarriage (especially if the procedure is performed before the eleventh week of pregnancy), fetal injury.

Anesthesia, General

Alternate names—n/a.

Purpose

To induce unconsciousness in a patient in order to perform a major procedure, usually surgery but also such procedures as electroconvulsive therapy.

Indications

As required for the procedure involved.

Anatomy

The respiratory, circulatory, and nervous systems are involved.

Pathology

n/a

Physiology

Various chemicals interact with the nervous system to affect consciousness.

Staffing

Anesthesiologist, possibly respiratory technologist.

Equipment and Supplies

Anesthetic machine, anesthetic gas agents, patient breathing circuits, endotracheal tube, physiological monitor, laryngoscope (Figure 27).

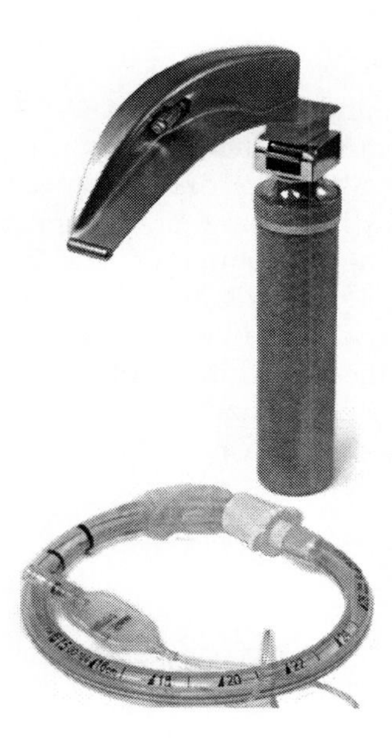

FIGURE 27
A laryngoscope and endotracheal tube. The scope blade keeps the patient's tongue from obscuring the pathway to the trachea. The tube is marked to indicate the depth of insertion.

Preparation

The surgical procedure, including anesthesia, is discussed with the patient, and an intravenous line is inserted. A sedative is administered via the IV line.

Procedure

Sedative dose is increased until the patient loses consciousness, a muscle relaxant is given, and an endotracheal tube is put in place.

The ventilator portion of the anesthetic machine begins to breathe for the patient, and the selected gas anesthetic agent is added to the inhaled air. Level of consciousness is monitored during the surgical procedure. After surgery is complete, the anesthetic agent delivery is stopped, and drugs may be given to counteract the sedatives and muscle relaxants used. The patient is monitored carefully until consciousness returns. During anesthesia, patients sometimes are aware of their surroundings, because the level of anesthetic was not sufficient; many hospitals now use a level-of-consciousness monitor to help avoid such unpleasant situations.

Expected Outcome and Follow-Up

Unconsciousness of a degree sufficient for performing surgery is attained and maintained during the procedure, while not allowing excessively deep unconsciousness that could have adverse effects. The patient is monitored for an hour or more after regaining consciousness to guard against any possible complications.

Complications

Generally complications during and after general anesthesia are more common or severe when the patient is in a fragile medical state initially, due to age (very young or very old), cardiac or respiratory problems, alcohol or drug abuse, or other disease conditions. Heart attack, stroke, severe respiratory distress, and brain damage can all occur, sometimes leading to death. Nausea and vomiting are common reactions to anesthesia, and patients must be watched closely for any signs of such distress because they may not be able to clear their own airways should they vomit while still drowsy. Headache, weakness, and drowsiness may persist for some time after the procedure. Malignant hyperthermia is a rare but severe complication of anesthesia, in which the patient's body temperature spikes and heart and breathing problems ensue. Brain damage can occur if temperature is not controlled quickly.

Anesthesia, Local

Alternate names—Local, spinal, epidural, regional.

Purpose

To reduce or eliminate the sensation of pain in a specific area of the body. Local anesthesia generally has fewer side effects and a shorter recovery time than general anesthesia.

Indications

Any condition or procedure in which significant pain might be experienced, such as minor surgery, dental work, or childbirth.

Anatomy

Nerve endings carry electrochemical signals from various body parts to the brain, where they are perceived as pain. Staff administering injected local anesthetics must be very familiar with the location of the nerves involved (Figure 28).

Pathology

n/a

Physiology

By interrupting the pain signals before they reach the brain, the sensation of pain can be reduced or eliminated, often for a long enough time that the situation producing the pain signals is removed. Different agents work on different parts of the nerve conduction pathway.

Staffing

Usually an MD, dentist, or nurse practitioner, though an imaging technologist may be needed in situations where needles have to be inserted into precise locations in the body.

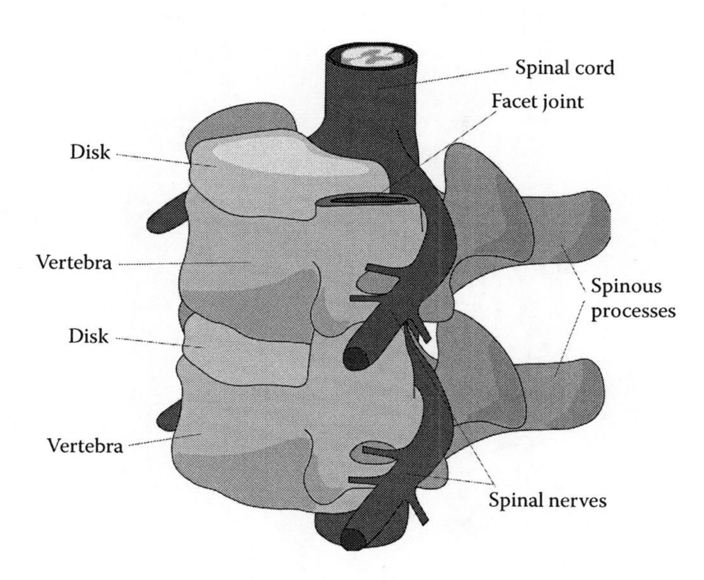

FIGURE 28
Vertebrae, spinal cord, and spinal nerves. The spinous processes are at the back of the spinal column, close to the skin.

In these cases an anesthesiologist may administer the drugs. Some local anesthetics are self-administered, such as the sprays or creams used for sunburns.

Equipment and Supplies

Anesthetic agent, delivery equipment such as needles and syringes or topical devices.

Preparation

The patient is informed of the reason for the procedure, and what sensations to expect, such as a "poke" when a needle is inserted. For injections, the skin in the area is cleaned. Sometimes a topical anesthetic is used to make needle insertion less painful.

Procedure

The agent is applied and left for some time to take full effect before the subsequent procedure is begun. Injected agents are delivered to the area near the nerves that supply the body area involved in order to block pain signals from that area. For dental work, this involves various nerves in the area of the mouth. For childbirth, the agent is injected between specific vertebrae and into the epidural space surrounding the spinal cord, thus the term "epidural" (Figure 29).

Spinal anesthetics are delivered past the epidural space, directly against the spinal cord.

Spinal anesthesia requires less agent and works more quickly than epidural, but requires more precision in administration and is more subject to side effects.

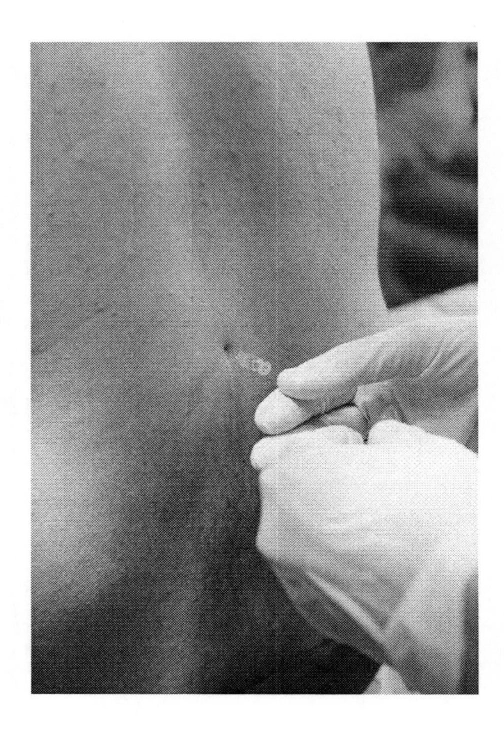

FIGURE 29
A needle inserted into the spinal canal for injection of epidural anesthetic. Spinal anesthetics are delivered past the epidural space, directly against the spinal cord.

Expected Outcome and Follow-Up

Adequate reduction or elimination of pain during the procedure is to be achieved. Patients, especially those undergoing spinal or epidural anesthesia must be monitored for some time afterward to ensure there are no complications. The numbness produced by the anesthetic may mean that the patient inadvertently causes damage because he cannot feel an accidental cut or burn.

Complications

Epidural anesthesia for childbirth may prolong labor, sometimes to the point where a caesarean section is required. Headaches may result from epidural or spinal injections, and paralysis is a rare occurrence.

Angiography

Alternate names—n/a.

Purpose

To study the condition of blood vessels in various parts of the body.

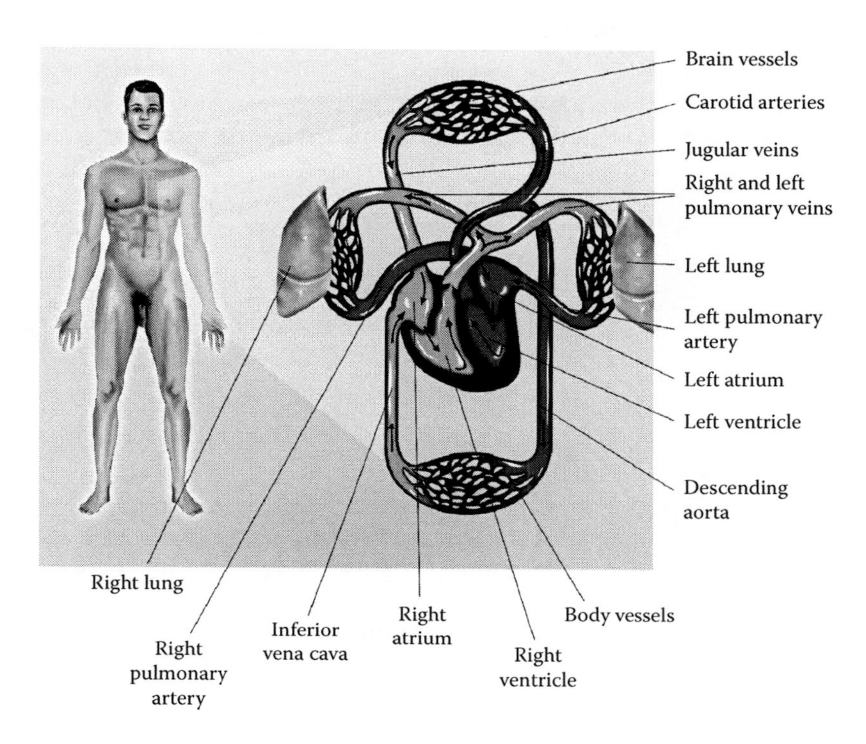

FIGURE 30
Schematic of the circulatory system.

Indications

Any situation where compromised blood flow may be a factor, or where detailed information about the blood supply of particular organs is required.

Anatomy

Blood vessels carry blood throughout the body (Figure 30). The blood supply to specific organs, such as the heart, brain, lungs, liver, or kidneys, is especially complex and critical.

Pathology

If blood vessels are damaged by injury or disease, blood flow can be reduced, potentially impairing function of the area or organ supplied.

Physiology

There is great variation in the physiology involved, depending on the organ or system being studied.

Staffing

Physician, nurse, imaging technologist, possibly a radiologist.

Equipment and Supplies

Angio injector, physiological monitor, X-ray/fluoroscopy machine, angiography catheter and associated insertion apparatus, and radiographic contrast medium, which is usually iodine based.

Preparation

The procedure is explained to the patient, and he is sedated but remains conscious, because he may have to respond to staff instructions or questions. The catheter insertion location is cleaned thoroughly.

Procedure

The catheter is inserted and guided into correct position with the aid of the X-ray unit. When in place, a small manually controlled injection of contrast medium may be given to confirm placement. The patient is instructed to remain very still, and large injection is made, usually via the power angio injector. A series of radiographic images are taken as the contrast medium progresses through the area of study; this must occur rapidly because the blood flow rate is usually high and the medium is carried away quickly. Various patient positions may be used, and multiple injections may take place. If the results are satisfactory, the catheter is withdrawn. Because the catheter was inserted into an artery, pressure on the insertion site must be firm and applied over a considerable time, 20 to 30 minutes, to ensure there is no bleeding. Patients are monitored carefully for several hours after the procedure.

Expected Outcome and Follow-Up

Adequate images of the area being studied are obtained and used by various physicians to determine diagnosis and treatment options.

Complications

Injection of the medium may cause a burning sensation, nausea, headache, irregular heartbeat, or chest pain. Asking the patient to cough at this time may alleviate some of the cardiac issues, which makes it important that the staff be aware of possible hearing impairments in the patient. Internal bleeding may occur, as well as heart attack or stroke. Cardiac arrhythmias and allergic reactions are also possible.

Angioplasty

Alternate names—Artery widening, artery reaming, balloon angioplasty, laser angioplasty.

Purpose

To restore adequate blood supply to cardiac muscles.

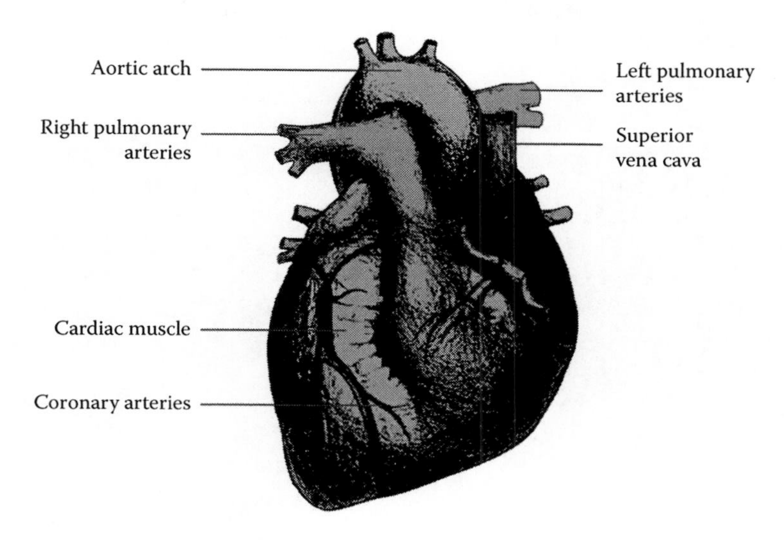

FIGURE 31
Heart and associated blood vessels.

Indications

Reduction in blood flow to the muscles of the heart because of blockage of one or more coronary arteries, as shown by myocardial infarction, ECG measurement, or more definitively, by diagnostic imaging such as angiography. Angiography is required to accurately locate the blockage.

Anatomy

The heart is a muscular organ consisting of left and right atria, left and right ventricles, four valves, and associated blood vessels (Figure 31). These blood vessels carry the blood pumped by the heart out to various body parts and also carry blood to the heart itself. The atria are thinner walled than the ventricles. Nerves and other pathways in the heart carry electrical signals to control pumping contractions.

The heart itself is supplied with blood by the coronary arteries, branches of which supply different areas.

Pathology

Arteriosclerosis (Figure 32) can cause narrowing of the coronary arteries to such an extent that muscle function, and therefore heart pumping effectiveness, is impaired (cardiac ischemia). If the blockage is severe enough, damage to or death of some of the cardiac muscle can result (myocardial infarction).

Physiology

The heart, being mostly muscle that is active all the time, requires an excellent blood supply to provide oxygen and nutrients and to remove wastes.

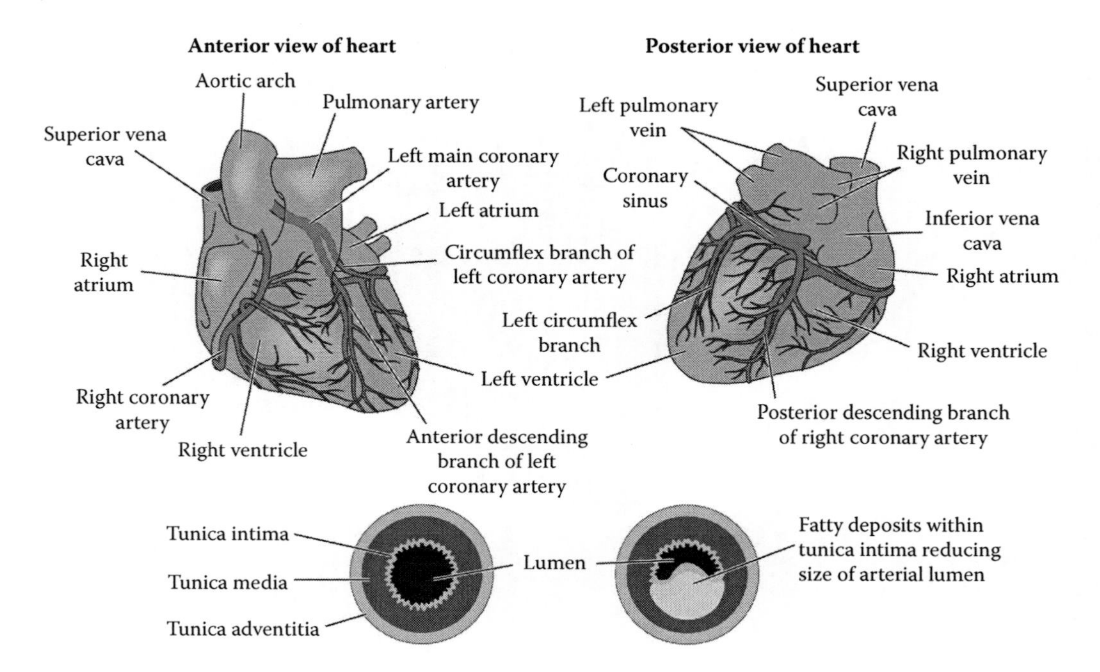

FIGURE 32
The blood vessel on the lower left is normal, while the one on the lower right shows the buildup of deposits within the artery. Such arteriosclerosis can occur in any artery, but when it is in the coronary arteries it has the potential of causing heart distress or heart attacks.

Signals from a center in the heart produce electrical signals that are carried throughout the heart to produce contractions. Various components provide delays to the signal so that contractions are coordinated and allow effective pumping.

Staffing

Normal surgical team; the surgeon is a cardiologist. Diagnostic imaging technologist.

Equipment and Supplies

Standard surgical plus angioplasty device and diagnostic imaging equipment. This device is a catheter, which has various components at its tip such as a balloon, a link for carrying a stent, a laser, or a mechanical grinder.

Preparation

Anticoagulant therapy is performed to help prevent clots during the procedure. The catheter insertion site is cleaned, and the patient is sedated. A local anesthetic is administered.

Procedure

An incision is made at the catheter insertion site, and the artery to be used is exposed. The femoral artery in the groin is commonly used. The catheter is inserted into the artery, and

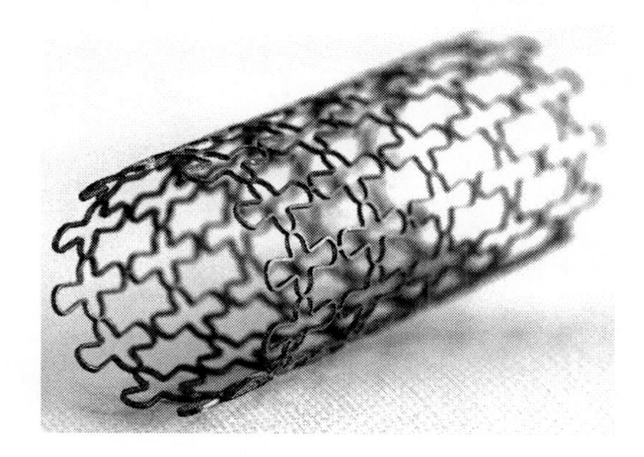

FIGURE 33
Stainless steel mesh stent.

with the aid of diagnostic imaging, it is guided to the aorta and into the target artery. The expansion or cleaning component is activated to increase the size of the arterial interior or lumen.

Expansion components may consist of a small balloon that can be inflated gradually in order to stretch the artery open. Inflations may be cycled, gradually increasing the peak pressure until the desired expansion is achieved. The catheter may also place a stent; this is a collapsed wire mesh tube that can be expanded and fixed in size to hold the artery open (Figure 33).

Cleaning components may be spinning blades or other cutting tools that break away the blockage, leaving the arterial lumen much more open. Alternatively, a laser tip may be used to burn away the blockage. In both cases, the device must be designed to produce very small particles that pose no danger of embolism.

Once the expansion or enlargement process is complete, the catheter is partially withdrawn, and the effectiveness of the procedure is evaluated with diagnostic imaging. The process is repeated, if necessary, and may be performed on another artery as required.

When all the work is complete, the catheter is withdrawn completely, and the artery insertion site checked for bleeding. Pressure is applied as necessary to stop bleeding, and when that is controlled, the incision is closed.

Expected Outcome and Follow-Up

Resumption of adequate blood supply to cardiac muscles and alleviation of previous symptoms. The patient is closely monitored. Normal activities are slowly resumed, with careful monitoring of each stage.

Complications

Puncture of the coronary artery by the catheter and associated components; heart attack or stroke caused by spasm of the coronary artery due to irritation from the procedure or from an embolism released from the site.

Aortic Aneurysm Graft

Alternate names—n/a.

Purpose

To repair an aortic aneurysm.

Indications

Many patients with aortic aneurysms are asymptomatic, especially if the aneurysm is in the chest. Those in the abdominal area may cause transient pain, and large aneurysms in any area can put pressure on adjacent organs, producing symptoms specific to that organ. Aneurysms are often discovered in the course of other medical investigations, using X-ray, CT, MRI, or ultrasound results.

Anatomy

The aorta is the major blood vessel carrying oxygenated blood from the heart to the rest of the body (Figure 34). It divides in two parts, called the ascending aorta and the descending aorta. Being close to the heart, the aorta is subject to the highest blood pressures in the circulatory system.

Pathology

If a weakness develops in the wall of the aorta a bulge, or aneurysm, can develop. Emboli can develop in areas of the aneurysm, and these can break loose and move to other parts of the

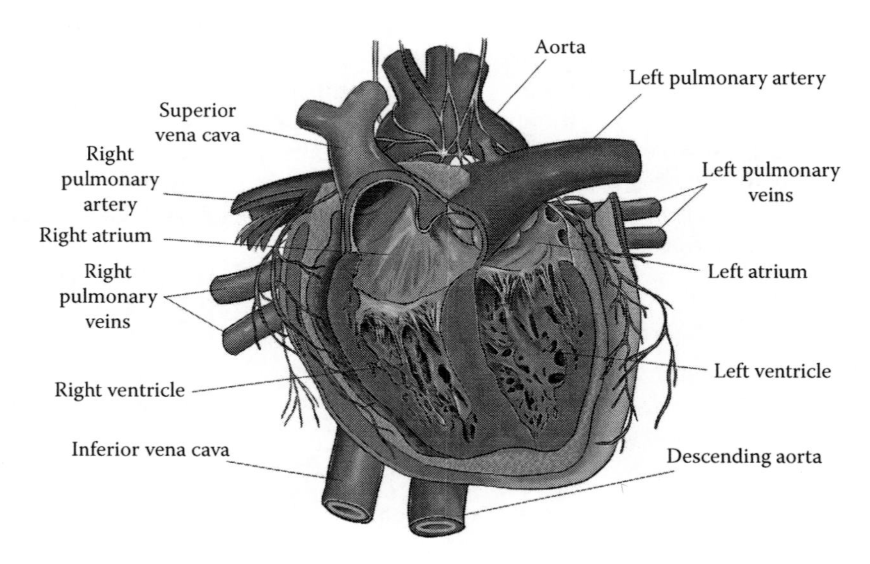

FIGURE 34
Heart and associated blood vessels, aorta at top.

body where they can block blood flow, causing severe problems. If the wall becomes weak enough, the aneurysm can rupture, resulting in critical internal bleeding, shock, or death.

Physiology

n/a

Staffing

Major surgical team, perfusionist.

Equipment and Supplies

Standard surgical supplies and equipment, heart–lung machine, graft material.

Preparation

Standard surgical preparation.

Procedure

An incision is made over the site of the aneurysm, and the aneurysm is exposed. Blood is diverted from the patient through a heart–lung machine. The aorta is clamped off above and below the aneurysm, and the weakened area is removed. A tube of artificial material is grafted in to replace the section that was removed, the heart–lung machine is disconnected, and the site is closed.

Expected Outcome and Follow-Up

Alleviation of symptoms and prevention of the major consequences of an untreated aneurysm. Full recovery may take 6 to 12 weeks.

Complications

The graft material may be rejected by the patient's system, though this is rare.

Appendectomy

Alternate names—n/a.

Purpose

To remove the diseased appendix from the patient.

Indications

Appendicitis is diagnosed from symptoms such as abdominal pain and tenderness on the right side, nausea, fever, and an elevated white blood cell count. X-rays, ultrasound, CT, or MRI scans may help confirm the diagnosis.

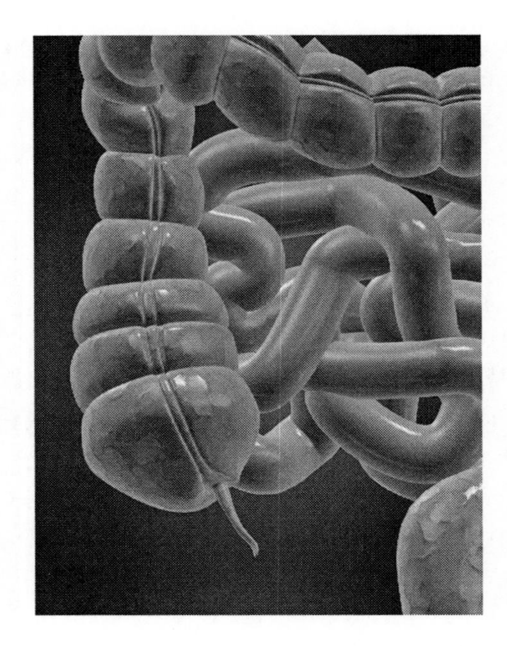

FIGURE 35
Intestinal organs, with appendix at lower left.

Anatomy

The appendix is a small sac or blind tube that projects off the intestine near the point where the small and large intestines join (Figure 35).

Pathology

Appendicitis is caused by a blockage of the appendix, leading to inflammation and infection of both the appendix itself and possibly of some surrounding tissue (Figure 36). If the infection is severe enough, it can weaken the wall of the appendix to the point where it ruptures. This is an emergency situation because the rupture can lead to potentially fatal peritonitis.

Physiology

The appendix is a remnant from human evolution, originally used to help digest some foods. It is now a minor part of the immune system.

Staffing

Normal surgical team.

Equipment and Supplies

Standard surgical.

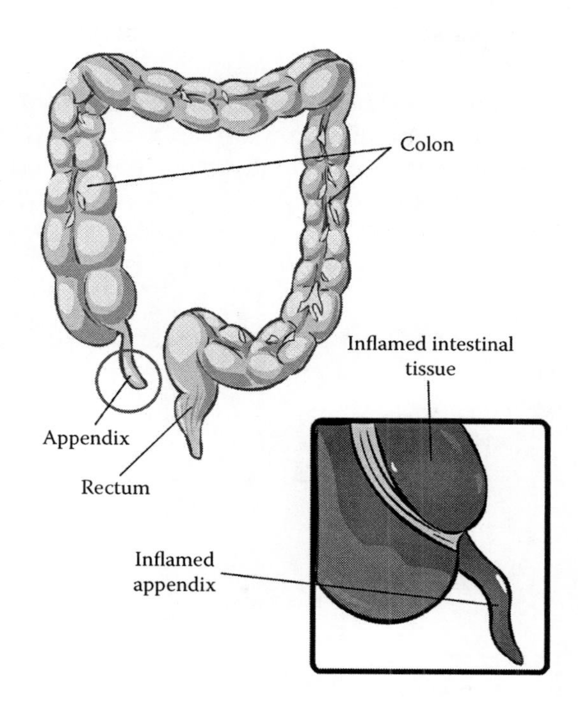

FIGURE 36
Appendicitis.

Preparation

Standard surgical.

Procedure

Appendix removal can be done either through open surgery or laparoscopically. Open surgery involves greater trauma with larger incisions, but laparoscopic surgery usually takes somewhat longer, resulting in greater exposure to anesthetics.

Open surgery: A relatively short incision is made in the lower right abdominal wall, and the appendix is located in relation to other nearby organs. The appendix is gently lifted up through the incision and is removed. The resulting opening is closed, and the intestine is returned to the abdomen, and the incision is sutured layer by layer.

Laparoscopic surgery: Several very small incisions are made at different points on the abdomen. These incisions are used to insert tubes for illumination and video viewing, for inflation (insufflation) of the abdomen to allow room to view and work, for suction of blood and other tissues and fluids, and for access with surgical tools. Once all the equipment is in place, the surgeon locates the appendix via the video image and removes it, being careful to keep it sealed so that its contents are not released into the abdomen.

The resulting opening in the intestine is closed, the appendix is withdrawn, the insufflating gas is released, and the small incisions are closed.

Expected Outcome and Follow-Up

Removal of the diseased appendix should result in alleviation of the appendicitis symptoms, and the patient returns to normal. Follow-up care is standard, checking for internal and surface infections. Because this is intestinal surgery, dietary restrictions are imposed for several days. If the appendix ruptured before removal, drain tubes will be placed and aggressive antibiotic therapy applied.

Complications

Adhesions of the surgical area to other abdominal organs or surfaces may occur. If the appendix ruptured, peritonitis will be seen. This is very serious and requires intensive monitoring and treatment.

Arthroscopy of Knee

Alternate names—ACL, PCL, MCL or LCL repair, knee repair.

Purpose

To repair damage to the knee joint resulting from injury or disease, or to alleviate pain from such damage.

Indications

Excessive and prolonged pain in the knee, with damage confirmed by X-ray, CT, or MRI scans, and/or by previous arthroscopic examination.

Anatomy

The knee joint consists of the lower end of the femur, the upper ends of the tibia and fibula, a cartilage cushion between the bones, supporting cartilage and ligament structures, the leg muscles that attach to the area, and associated blood vessels and nerves (Figure 37). There are four main ligaments that help stabilize the knee: the anterior cruciate ligament (ACL), the posterior cruciate ligament (PCL), the medial collateral ligament (MCL) and the lateral collateral ligament (LCL). The patella, or kneecap, helps protect the knee joint and also provides extra leverage to the muscles used in extending the knee. Several fluid-filled sacs, or bursae, surround the joint and provide cushioning.

Pathology

Sporting and other accidents can damage the cartilage and ligaments of the knee through physical tearing or crushing. Arthritis can result in the thinning or complete loss of the cartilage between the various bone surfaces, as well as roughening of the bone surfaces. Ligament and other damage can cause the patella to become misplaced or misaligned.

Physiology

n/a

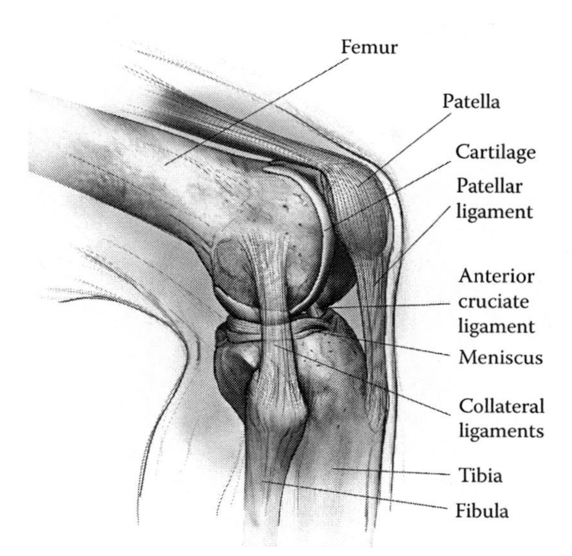

FIGURE 37
Knee joint anatomy.

Staffing

Normal surgical team.

Equipment and Supplies

Standard surgical.

Preparation

Standard surgical.

Procedure

This surgery may be done under local or general anesthesia.

Several small incisions are made into the knee joint, and arthroscopic instruments are inserted. The joint area is insufflated to allow for viewing and working (Figure 38).

The remaining process is determined by the condition of the knee joint.

Cartilage damage due to arthritis can be alleviated by removal of some of the cartilage; detached portions are washed and suctioned out, while damaged portions may be cut away and then washed and suctioned. Some researchers are working on methods of transplanting cartilage into the joint, which involves removing some of the old tissue and replacing it with cartilage either taken from another part of the body or grown from the patient's cells in a lab culture. Rough bone surfaces can be smoothed out, improving mechanical function and reducing pain.

Cartilage damage due to injury is treated in much the same manner as that due to arthritis, though the structural changes are different.

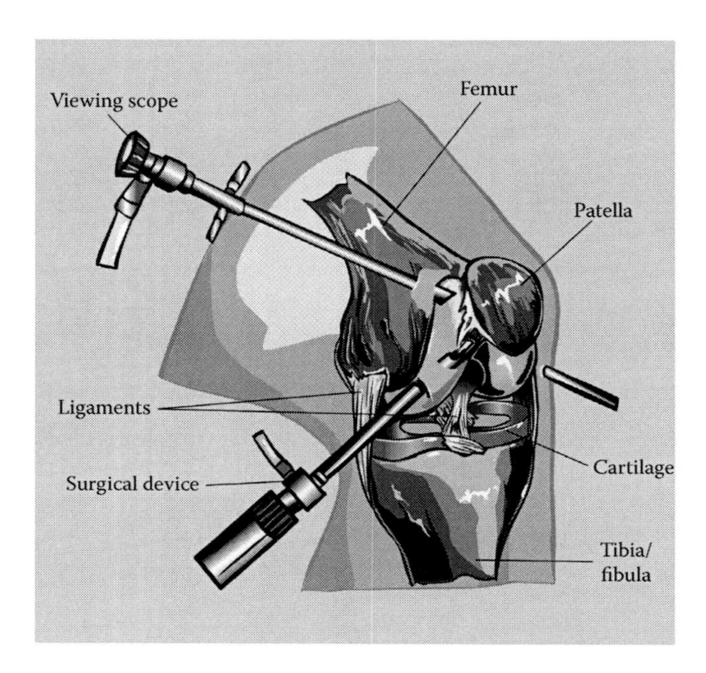

FIGURE 38
Arthroscopic procedure for knee repairs.

Ligament damage can be repaired by realigning ligaments, rejoining and strengthening torn ligaments, or grafting new sections of ligament into the damaged area.

Patellar misalignment can be corrected by adjusting and repairing the ligaments that normally hold it in the correct position.

Expected Outcome and Follow-Up

Alleviation of pain and improvement in mobility and function are the goals of this surgery. Physiotherapy treatments and specific exercises following surgery can greatly increase the chances of success, though the joint may need to be immobilized for some time immediately following surgery to allow tissues to heal.

Complications

In some cases, the pain may not be alleviated or may in fact worsen.

Bariatric Surgery

Alternate names—Gastric bypass, stomach banding, stomach stapling.

Purpose

To reduce the physical capacity of the digestive system in order to induce weight loss.

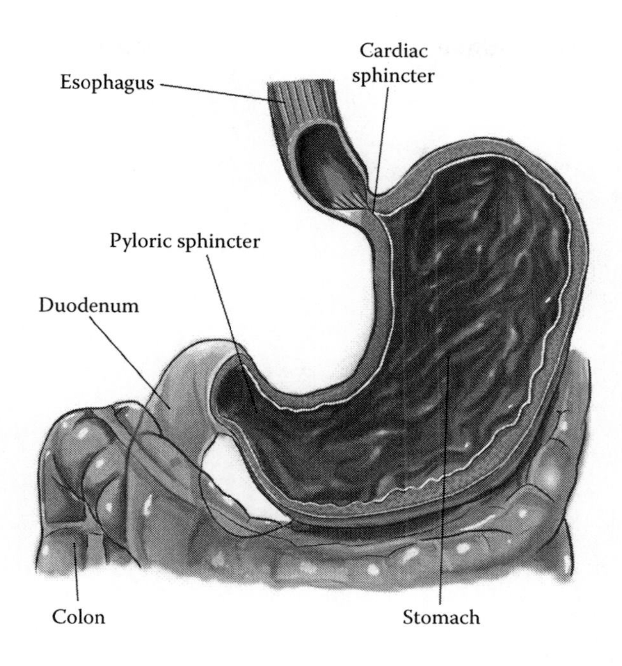

FIGURE 39
Stomach anatomy.

Indications

Bariatric surgery is usually only considered for severely or morbidly obese patients for whom other weight loss tactics have not worked.

Anatomy

The stomach is a pouch where the main digestion of food begins (Figure 39). Food enters the stomach from the mouth and esophagus via the esophageal sphincter, a ring of muscle that helps prevent food from being pushed back up the esophagus. Another sphincter, the pyloric, keeps food in the stomach until it is ready to be passed into the small intestine. The stomach walls are muscular to help move food around inside and promote digestion, but there is little absorption of nutrients in the stomach. From the stomach, food passes into the small intestine, which consists of three main sections: the duodenum, the jejunum (the longest section), and the ileum.

Pathology

Obesity can contribute to a wide variety of health problems including diabetes, heart disease, cancer, and joint and mobility problems.

Physiology

Glands in the lining of the stomach produce digestive enzymes and hydrochloric acid to start digestion of food. They also produce mucous material to prevent the enzymes and acid from digesting the stomach itself. Further digestive enzymes are added in the small intestine, and digestion continues. Nutrients from digested food are absorbed into the bloodstream from the small intestine.

Staffing

Normal surgical team.

Equipment and Supplies

Standard surgical.

Preparation

Standard surgical. A nasogastric tube is inserted to remove gastric fluids from the surgical area during and for a day or so after surgery.

Procedure

There are a number of different types of bariatric surgery. They are generally divided into two groups, malabsorptive procedures and restrictive procedures, though some techniques combine the two. Malabsorptive procedures aim to reduce the absorption of nutrients by the digestive system, while restrictive procedures reduce the capacity of the system, thereby reducing food intake.

Most of the procedures can be performed either with open surgery or laparoscopically.

Malabsorptive procedures involve resecting the stomach somewhat to reduce its capacity, and then bypassing the duodenum and jejunum so that food passes from the stomach right into the ileum. This allows for a relatively normal diet, but since most of the small intestine is bypassed, there is a greatly reduced absorption of nutrients.

Restrictive procedures involve reducing the capacity of the stomach to a large degree. A section of the stomach may be removed to produce a smaller pouch, or part of the stomach may simply be closed off to form one section with no openings and another with normal openings but greatly reduced capacity. The section may be closed off by suturing or stapling. Another restrictive procedure involves the use of a band placed around part of the stomach to create a small "pre-stomach," which restricts food intake capacity (Figure 40).

Food then passes from this pre-stomach to the main stomach where it is digested normally. The band may be adjustable to allow for individual variations and postoperative changes.

Expected Outcome and Follow-Up

Reduced food intake and/or reduced nutrient absorption will result in weight loss. The patient will not be able to eat for a day or two following surgery, and easily digested foods will be introduced as tolerated. Dietary modifications will be necessary to accommodate the changes in the digestive system. Patients must be monitored to ensure adequate nutrition and to prevent excessive weight loss.

Complications

Normal surgical complications may ensue. In some cases, the new openings in the digestive system may become reduced in size and cause problems with the movement of food,

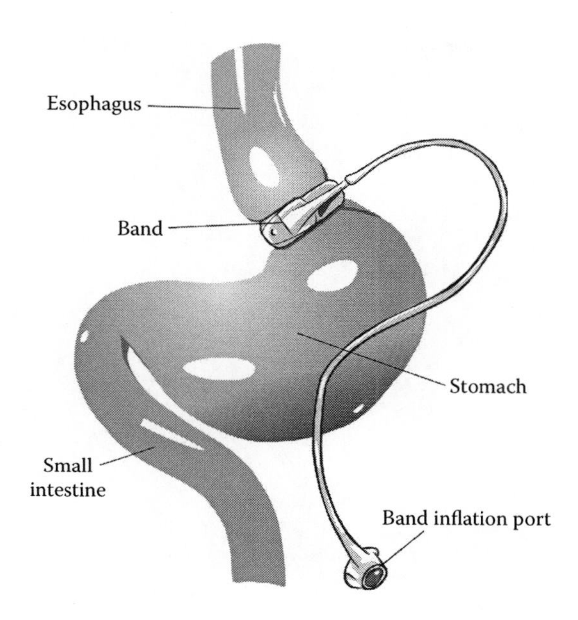

FIGURE 40
Stomach banding.

possibly resulting in vomiting. Gallstones are more likely to form following bariatric surgery.

Biopsy, Bone Marrow

Alternate names—n/a.

Purpose

To obtain a diagnostic sample of bone marrow.

Indications

Abnormal blood test results or bone lesions visible on X-ray images, chemotherapy monitoring, bone marrow transplant monitoring.

Anatomy

Bone marrow is present in cavities inside larger bones. The most accessible areas for this procedure are the upper, outer portion of the hip bone, the sternum, certain vertebrae, and the tibia (Figure 41).

Pathology

Various blood disorders are a result of abnormalities in the production of blood cells, either due to disease of the marrow or increased demand for certain cell types because of

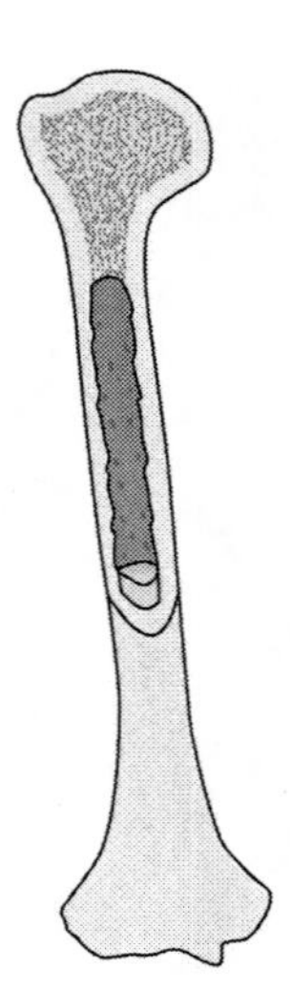

FIGURE 41
Cutaway of femur showing marrow inside.

disease processes outside the marrow. Some anemias and cancers such as leukemia and non-Hodgkin's lymphoma are a result of abnormal blood cell production in the marrow.

Physiology

Bone marrow is well vascularized. It produces most of the red blood cells and platelets in the body, as well as many of the white blood cells.

Staffing

Physician (usually a pathologist or oncologist) and perhaps an assistant.

Equipment and Supplies

Minor surgery supplies, special biopsy needle assembly and syringe.

Preparation

The procedure is discussed with the patient, and the importance of remaining still during needle insertion is stressed. The site is cleaned and antiseptic applied; then a local anesthetic is given, sometimes in two stages, surface and deeper.

Procedure

A small incision is made over the target area, and the needle is inserted until it contacts bone. Pressure is applied until the needle is pushed through the bone and into the marrow cavity, at which point the pressure required decreases greatly (Figure 42).

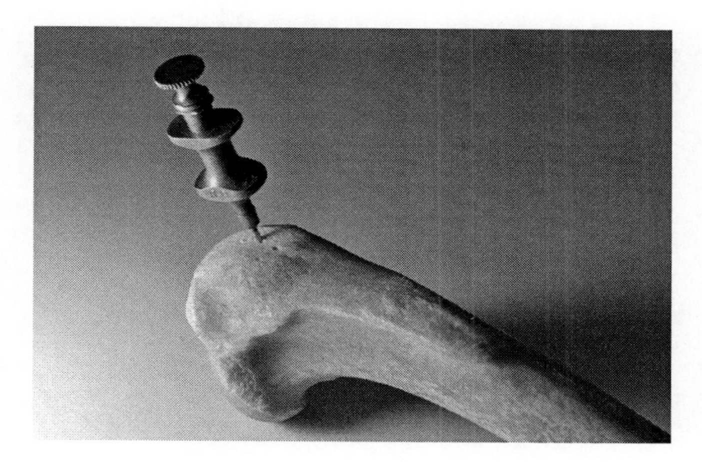

FIGURE 42
Representation of bone puncture for marrow aspiration.

Samples are aspirated and stored for lab analysis, and then a special cutting blade is inserted through the needle and into the marrow. A rotating action captures a small, intact portion of marrow inside the needle, which is then withdrawn. The sample is extruded and captured for analysis. The incision is closed and pressure applied for an hour to ensure the bleeding has stopped.

Expected Outcome and Follow-Up

Obtaining a suitable sample that can be analyzed in the lab will provide important diagnostic information. The patient is monitored for an hour or so to be sure that there is no continued bleeding from the site and that vital signs are normal.

Complications

n/a

Biopsy, Breast

Alternate names—n/a.

Purpose

To obtain a sample of breast tissue for lab examination.

Indications

Abnormalities in breast tissue detected by palpation, mammography, ultrasound, or other imaging techniques. These may include lumps or other masses, or changes in tissue when compared to earlier, baseline images.

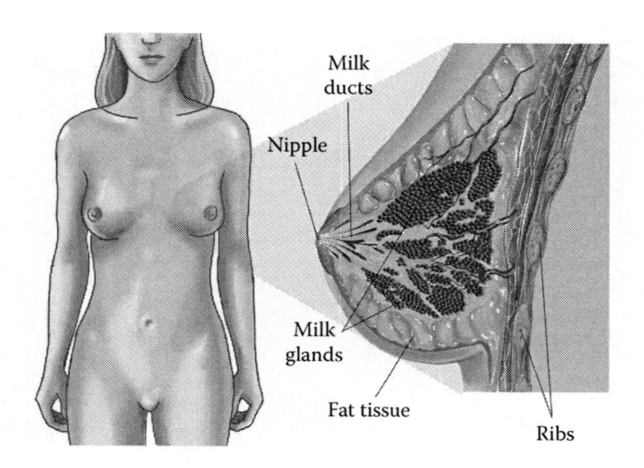

FIGURE 43
Breast anatomy.

Anatomy

The breast consists mainly of fat cells and milk-producing glands (Figure 43). Breast tissue can have a wide range of textures.

Pathology

Breast cancer and other diseases of the breast can produce abnormal masses within the breast.

Physiology

Breasts produce milk following pregnancy; when not lactating, they are essentially dormant.

Staffing

Basic surgical, possibly an ultrasound or other imaging technologist, for needle biopsies. For surgical biopsies, minor surgical team.

Equipment and Supplies

Biopsy needles, ultrasound or other imaging device for needle biopsies, standard surgical equipment and supplies for surgical biopsies. A device for physically stabilizing the breast is used.

Preparation

The procedure is discussed with the patient. Antiseptic is applied to the site. Local anesthetic may or may not be administered for needle biopsies, depending on the size of needle used. Surgical biopsies are usually performed with local anesthesia, though general may also be used. A sedative may be given for either type.

Procedure

Two types of biopsies may be performed, needle or surgical.

In a needle biopsy, a needle is inserted into the target tissue, guided either by direct feel or by diagnostic imaging if the site is small and/or less accessible. Depending on the requirements of the test, either a small or large needle is used. Small, fine needles produce very little pain sensation and probably do not need even local anesthesia, but larger needles can be painful, and local is used. Once the needle tip is in the correct location, a sample is aspirated, and the needle is removed. Pressure is applied to the insertion site to stop bleeding, and the sample is sent to the lab for analysis.

For a surgical biopsy, an incision is made over the target site and is extended until the target is reached. All or a portion of the suspect mass is then removed and sent to the lab. The incision is closed.

Expected Outcome and Follow-Up

An adequate sample will provide diagnosis of the disease process involved, or may eliminate the possibility of disease. Patients undergoing needle biopsy will likely have few if any aftereffects; those having surgical biopsies can expect normal effects following minor surgery.

Complications

There is a small risk of hematoma.

Biopsy, Liver

Alternate names—Percutaneous liver biopsy.

Purpose

To obtain a sample of liver tissue for lab examination.

Indications

A wide range of liver disorders can be tested for using biopsy results. Some of these disorders include jaundice, cirrhosis, hepatitis, various cancers of the liver, and unexplained enlargement of the liver. Liver biopsies may also be used to help determine if a transplanted liver is undergoing rejection.

Anatomy

The liver is a large abdominal organ, most of which is on the right side of the body just under the diaphragm. It is very well vascularized, and is connected to the small intestine via the common bile duct, which also connects to the gall bladder (Figure 44).

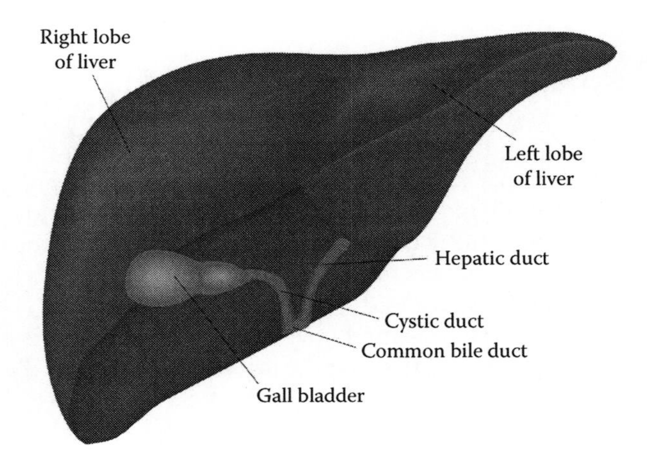

FIGURE 44
A color version of this figure follows page **176.** Liver anatomy.

Pathology

Disease and chemical or physical trauma can all affect liver function and eventually alter its structure. Alcohol abuse, poisoning by various chemicals and drugs, and metabolic disorders such as hemochromatosis can all affect liver structure and function. Diseases such as jaundice (a buildup of bilirubin in the body), cirrhosis (the development of scar tissue and nodules in the liver), hepatitis, various cancers of the liver, and tuberculosis may also compromise the liver.

Physiology

The liver performs a variety of functions, including the production of bile, which is used in the digestion process, the breakdown and excretion of toxic substances from the blood, and production of certain amino acids. The liver can store excess glucose from digestion in the form of glycogen, and it also produces cholesterol, which is an important metabolic compound.

Staffing

Basic surgical; the physician may be a radiologist.

Equipment and Supplies

Basic surgical plus special needle and syringe.

Preparation

Basic surgical. A sedative may be given.

Procedure

A local anesthetic is administered, and the biopsy needle is inserted. Because the liver is a large organ and the sample to be obtained can usually come from almost any area of the liver, needle placement is relatively straightforward. If a specific target area is identified, the needle placement may be guided by diagnostic imaging. Once the target is reached, a sample is aspirated and the needle removed. A dressing is applied to the insertion site. The sample is sent to the lab.

Expected Outcome and Follow-Up

Definitive diagnosis of specific diseases or clear indications of liver function. The patient rests for a few hours following the procedure, and normal activities are resumed after a few days.

Complications

n/a

Biopsy, Lung

Alternate names—n/a.

Purpose

To obtain a sample of lung tissue for lab examination.

Indications

Suspected or known abnormalities of the lungs, including cancers, pneumonia, or tuberculosis, as shown by diagnostic images or lab results.

Anatomy

The lungs consist of several lobes, three on the right and two on the left (Figure 45). These are made up of small sacs called alveoli, which are connected together by small tubes called bronchioles (Figure 46). The bronchioles collect together to larger bronchial tubes, which then join to form the trachea, the tube that connects to the mouth. A dome-shaped muscle (the diaphragm) under the thoracic cavity causes air to be drawn into the lungs when it contracts. Small blood vessels surround the alveolar sacs.

Pathology

Lung tissue can be damaged by diseases such as cancer (Figure 47), emphysema, tuberculosis, and cystic fibrosis. Various inhaled substances can also cause damage, including smoke, dust particles, and toxic gases. Damaged tissue results in less than optimal lung function.

Physiology

On inspiration, air enters the body through the mouth or nose and passes down into the lungs via the trachea, bronchial tubes, and bronchioles, ending in the alveoli. There carbon dioxide from the blood is passed into the alveoli and oxygen from the air moves into the bloodstream. The carbon dioxide is then expelled from the lungs during exhalation.

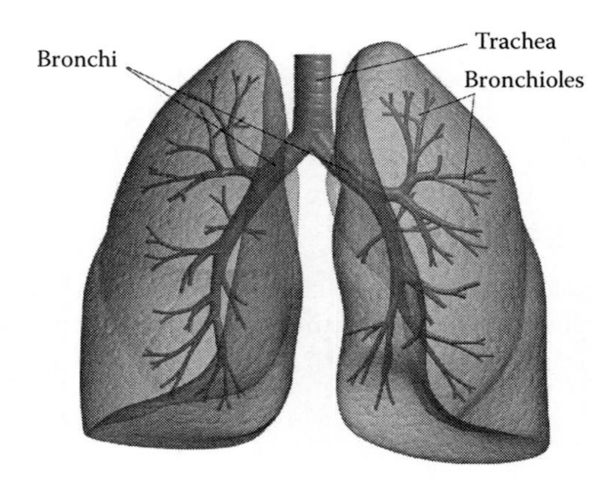

FIGURE 45
Lung anatomy.

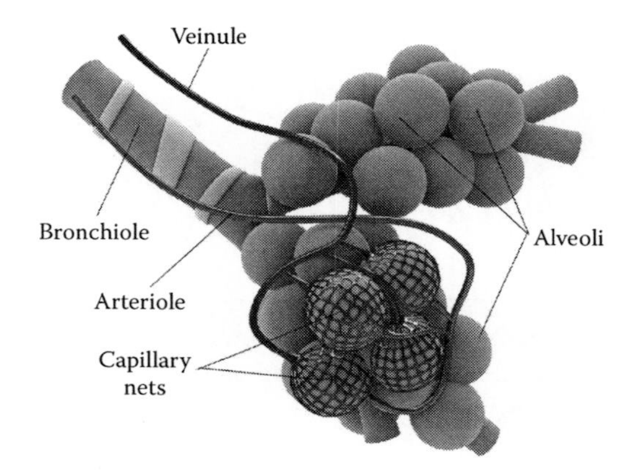

FIGURE 46
Bronchioles and alveoli.

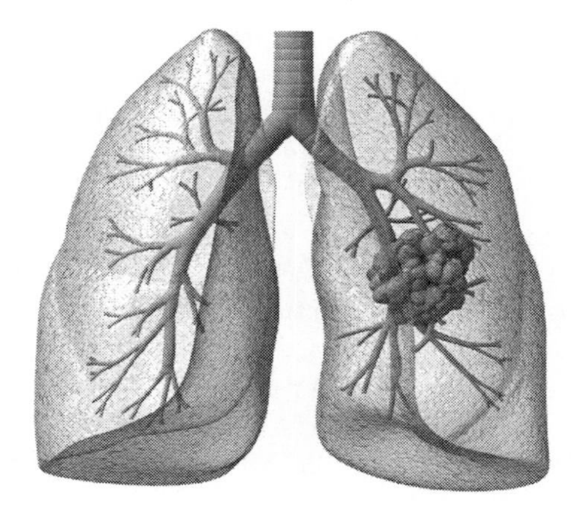

FIGURE 47
Diagram of lungs showing a lung cancer tumor.

Staffing

Physician and assistant, or minor surgical team if a surgical biopsy is to be performed.

Equipment and Supplies

Depending on the type of biopsy to be performed: bronchoscopy equipment, tissue sampling device, sampling needle and diagnostic imaging device, or minor surgical equipment and supplies.

Preparation

The details of the procedure are discussed with the patient. Because some patients may find the bronchoscopy disturbing due to the gagging sensation when the scope is inserted, an amnesiac sedative may be given. Needle biopsies use local anesthesia, while surgical biopsies are usually done with general anesthesia.

Procedure

Three different types of lung biopsy may be performed: bronchoscopic, needle, or surgical.

Bronchoscopic biopsy involves the insertion of a thin, flexible bronchoscope tube down the patient's trachea and into the bronchial tubes as far as necessary and possible. The scope has illumination and video capabilities, and the physician operating the equipment guides the scope according to the anatomical structures viewed on a video monitor. Samples may be drawn through the scope passages via simple suction, or small cutting instruments may be directed through the scope to help obtain samples. These are then aspirated with suction. Still photo or video images can be recorded during the procedure for future reference. Once the desired samples have been obtained, the scope is withdrawn.

Needle biopsy is performed if the lesion in question is close to the thoracic wall. A needle is inserted, guided by diagnostic imaging, and samples aspirated. A cutting device may be used.

Larger or less accessible lesions may have to be accessed via surgery. An incision is made over the target and a sample removed. The incision is then closed.

Expected Outcome and Follow-Up

Definitive diagnosis of specific lung diseases or conditions. The patient rests for a few hours following a surgical procedure, and normal activities are resumed after a few days. Recovery from bronchoscopic or needle biopsies is much shorter.

Complications

Surgical biopsies carry the greatest risk of complications, the most serious of which is pneumothorax, in which the lung collapses due to a leak from the lung into the thoracic cavity, possibly caused by puncture from the surgical biopsy procedure. Pneumothorax may also occur less frequently from puncture by a biopsy needle. Pneumothorax requires immediate medical intervention.

Biopsy, Prostate

Alternate names—n/a.

Purpose

To obtain a sample of prostate tissue for lab examination.

Indications

Suspected prostate cancer as indicated by elevated blood prostate-specific antigen (PSA) levels, direct physical examination, or diagnostic imaging results.

Anatomy

The prostate is a gland that surrounds the urethra just below the bladder in males (Figure 48).

Pathology

Normally the prostate is small, but disease processes can cause it to become enlarged, which can cause problems with urination and sexual function. Prostate cancer is potentially fatal and must be detected and treated early for the best chance of survival (Figure 49).

Physiology

The prostate contributes components of seminal fluid.

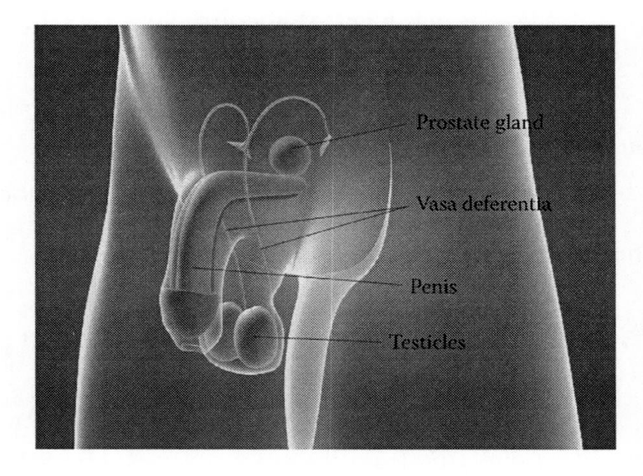

FIGURE 48
A color version of this figure follows page **176.** Male anatomy schematic showing prostate gland.

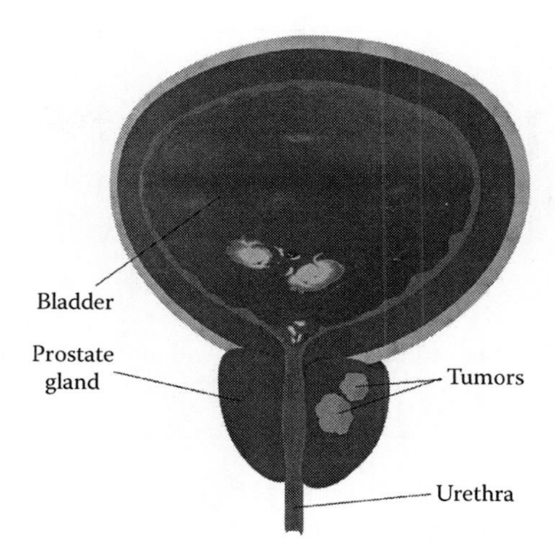

FIGURE 49
A color version of this figure follows page **176.** Prostate tumors.

Staffing

Basic surgical team.

Equipment and Supplies

Basic surgical, diagnostic imaging device, biopsy needle or gun.

Preparation

Basic surgical. A sedative may be given.

Procedure

There are several different methods of taking a prostate biopsy, depending on the method of access. The most common method uses a needle inserted either through the urethra, through the rectum, or through the perineum. Local anesthesia is used for needle biopsies. For perineal sampling, a small incision is made to allow access, while transrectal sampling uses only a biopsy needle. Transurethral sampling may use a thin scope called a cystoscope that allows direct viewing of the prostate during the procedure.

Expected Outcome and Follow-Up

Positive or negative diagnosis of prostate cancer. The patient rests for a few hours following the procedure, and normal activities are resumed after a few days.

Complications

n/a

Bladder Suspension

Alternate names—Retropubic suspension, Burch procedure, also known as retropubic urethropexy procedure or Burch colosuspension and Marshall-Marchetti-Krantz procedure (MMK).

Purpose

To alleviate stress incontinence due to lack of muscle support of the urethra.

Indications

Stress incontinence, with the underlying cause confirmed by diagnostic imaging and possibly measurement of bladder pressures that induce incontinence.

Anatomy

The bladder collects urine from the kidneys and holds it until it can be released via urination. Sphincter muscles are at the exit of the bladder, and the urethra leads from the bladder to the outside (Figure 50). The bladder and urethra are held in place by a set of muscles and ligaments.

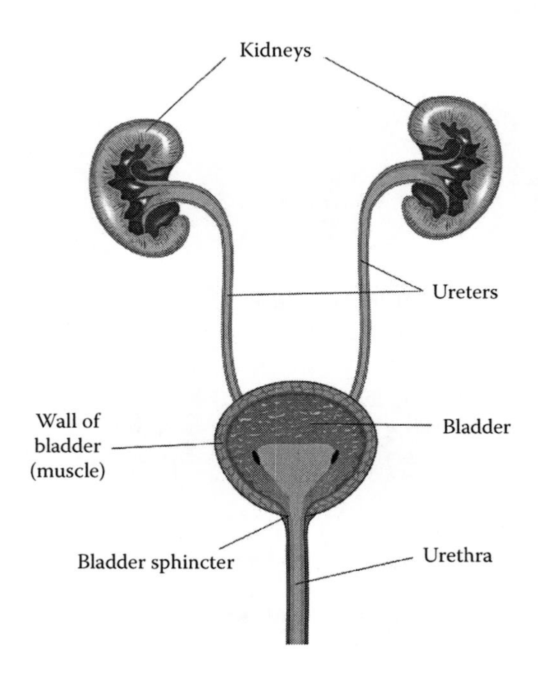

FIGURE 50
Urinary system.

Pathology

If the bladder sphincter muscles are weakened, or if the muscles and ligaments supporting the bladder and urethra are weakened, involuntary urine flow (incontinence) may occur, especially at times of stress or pressure such as lifting, sneezing, coughing, exercising, or laughing.

Physiology

The bladder sphincter holds urine in the bladder until urination is begun voluntarily.

Staffing

Normal surgical.

Equipment and Supplies

Standard surgical.

Preparation

Standard surgical.

Procedure

This procedure may be done with open surgery, in which case the patient will be given general anesthesia, or laparoscopically, in which case anesthesia may be general or (more commonly) local. In either case, the actual suspension procedure is the same.

Patients who are undergoing other abdominal surgery, such as a hysterectomy or urethral reconstruction, may have bladder suspension done at the same time to avoid repeated surgeries.

Open surgery involves an abdominal incision, usually at the bikini line, or a transvaginal incision. Laparoscopic surgery uses small incisions in the area of the navel and the pubic hairline.

In the Burch procedure, the urethra and bladder are attached to pelvic muscles, while in the MKK procedure, they are attached to the pelvic cartilage.

Expected Outcome and Follow-Up

Reduced or eliminated stress incontinence. A catheter may be inserted into the bladder, exiting through the abdominal wall, in order to drain the bladder for the first few days following surgery. Strenuous activity is to be avoided for a few months.

Complications

If the suspension is too tight, urination may be difficult; this may require a corrective surgery. Vaginal prolapse may occur.

Blood Pressure Measurement

Alternate names—BP measurement, NIBP/NBP (noninvasive blood pressure).

Purpose

To determine the systolic and diastolic values of blood pressure, usually in the arterial system but sometimes in other parts of the circulatory system.

Indications

Standard vital sign measurement.

Anatomy

Arteries and veins carry blood throughout the body.

Pathology

Excessively high blood pressure (hypertension) can lead to strokes or heart attacks; excessively low pressure (hypotension) can lead to fainting, coma, or death. Variations in blood pressure beyond normal limits can be caused by various disease processes, trauma, or drug or chemical reactions. High blood pressure is often associated with hardening of the arteries (arteriosclerosis), while low blood pressure may be related to congestive heart failure, myocardial infarction, blood loss, or dehydration.

Physiology

The heart beats in a two-stage cycle, with the atria contracting first, followed by the more powerful ventricles. This cycle produces a typical pressure waveform in the blood as it passes through the circulatory system (Figure 51). The pressure is highest at contraction (systolic pressure) and lowest at relaxation of the heart (diastolic pressure). See Appendix C for normal blood pressure values. These measurements are highest in arteries closest to the heart and decrease further away from the heart and as arteries get smaller. After blood passes through capillaries and into the venous system, pulsatile variations are mostly removed, and the remaining pressure is sufficient to return the blood to the heart. Pressure may be measured indirectly by determining the external pressure required to occlude arterial blood flow, or internally by use of a catheter inserted into a target blood vessel and connected to a pressure transducer and monitor.

If arterial blood flow is completely occluded in NIBP measurement (that is, cuff pressure is above systolic pressure), the arterial wall does not move in time with the pulse. Between systolic and diastolic points, the artery opens and closes with each pulse, making a typical physical waveform in the cuff area and producing a sound (Korotkoff sound) that can be discerned with a stethoscope. When cuff pressure drops below diastolic, blood flow is smooth and the pressure waveforms and sounds essentially disappear.

Staffing

Physician or nurse. Special training is required to insert catheters for invasive blood pressure measurement.

Equipment and Supplies

Noninvasive blood pressure apparatus, or catheter and associated pressure transducers and monitoring equipment for invasive pressure monitoring, possibly diagnostic imaging equipment.

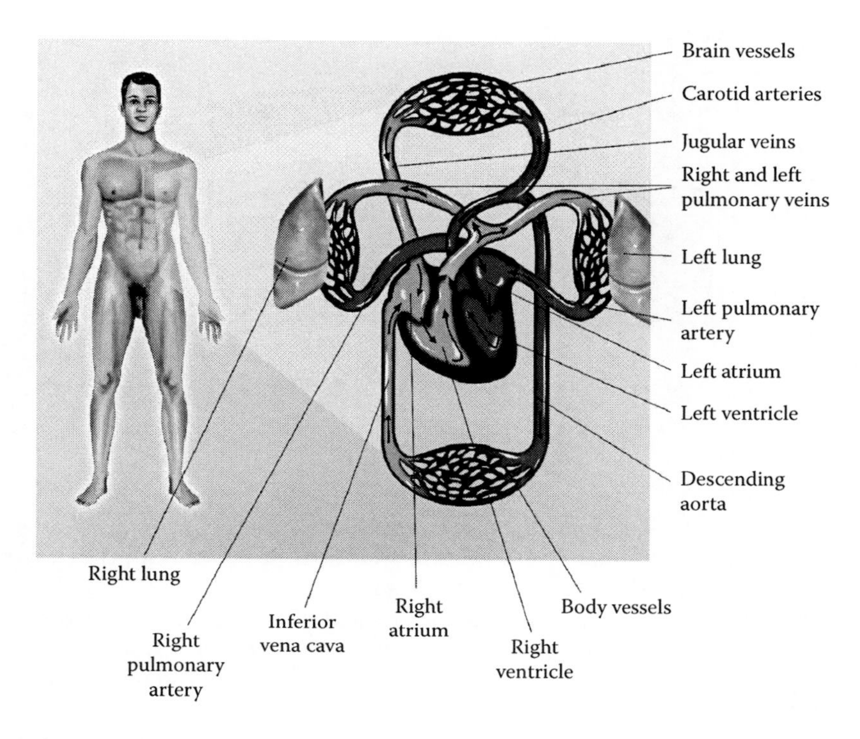

FIGURE 51
Circulatory system schematic.

Preparation

For NIBP, a cuff is placed around the patient's limb, usually the upper arm but sometimes the thigh or other locations. The patient is told that the cuff may become somewhat uncomfortable as it inflates.

For invasive measurements, the catheter insertion site is cleaned.

Procedure

NIBP can be done manually or automatically; in either case the cuff is inflated until arterial blood flow stops and then released until it resumes, first intermittently just below the systolic point, and then smoothly at the diastolic point.

For manual, or auscultatory, measurements, a stethoscope is placed below the cuff at a point where a large artery is relatively near the surface, allowing the practitioner to hear the Korotkoff sounds (Figure 52). The cuff is inflated to a point above systolic pressure, and then the pressure is slowly released. When the Korotkoff sounds are first heard, the practitioner notes the pressure reading as the systolic value. When the Korotkoff sounds cease, the pressure is noted as diastolic. After measurement is complete the cuff is removed.

Automatic blood pressure devices sometimes use a similar method to the manual one described above, with a microphone and analysis circuitry in place of the stethoscope and the practitioner's ears. The machine automatically inflates and deflates the cuff, analyzing the sounds to determine systolic and diastolic values. These are then displayed on a screen and/or printed out on a recorder.

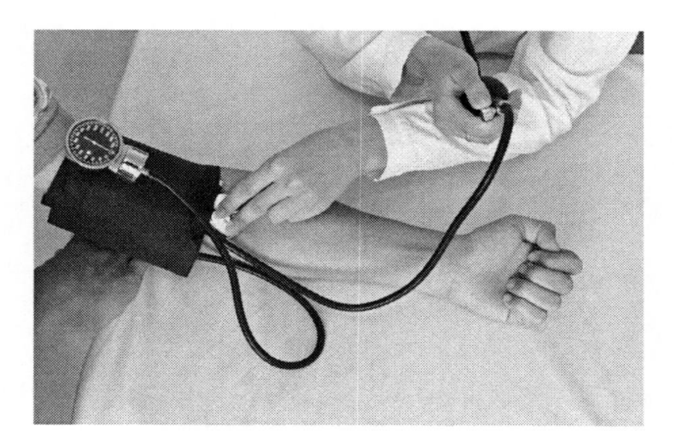

FIGURE 52
Manual blood pressure measurement.

Other automated devices measure variations in the air pressure within the cuff and connecting tubes to determine blood pressure values.

Invasive pressure monitoring involves the insertion of a catheter into the circulatory system. Since catheter placement is not as restricted physically as cuff placement, measurements can be taken at various points in the circulatory system, including in vessels near the heart and in veins. The catheter is placed, usually by pushing it a specific distance beyond the insertion point, though for precise placement near the heart, diagnostic imaging equipment may be used. Once placed, the system is zeroed to account for variations in atmospheric pressure, and then the catheter and pressure transducer are connected. The system measures blood pressure values in real time, and the waveform can be displayed on a screen. Analysis circuitry determines systolic and diastolic values. This system can be used continuously for long-term monitoring of blood pressure.

Since the different methods of blood pressure measurement use different techniques, they should not be expected to exactly coincide. Only a properly used invasive pressure setup can provide truly accurate pressure values.

Expected Outcome and Follow-Up

A useful measurement of blood pressure.

Complications

n/a

Blood Transfusion

Alternate names—n/a.

Purpose

To increase the amount of blood in a patient's system to levels adequate for health.

Indications

Low blood volume.

Anatomy

Blood fills the circulatory system.

Pathology

Low blood volume may be due to blood loss in trauma, disease, or surgery or to inadequate production of blood in the body.

Physiology

Blood is the fluid that carries oxygen, carbon dioxide, nutrients, and waste products throughout the body.

Staffing

Physician or nurse.

Equipment and Supplies

Whole blood or blood products, insertion needles or catheters. If larger quantities of blood may be required, or if the patient's body temperature is not adequately controlled, a blood warming unit may be used. An infusion pump may be used to provide even flow rates during delivery.

Preparation

The patient's blood type is determined by lab testing, confirmed and clearly indicated, because mismatched blood types can cause serious problems or even death. An indication of relative blood volume is made by various means in order to determine how much blood might need to be infused. The infusion site is cleaned.

Procedure

A needle or catheter is inserted into a vein, and the blood supply is connected, possibly through a blood warmer and/or infusion pump. Blood is infused until adequate levels are attained, after which the needle/catheter is removed.

Expected Outcome and Follow-Up

Adequate blood supply for normal life functions. The patient is monitored, and the insertion site is checked for bleeding. Because an anticoagulant agent is used in the donated blood to prevent its clotting, the patient is checked for excess bleeding for some time following the transfusion.

Complications

If blood matching was not done properly, the patient may suffer from a hemolytic reaction in which their body destroys the "foreign" blood cells, which can be fatal. Other complications may include an allergic reaction to a component of the process, or infection.

Breast Augmentation Surgery

Alternate names—Breast implants, breast job, breast enlargement, mammoplasty.

Purpose

To increase the size of the breast, either for cosmetic purposes or to compensate for loss or imbalance of breast tissue due to disease or anatomy.

Indications

Inadequate breast size, as determined by the patient.

Anatomy

The breasts consist of fat tissue and milk glands and ducts, covered by skin (Figure 53).

Pathology

Breast tissue may be lost due to injury or to diseases such as breast cancer and resulting mastectomy.

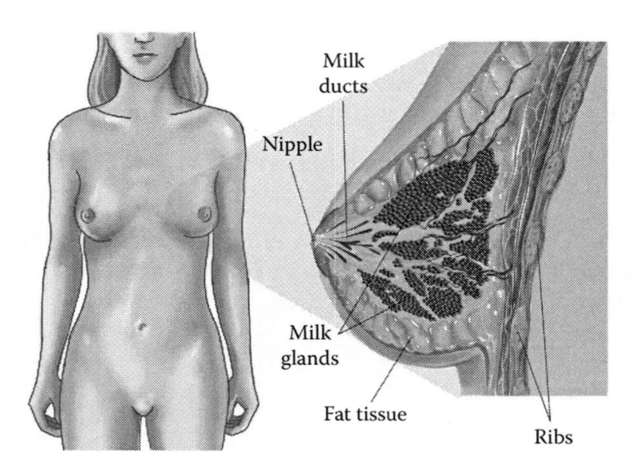

FIGURE 53
Breast anatomy.

Physiology

n/a

Staffing

Normal surgical team.

Equipment and Supplies

Standard surgical, plus implant material. Implants are silicone sacs filled with saline.

Preparation

The procedure is discussed with the patient, having previously determined the size and general shape of implant required.

Procedure

Surgery may be performed with either local or general anesthesia. An incision is made either along the side of the breast stating from the axilla, along the fold under the breast, or around the areola. The implant area may be either under the muscles underlying the breast, or above the muscles, between the muscles and the breast itself. The implant position is adjusted and the incisions closed (Figure 54).

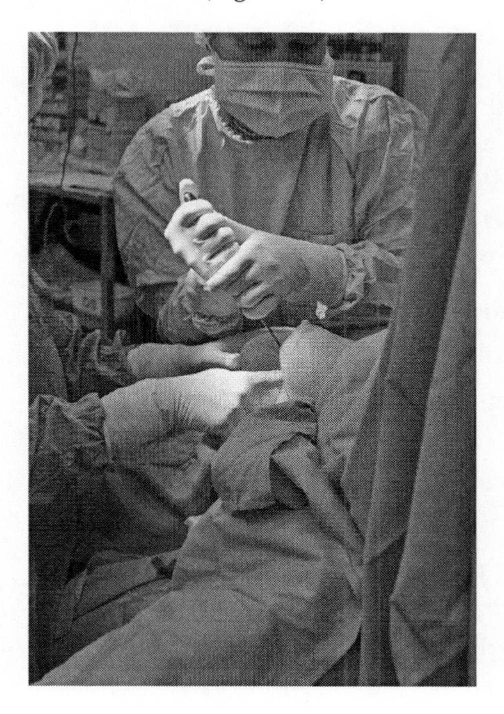

FIGURE 54
Surgeon inserting breast implant. This image shows a breast augmentation procedure, but this aspect of reconstruction surgery is similar.

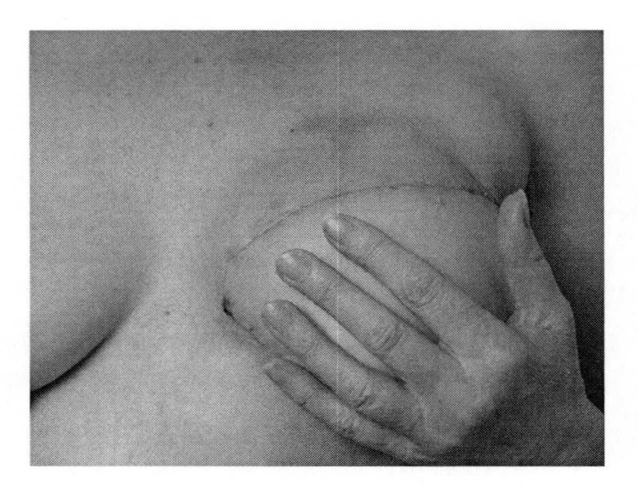

FIGURE 55
Reconstructed breast.

Expected Outcome and Follow-Up

Satisfactory breast size and shape (Figure 55, Figure 56). Normal activities are slowly resumed. Incisions are checked for inflammation.

Complications

Sensation in the breast may be reduced, and breast-feeding may be more difficult or impossible. The implant may move, or it may rupture, and scar tissue may form around the implant.

Breast Reduction Surgery

Alternate names—Mammoplasty, reduction mammoplasty.

Purpose

To reduce breast size for cosmetic or physical reasons.

Indications

Excessive breast size, which can result in severe physical and/or psychological discomfort for the patient; imbalanced breast size.

Anatomy

The breasts consist of fat tissue and milk glands and ducts, covered by skin.

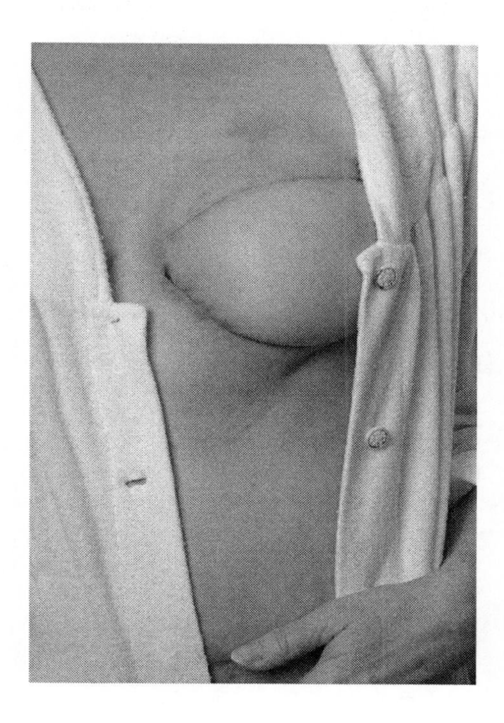

FIGURE 56
Another perspective.

Pathology

Large breasts may simply be a result of normal body growth rather than any pathological process, though some diseases and certain chemicals may cause excess growth of breast tissue. Large breasts can cause physical discomfort such as back strain or shoulder pain, and they may interfere with some physical activities. The patient may suffer psychologically as well, by feeling "different" or as a result of unfeeling comments or looks from others.

Breast reduction surgery may also be required in order to balance breast size and appearance if the two breasts are substantially different from each other or if one breast has had to be altered due to breast cancer surgery, accident, or other causes.

Physiology

n/a

Staffing

Normal surgical team.

Equipment and Supplies

Standard surgical.

Preparation

The procedure is discussed with the patient, having previously determined the size and general shape of breasts desired.

Procedure

An incision is made around the areola and down the underside of the breast. Excess breast tissue and skin is removed to provide the desired size and shape of breast, and the nipple/areola is repositioned and reattached. The incision is closed. The procedure is repeated for each breast as necessary, with symmetry as a goal.

Expected Outcome and Follow-Up

Satisfactory breast size and shape. Normal activities are slowly resumed. Incisions are checked for inflammation.

Complications

Reduced sensation, reduction or elimination of the ability to breastfeed.

Caesarean Section

Alternate names—C-section, caesarean, caesarean delivery, caesarean birth.

Purpose

To deliver a fetus or fetuses that cannot be delivered vaginally.

Indications

Vaginal birth may not be possible or inadvisable due to a number of factors. The most common factor is that the mother has had a previous caesarean, especially one with a vertical or "classical" incision in the uterus. Nonprogression of labor is another major reason for performing C-sections, as well as the fetus being in a breech position, fetal distress, umbilical cord prolapse, multiple fetuses, placenta previa, and abruptio placentia. Some mothers may choose to have a C-section for their own needs, in consultation with their physician.

Anatomy

The fetus develops within the mother's uterus, a muscle-walled organ that is pear shaped before conception but expands as the fetus grows (Figure 57). The fetus is contained by a thin-walled structure called the amniotic sac within the uterus; this sac is filled with fluid that helps protect the developing fetus. The fetus is connected to the mother via its umbilical cord, which carries blood in both directions between the fetus and the placenta. The placenta is a highly vascularized organ that is fetal tissue and has extensive connections to the wall of the uterus. The birth canal consists of the exit to the uterus (the cervix) and the vagina, which are surrounded by the bones of the pelvis.

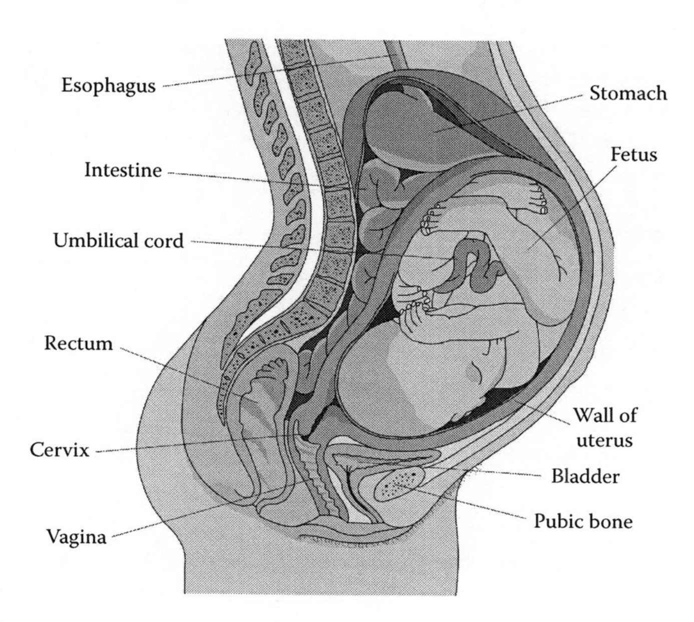

FIGURE 57
Full-term fetus in pregnancy.

Pathology

A vertical or "classical" incision in the uterus in a previous caesarean section results in uterine walls that are not strong enough to withstand the stresses of labor contractions.

Nonprogression of labor (dystocia) may occur for various reasons including abnormalities in the birth canal such as a narrow pelvic opening in relation to the size of the fetus, improper positioning of the fetus, and inadequate strength of contractions.

If during labor the placenta separates from the uterine wall prematurely (abruptio placentia) hemorrhaging will occur, endangering both fetus and mother.

Sometimes the placenta develops in a location that partially or completely blocks the cervical opening (placenta previa), a condition that prevents normal delivery from taking place.

If the umbilical cord comes between the fetus and the cervix, it can be pushed out of the uterus before the fetus (umbilical prolapse), which compresses the cord and can result in the blood supply to the fetus being reduced or stopped, which will cause major damage to the fetus if not corrected very quickly. Also, the cord may become wrapped around parts of the fetus in such a way that it is compressed within the uterus during contractions, again compromising fetal blood supply.

The birth canal is designed to accommodate a fetus in the head-down position. If the fetus is in other positions, such as breech with the buttocks down or sideways with the shoulder and arm closest to the cervix, normal progression through the birth canal may be difficult or impossible. Labor may be blocked, or continued labor may cause damage to the fetus.

Some women may have smaller than normal pelvic openings, or the fetus may be larger than normal, sometimes due to gestational diabetes. In these situations, normal progression of labor is difficult or impossible.

Following birth, the status of the infant may be evaluated at specific time intervals using the APGAR scoring method. Though named after its originator, Dr. Virginia Apgar, the term has been developed into an acronym for the five factors that are considered. Each factor is rated as either 0, 1, or 2, and the total gives a good indication of overall infant condition. The five factors are:

A—activity, or muscle tone
P—pulse presence and rate
G—grimace, or reflex response
A—appearance, mainly skin color
R—respiration quality

The total APGAR score can range from 0 to 10.

Physiology

When the fetus reaches full term, hormonal changes cause the uterus to begin contracting in cycles. The contractions gradually become stronger and more frequent, and the pressure of the contractions causes the cervix to dilate. When the cervix is dilated to the maximum extent, further uterine contractions and "pushing" by the mother move the fetus through the birth canal and into the outside world. Normally, the placenta separates from the wall of the uterus at this time and is pushed out of the uterus by residual contractions. The now-empty uterus contracts enough to prevent significant bleeding from the placental site.

Staffing

Normal surgical.

Equipment and Supplies

Standard surgical, fetal monitor, infant resuscitation equipment.

Preparation

Standard surgical. C-sections may be done on a scheduled basis, or they may be done in an emergency. Emergency sections are usually done using general anesthesia, while scheduled sections may be done with either regional or general anesthesia. If the procedure is to be performed using epidural or spinal anesthesia, the line for anesthetic agent infusion is established.

Procedure

An incision is made at the bikini line, opening the abdomen. Access to the uterus is established, and an incision into the uterus is made, usually laterally and low on the uterus to allow for stronger uterine walls after healing (Figure 58). The amniotic sac is opened, and the fetus is delivered.

The umbilical cord is clamped and cut, and the placenta is removed. The uterine incision is closed, followed by the abdominal incision (Figure 59).

In emergency situations, a vertical incision in both the abdomen and uterus may be used, as they provide better, quicker access to the fetus.

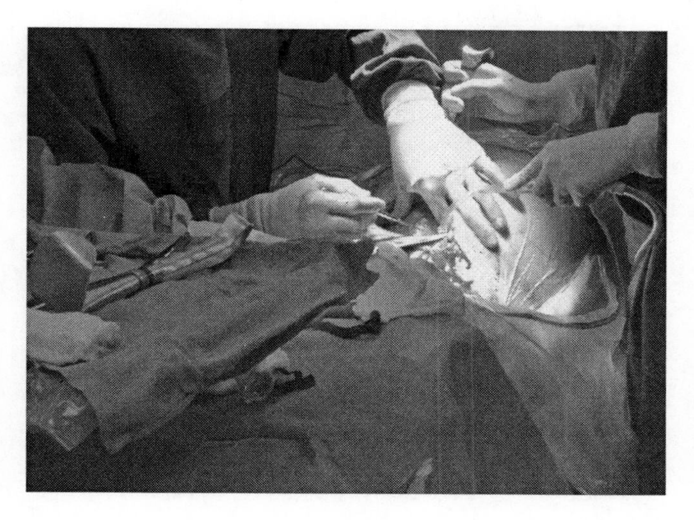

FIGURE 58
A color version of this figure follows page **176.** Caesarean section in progress.

If general anesthesia is used, the procedure usually is performed relatively quickly after induction of anesthesia because the agents can pass from the mother to the fetus, potentially causing problems. Even with a quick delivery, the newborn must be checked carefully for any issues related to anesthesia.

After delivery, the infant is checked for clear breathing passages and suctioned if necessary. Further checks to determine the overall level of health are performed and corrective action taken as necessary (Figure 60).

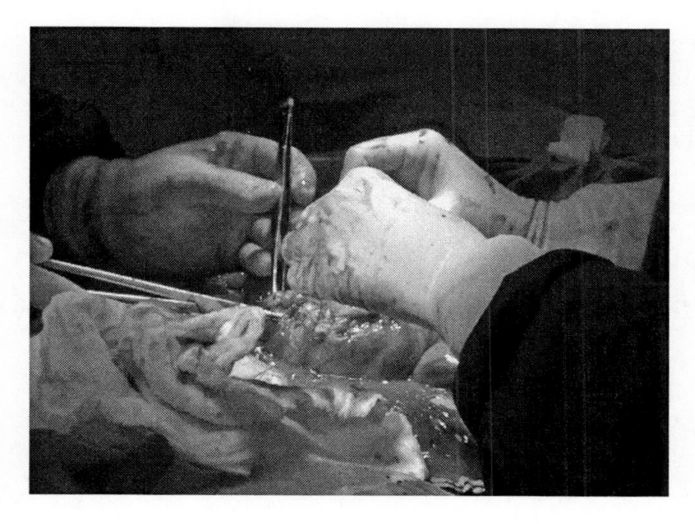

FIGURE 59
A color version of this figure follows page **176.** Suturing uterus following caesarean delivery.

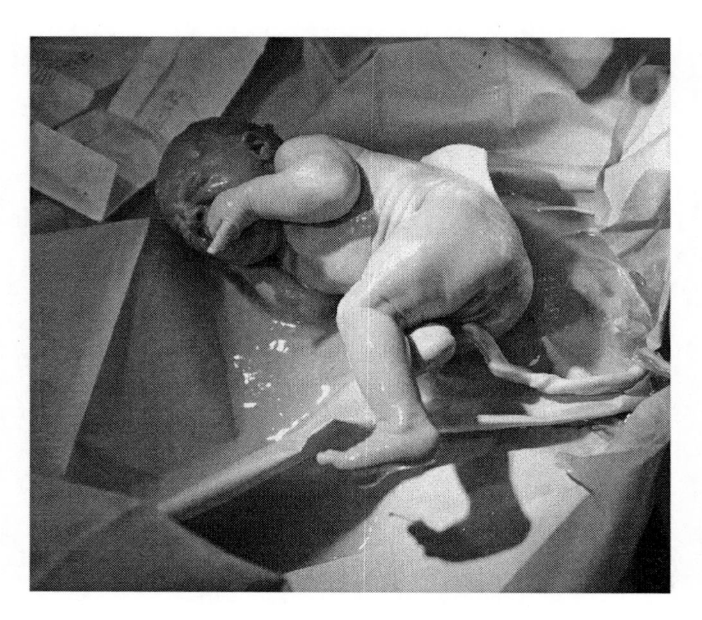

FIGURE 60
The baby, immediately following caesarean delivery. Umbilical cord still attached.

Expected Outcome and Follow-Up

Successful delivery of normal, healthy infant. The mother is monitored for bleeding and changes in blood pressure. Normal activities are resumed slowly. If the incisions heal well, the uterus may be determined to be sound enough for the mother to attempt vaginal birth (VBAC, vaginal birth after caesarean) in subsequent pregnancies if she should so chose. The majority of mothers opt for another caesarean after the first one.

Complications

Excessive bleeding, damage to surrounding organs such as the bladder or intestines, especially after emergency surgery. Hematomas and blood clots, possibly leading to a pulmonary embolism, may occur.

Cardiac Output Measurement

Alternate names—Thermodilution measurement, Swan-Ganz, Q measurement.

Purpose

To determine the effectiveness of cardiac pumping, thus helping diagnose certain heart problems.

Indications

Known or suspected heart problems.

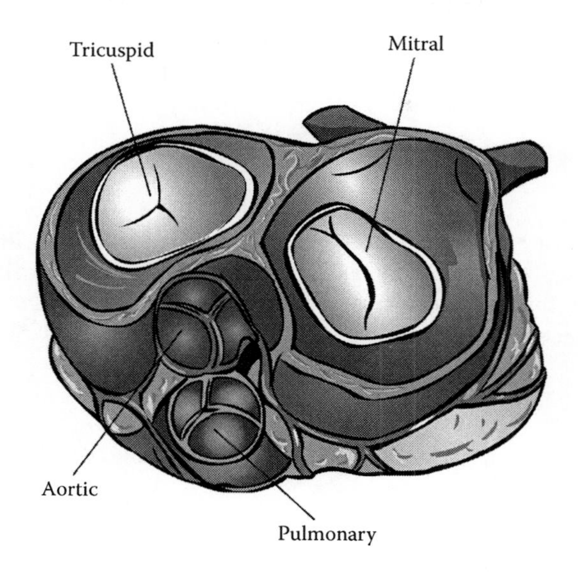

FIGURE 61
A color version of this figure follows page **176.** Heart valves.

Anatomy

The heart is a muscular organ consisting of left and right atria, left and right ventricles, four valves, and associated blood vessels (Figure 61). These blood vessels carry the blood pumped by the heart out to various body parts and also carry blood to the heart itself. The atria are thinner walled than the ventricles. Nerves and other pathways in the heart carry electrical signals to control pumping contractions.

Pathology

Heart pumping effectiveness can be compromised by various problems within the heart. The muscles of the atria and especially the ventricles can be damaged by disease or lack of blood supply (cardiac infarction). The cardiac valves can be damaged by disease or age, thus allowing backflow of blood. Finally, the conductive pathways that help produce and coordinate contractions may be damaged, so that contractions are either weaker or less coordinated, or both.

Physiology

The heart pumps blood throughout the body to carry nutrients and wastes for use or disposal. Blood is pumped when the muscles of the heart wall contract, expelling blood from the interior chamber. Valves between the various parts of the heart keep blood flowing in the correct direction by preventing backflow. Flow proceeds from the venae cava (the main veins bringing blood back from the body to the heart) into the right atrium, through the right atrioventricular (or tricuspid) valve, into the right ventricle, through the pulmonary semilunar valve, through the pulmonary artery to the lungs, back via the pulmonary veins to the left atrium, through the left atrioventricular (or bicuspid) valve, into the left ventricle, out through the aortic valve, the aorta, and to the rest of the body.

Signals from a center in the heart produce electrical signals that are carried throughout the heart to produce contractions. Various components provide delays to the signal so that contractions are coordinated and allow effective pumping.

Definition of Q

Q is a term used to represent cardiac output. It is defined as the volume of blood pumped by the heart in one minute, and is measured in units of liters per minute. Q is equivalent to the stroke volume (the volume of blood pumped by the heart in one beat cycle) in liters times rate in beats per minute.

Staffing

Physician (usually a cardiologist or internist), assistant, diagnostic imaging technologist.

Equipment and Supplies

Different techniques may be used to measure cardiac output, each requiring specific equipment. These are described in the procedure section, below.

Preparation

This again varies depending on the method used, described below.

Procedure

Cardiac output can be measured invasively or noninvasively.

Invasive techniques usually involve the insertion of a catheter into the venous system, usually in the groin. The catheter has a small balloon at its tip, which is carried through the veins by blood flow. It passes through the heart and lodges in one of the pulmonary arteries, at which point the balloon is deflated. The catheter also has a measuring device that sits a specific distance past the balloon; this may be a temperature probe or an optical probe. Once all is in place, a bolus of material is injected through the catheter and out an opening near the balloon. The material may be cold saline, or a dye solution. The measuring device at the far end of the catheter detects the flow of the cold saline or dye, and the volumes, measurement changes, and other factors are combined and analyzed to provide a value for cardiac output. Measurements may be repeated two or three times to improve accuracy. These methods have a number of problems and are used less commonly than noninvasive methods.

Noninvasive methods for measuring cardiac output include:

- Doppler ultrasound, in which sound waves are directed at the blood flowing out of the heart. Moving blood cells reflect some of the sound waves back to the transducer, and the change in frequency of the reflected waves can be used to calculate the velocity of the blood. The diameter of the vessel being examined has been previously determined, and these values can be combined to provide an accurate measure of Q.
- MRI can also provide flow velocity values plus vessel diameter, allowing quick calculation of Q.

Expected Outcome and Follow-Up

An accurate and clinically useful measurement of cardiac output. Noninvasive methods require no follow-up for the procedure itself, while dilution techniques require that the patient be monitored for complications.

Complications

Embolisms, for invasive methods.

Carotid Endarterectomy

Alternate names—n/a.

Purpose

To remove plaque deposits from one or both carotid arteries.

Indications

Symptoms of decreased blood flow to the brain, including dizziness, confusion, lack of coordination, and blackouts, when other possible causes have been eliminated. Such symptoms may occur periodically, and if caused by carotid plaques, are referred to as transient ischemic attacks, or TIAs. Diagnosis can be confirmed by various diagnostic imaging techniques.

Anatomy

The carotid arteries are on either side of the trachea in the neck, leading from the thorax to the head and brain.

Pathology

A buildup of cholesterol or plaque in the carotid arteries can restrict blood flow to the brain, causing various neurological symptoms and possibly a stroke.

Physiology

The carotid arteries carry blood from the heart to the head, especially the brain.

Staffing

Normal surgical; the surgeon is usually a vascular surgeon or neurosurgeon.

Equipment and Supplies

Standard surgical, plus an arterial bypass shunt.

Preparation

Standard surgical; the procedure is usually performed under general anesthesia.

Procedure

An incision is made in the neck in order to expose the carotid artery. A temporary shunt is connected to carry blood around the area of blockage. The artery is cut open longitudinally, and the plaque material is removed, if possible, and then shunt is removed and the artery is sutured up. If removal is not possible or if the artery is damaged, a permanent bypass may be put in place using a vein from the patient's leg. The incision in the neck is then closed.

Expected Outcome and Follow-Up

Increased blood flow to the brain, elimination of the TIA symptoms, and reduced risk of stroke. The patient is monitored closely for a day after surgery, and then normal activities are gradually resumed. Lifestyle modifications to help avoid future plaque buildup may be suggested.

Complications

Stroke, high blood pressure, heart attack.

Cataract Removal

Alternate names—Phacoemulsification, phaco, lens replacement, intraocular lens (IOL) insertion.

Purpose

To remove a lens clouded by cataract from the eye and replace it with a plastic lens, in order to restore more normal vision.

Indications

Cloudy vision, the presence of cataracts visible through the pupil of the eye.

Anatomy

The eye focuses light from outside using an ovoid lens that lies just behind the iris/pupil. The lens is enclosed by a fibrous capsule. The front of the eye is covered by a clear tissue called the cornea (Figure 62).

Pathology

Lenses can become clouded when the proteins that make up their structure become denatured, thus changing from a transparent state to a translucent or opaque state. This denaturing can be caused by excess exposure to ultraviolet light, infrared or microwave radiation, ionizing radiation, diabetes, high blood pressure, or simply age. The reduction in transparency of the lens means that light is not focused on the retina properly, and vision is impaired, possibly to the point of complete blindness.

Physiology

n/a

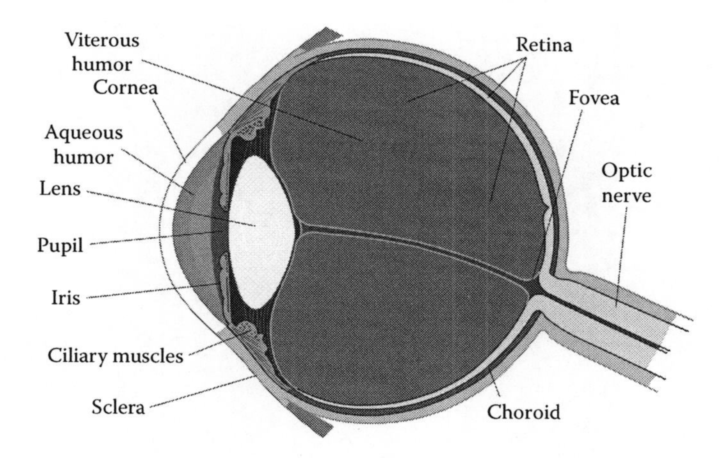

FIGURE 62
Eye anatomy.

Staffing

Minor surgical; the surgeon is an ophthalmologist.

Equipment and Supplies

Minor surgical, plus a phacoemulsifier machine (Figure 63) and an operating microscope (Figure 64). Plastic lens.

Preparation

The patient is sedated, and eye drops are given that dilate the pupil. Local anesthetic is administered to the eye (Figure 65).

Procedure

A very small incision is made in the cornea, and an even smaller incision is made through the lens capsule (Figure 66). Water is injected into the capsule to separate the middle part of the lens (the part that is clouded by cataract) from the capsule (Figure 67). The tip of the phacoemulsifier is inserted into the lens and ultrasound energy is applied, which breaks apart (emulsifies) the lens core.

Suction is applied through a small passage in the phaco tip, and the emulsified lens is removed, leaving the capsule in place. This is done in several stages. After the cataract is removed, the intraocular lens is inserted. The IOL is made of thin, flexible material that can be rolled into a thin tube, allowing it to be inserted through a very small incision.

Once in the capsule, the IOL unrolls and is held in position by the remains of the capsule (Figure 68). The phaco probe is removed, and because the incisions are so small, no sutures are necessary. The entire procedure may be completed in as little as 20 minutes.

Expected Outcome and Follow-Up

Restoration of near-normal vision. Antibiotic drops are used for some time following surgery to help prevent infection. Light sensitivity may be experienced, so exposure to bright

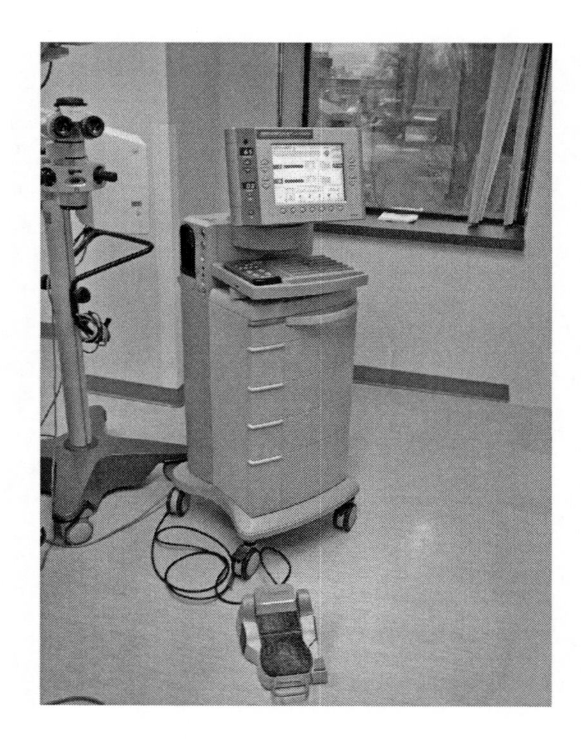

FIGURE 63
Phacoemulsifier unit.

lights should be avoided. Follow-up checks are performed several times in the weeks following surgery.

Complications

Retinal detachment.

Chemotherapy

Alternate names—Chemo.

Purpose

To destroy cancerous cells in the body in order to prevent recurrence of cancer or reduce the symptoms of cancer that cannot be cured. Chemotherapy may also be used to shrink tumors in order to make their surgical removal or treatment by radiation therapy easier or more effective.

Indications

Cancer.

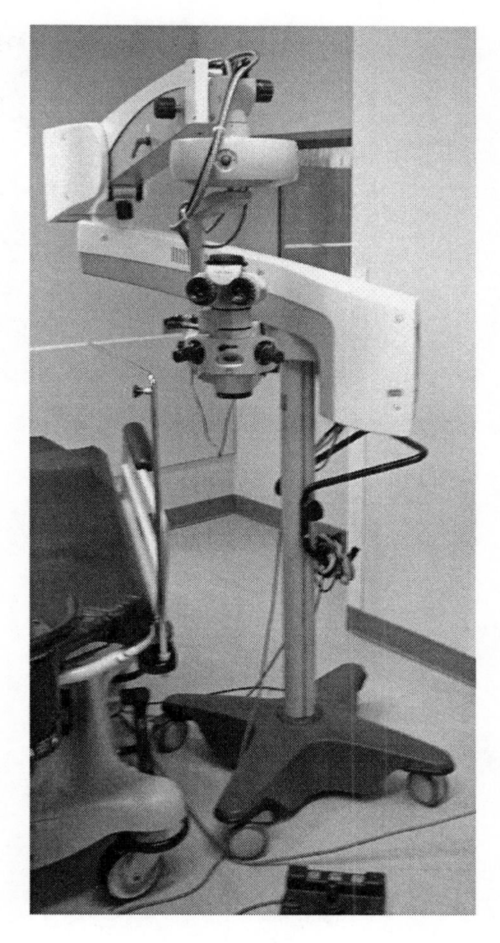

FIGURE 64
Operating microscope for cataract surgery.

Anatomy

Cancer can occur in almost any part of the body.

Pathology

Cancer is the uncontrolled growth of cells in certain tissues in the body. These cells may separate from the initial site and travel to other parts of the body where they can start to grow a new tumor, a process called metastasization. The tumor growth may damage vital organs, or it may simply draw more and more of the body's resources, leaving less and less for normal functions. In either case, death may result. Cancer may be the result of spontaneous changes in the genetic structure of cells, or these changes may be initiated by radiation or chemicals. Cancers are different depending on the type of tissue in which they originate.

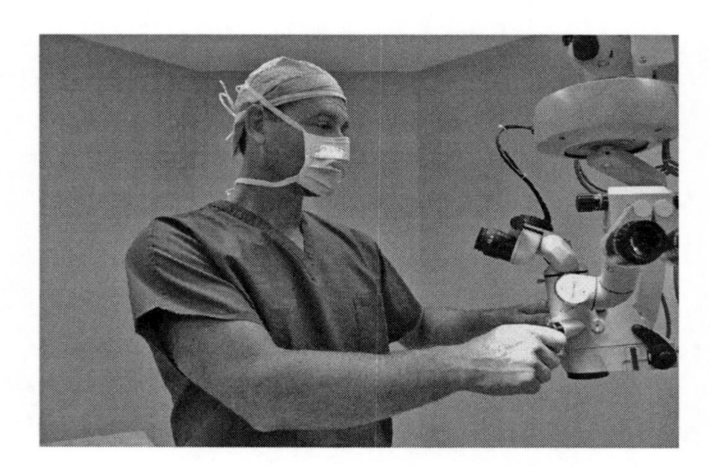

FIGURE 65
Surgeon preparing to use operating microscope for eye surgery.

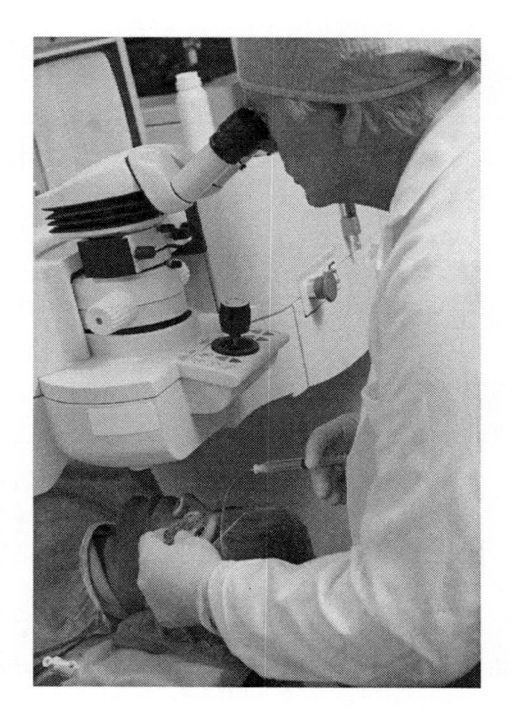

FIGURE 66
A color version of this figure follows page **176.** Surgeon performing eye surgery using an operating microscope.

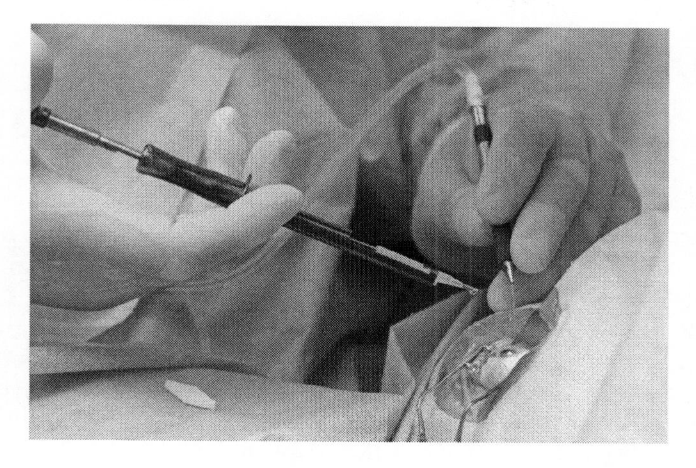

FIGURE 67
A color version of this figure follows page **176.** Cataract surgery.

Physiology

n/a

Staffing

Nurse.

Equipment and Supplies

Chemotherapy drugs, IV pump.

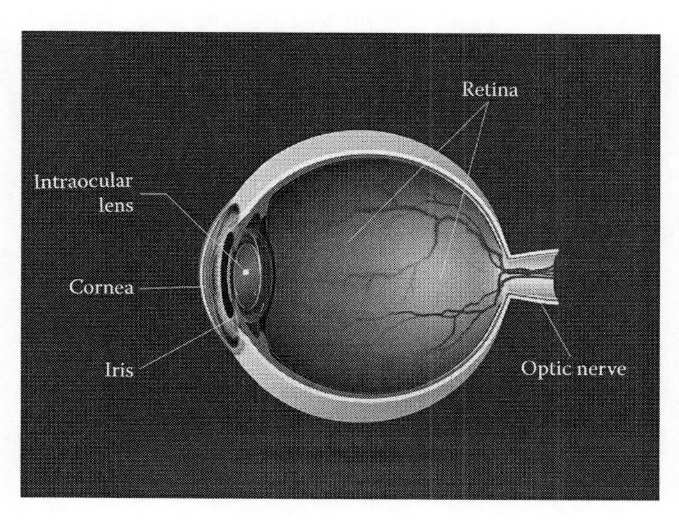

FIGURE 68
Cross section of eye showing implanted lens.

Preparation

An intravenous line is inserted. Antinausea medications may be given.

Procedure

Most chemotherapy drugs are administered intravenously, though some are given orally. The IV line is used to deliver saline solution at first, and then the chemo drugs are added. The type of cancer and the tolerance of the patient to the drugs determine the rate and amount of drugs delivered in any one session.

Because of their rapid growth, cancer cells are often susceptible to damage by specific chemicals. These chemicals are usually toxic to the body as a whole, but since their toxicity level varies with rate of cell growth, controlled doses of the chemicals can cause cancer cell deaths while allowing most normal cells to survive.

Cancer cells are not the only ones in the body that divide relatively rapidly, though other types of rapidly dividing cells do not approach the rate of division of cancer cells. However, this means that these normal cells may be destroyed by the chemo as well. The cells producing hair in the skin are one example of such cells, which is why patients undergoing chemo usually lose their hair.

Chemo treatments are repeated in accordance to the type of tumor and the tolerance of the patient. Most sessions are repeated weekly for several weeks.

Expected Outcome and Follow-Up

Reduction or elimination of cancer cells in the body. Patients are monitored to try to determine how effective the treatments have been in reducing the number of cancer cells. This measure is used to help decide on the frequency and level of subsequent treatments.

Complications

Nausea, physical exhaustion, digestive upsets, anemia, malnutrition, hair loss, organ damage, and immune system depression which may lead to infections and the possibility that secondary cancers may be initiated.

Cholecystectomy

Alternate names—Laparoscopic cholecystectomy, lap-chole, gall bladder surgery.

Purpose

To remove the gall bladder.

Indications

Gallstones that cannot be removed by other means, inflammation of the gall bladder. Definitive diagnosis is made on the basis of lab blood analysis and diagnostic imaging, usually using ultrasound.

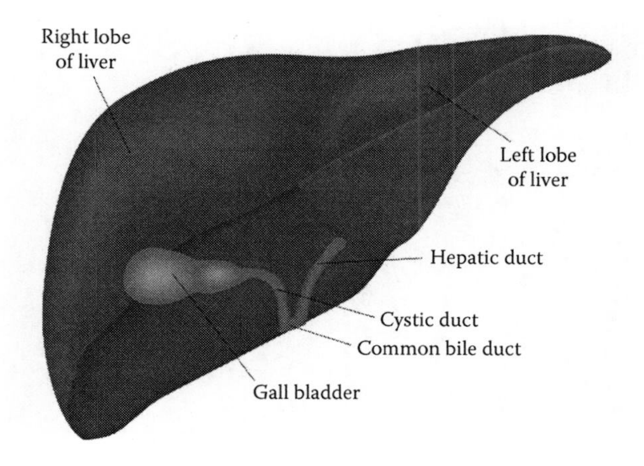

FIGURE 69
Liver and gall bladder.

Anatomy

The gall bladder is a sac that connects to the common bile duct. The duct connects the liver and the small intestine just below the stomach (Figure 69).

Pathology

Bile and/or other substances stored in the gall bladder can become concentrated enough to condense into hard nodules called gallstones or choleliths. Gallstones usually do not cause any problems, but if they are large or if they cause irritation of the gall bladder, significant pain can result. Gall bladder irritation or inflammation, called cholecystitis, can also cause nausea and vomiting.

Gallstone production is increased by certain factors, including age (more common between 40 and 60), gender (more common in females), obesity, genetic factors, and diabetes. High-fat, low-fiber diets increase the chances of gallstone formation.

Physiology

Bile is produced by the liver and helps digest fats in the intestine. Some excess bile can be stored in the gall bladder until needed.

Staffing

Normal surgical.

Equipment and Supplies

Standard surgical, laparoscopic equipment.

Preparation

Standard surgical. A nasogastric tube may be placed.

Procedure

Cholecystectomy may be performed with open surgery, but most cases are done using laparoscopy.

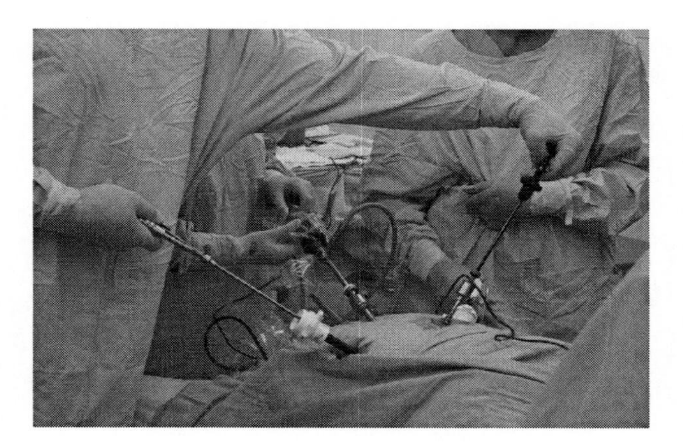

FIGURE 70
A color version of this figure follows page **176.** Laparoscopic surgery procedure.

In open surgery, an incision is made just below the ribs on the right side of the abdomen. Access to the gall bladder is established, and the organ is separated from surrounding tissues. The neck of the gall bladder, where it connects to the common bile duct, is sealed off, usually using electrocautery. A cut is made in the closed section so that both the bile duct end and the gall bladder end remain closed. The gall bladder is removed, and the incision is closed. A drainage tube may be placed to remove excess fluid for some time following surgery.

In laparoscopic cholecystectomies, several small incisions are made in the abdomen to allow access of the laparoscopic tubes. The abdomen is insufflated to allow clearance for the procedure and to permit visualization of the field (Figure 70).

The gall bladder is isolated in much the same way as in open surgery and is removed through one of the lap incisions. The site is checked for bleeding or other problems, the tubes are removed and the incisions closed. A drainage tube may be placed.

Expected Outcome and Follow-Up

Relief from symptoms. No food or liquid intake is permitted for 1 or 2 days (hydration is provided intravenously), and soft foods are introduced gradually after that.

Complications

Bile duct damage may occur, resulting in either blockage of bile or leaking of bile into the abdominal cavity.

Circumcision

Alternate names—n/a.

Purpose

Removal of the foreskin of the penis.

Indications

Circumcision is often performed for cultural or religious reasons. Recurrent infections under the foreskin, concerns regarding sexually transmitted diseases, reduction of urinary tract infections, and phimosis (a tightening of the foreskin around the penis) are some for the medical reasons for circumcision.

Anatomy

The penis has a fold of skin at the tip that normally covers most or all of the end of the penis (the glans).

Pathology

Because the foreskin encloses an area at the end of the penis, moisture can be trapped there, leading to growth of microorganisms if proper hygiene practices are not followed. Some organisms related to sexually transmitted diseases may be carried under the foreskin, making transmission to sexual partners more likely.

Physiology

The foreskin serves to protect the sensitive glans of the penis, and also to help prevent foreign matter from entering the urinary tract through the urethral opening. During sexual arousal, the foreskin is pulled back, exposing the glans.

Staffing

Individual physician, possibly an assistant.

Equipment and Supplies

Minor surgical, circumcision clamp.

Preparation

Local anesthesia may be administered.

Procedure

For infants, a circumcision clamp is applied that pulls the foreskin away from the glans. An incision is then made around the foreskin above the glans, and the foreskin is removed. The incision may require sutures, but usually the clamp has prevented blood flow before the incision, and there is little bleeding following the procedure. For older males, an incision is made around the foreskin, which is then pulled away and cut away from the penis. Sutures are usually required to close the incision.

Expected Outcome and Follow-Up

Successful removal of the foreskin if done for cultural or religious reasons. Reduction in infections or elimination of phimosis pain if done for medical reasons.

Complications

Reduced sensitivity of the penis during sex.

Colonoscopy

Alternate names—n/a.

Purpose

To examine the interior of the colon for abnormalities. Some lesions can be removed during the procedure.

Indications

Rectal bleeding, unexplained changes in bowel function, abdominal discomfort when other causes have been eliminated.

Anatomy

The colon, or large intestine, is shaped like an inverted U. It connects the lower end of the small intestine or ileum, and the rectum. It has a relatively large diameter, and its walls are composed of smooth muscle and a lining with mucous glands and structures for absorbing liquid from the digestive tract (Figure 71).

Pathology

A wide variety of problems can occur within the colon. Cancer (Figure 72), Crohn's disease, colitis and ulcerative colitis, polyps (Figure 73), diverticulitis/diverticulosis, Hirschsprung's disease, and irritable bowel syndrome can all cause bleeding and/or bowel malfunctions.

Physiology

The main functions of the colon are reabsorption of water and some electrolytes from the material passed into the colon from the small intestine and storage of waste products until they can be eliminated.

Staffing

Physician and assistant.

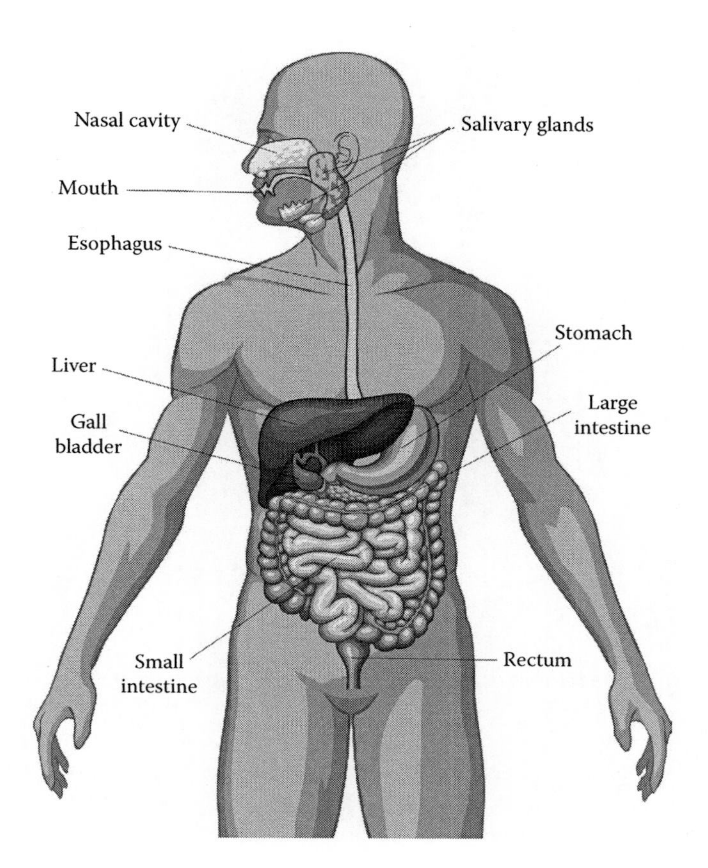

FIGURE 71
A color version of this figure follows page **176.** Digestive system.

Equipment and Supplies

Colonoscope and associated apparatus. The colonoscope is a long, thin tube with a number of interior passageways. These passageways can carry illumination, video images, gas supplies for inflation, water for irrigation, and suction for aspiration. Mechanisms allow the end of the scope to turn in various directions, and a cutting tool is available to help obtain tissue samples. The other end of the scope attaches to equipment that provides light, video, water, and suction, and has arms and levers to control pointing and cutting.

Preparation

Extensive measures are taken before the procedure to ensure that the colon is completely empty because any fecal material present can obscure the target sites. The patient is sedated. The most common position for colonoscopy is for the patient to lie on his side with his knees drawn up toward the chest.

Procedure

The colonoscope is lubricated and inserted slowly through the anus. Video images provide a continuous view of the interior of the colon, allowing the operator to guide the instrument to areas of interest. Images may be taken continuously, or still photos may be

FIGURE 72
A color version of this figure follows page **176.** Cutaway view of a colon tumor.

taken of specific sites and structures. Tissue samples can be obtained with the use of the cutting tool on the scope; these can then be aspirated up the scope and collected for later lab examination. Some structures such as polyps or small tumors can be excised during colonoscopy. When sufficient images and/or tissue samples have been obtained, the scope is withdrawn.

Expected Outcome and Follow-Up

Adequate images and tissue samples to aid in diagnosing the presenting problems. The patient is monitored until the sedative and analgesic have worn off. Normal activities can be resumed a few hours after the procedure.

Complications

Bowel perforation may occur rarely. If tissue samples are taken or structures such as polyps removed, bleeding may occur at the site.

Colostomy

Alternate names—n/a.

Purpose

To create an opening for the colon thought the abdominal wall following permanent or temporary severing of part of the colon.

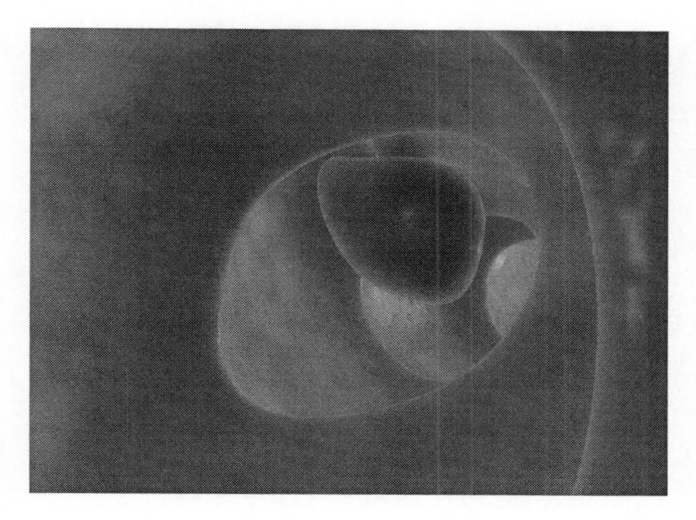

FIGURE 73
A color version of this figure follows page **176.** A large colon polyp.

Indications

Diseases of the colon or trauma that require either removal of much of the colon, or bypassing the colon to allow for healing. These conditions are diagnosed from physical examination, lab analysis of stool and blood, diagnostic imaging, and colonoscopy.

Anatomy

The colon, or large intestine, is shaped like an inverted U. It connects the lower end of the small intestine or ileum, and the rectum. It has a relatively large diameter, and its walls are composed of smooth muscle and a lining with mucous glands and structures for absorbing liquid from the digestive tract (Figure 74).

Pathology

A wide variety of problems can occur within the colon. Cancer, Crohn's disease, colitis and ulcerative colitis, polyps, diverticulitis/diverticulosis, Hirschsprung's disease, and irritable bowel syndrome can all cause bleeding and/or bowel malfunctions.

Physiology

The main functions of the colon are reabsorption of water and some electrolytes from the material passed into the colon from the small intestine and storage of waste products until they can be eliminated.

Staffing

Normal surgical.

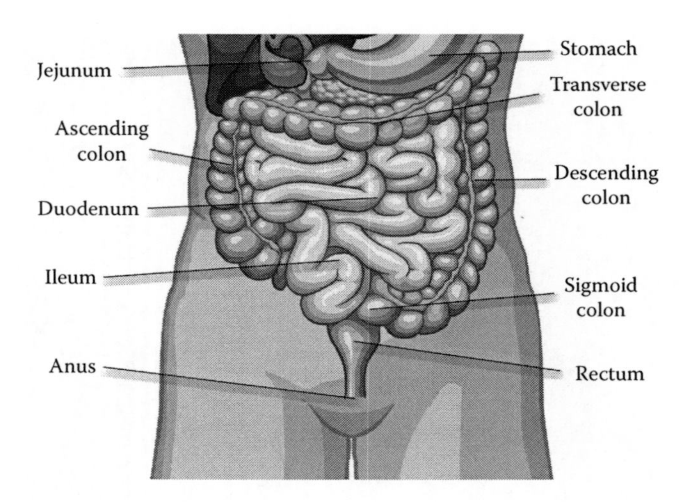

FIGURE 74
A color version of this figure follows page **176.** Intestinal organs.

Equipment and Supplies

Standard surgical.

Preparation

Extensive measures are taken before the procedure to ensure that the colon is completely empty. The patient is sedated, and general anesthesia is induced.

Procedure

A large incision is made vertically in the abdomen, usually just to one side of the navel. If the colon is to be removed, it is separated from supporting structures and associated arteries, and veins are closed off. The distal end of the portion to be removed is separated and closed off, the proximal end is separated, and the diseased or damaged section is removed from the abdomen. If the colon or part of it is to remain in place, the two parts are separated, and the section that will be allowed to rest is closed off.

A small opening is made in the abdominal wall on the side closest to the end of the colon that will remain connected to the small intestine. This end of the colon is drawn out through the opening and is everted, forming a "cuff" which has the inner lining of the colon facing outward. The edges of the colon are then sutured to the skin of the abdomen to form a stoma. Fecal material will now exit the body through the stoma, so a pouch is placed over the stoma to catch the material (Figure 75).

The abdominal incision is closed with drainage tubes in place.

Expected Outcome and Follow-Up

Elimination of the symptoms leading to the surgery, or provision of time for the bypassed section of the colon to rest and heal without the passage of fecal material.

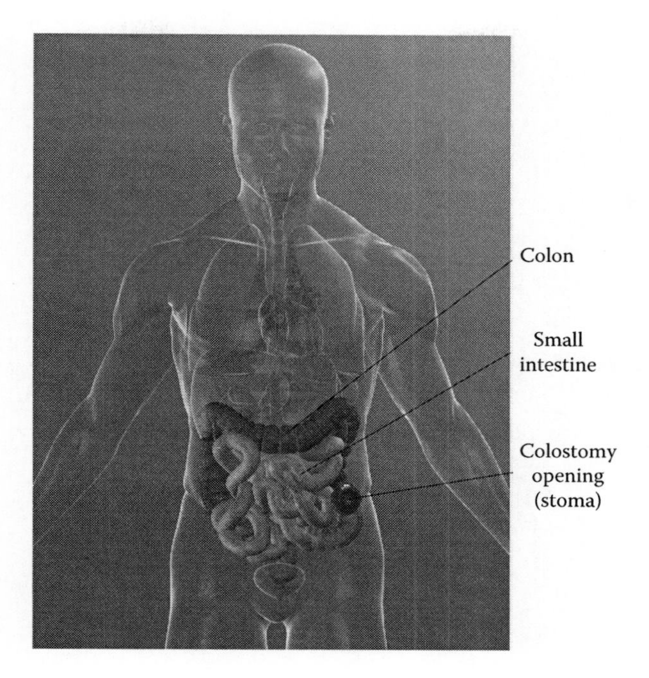

FIGURE 75
A color version of this figure follows page 176. Colostomy formed at the lower end of the descending colon.

Fluid and food intake is restricted for two or three days after surgery (hydration is provided intravenously), and then soft foods are gradually introduced.

The patient is educated concerning care of the colostomy. Some patients will require a pouch to be in place at all times, while others may only need a dressing over the stoma most of the time. These patients will irrigate the stoma to allow for controlled release of fecal material.

Dietary modifications will be required to allow for the change in the digestive system.

If the colostomy is temporary, after sufficient time has passed to allow healing of the damaged or diseased portion, the abdomen is reopened and the two ends of the colon are reattached.

Complications

Blood clots. Some patients may suffer from psychological trauma.

Cornea Transplant

Alternate names—Keratoplasty, lamellar keratoplasty, LK, penetrating keratoplasty, PK.

Purpose

To replace a cornea damaged by disease or trauma.

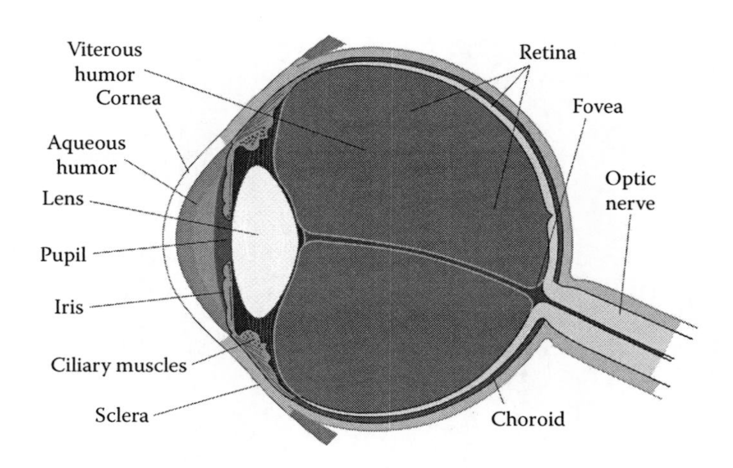

FIGURE 76
Cross section of eye.

Indications

Damage to the cornea preventing normal vision, not treatable by other means.

Anatomy

The cornea is a clear structure forming the outer surface of the front of the eye (Figure 76).

Pathology

A number of diseases can cause damage to the cornea, including herpes and other infections. Chemical and heat burns as well as other physical trauma can produce irreparable damage.

Physiology

The cornea, though nonvascularized, is living tissue and must be supplied with oxygen through continuous bathing with eye fluids. The lack of vascularization means that donated tissue is less likely to be rejected than with other types of transplants.

Staffing

Normal surgical; the surgeon is an ophthalmologist specializing in such procedures.

Equipment and Supplies

Normal surgical, donated cornea.

Preparation

The patient is sedated, and local or general anesthesia is induced.

Procedure

Either the outer layer of the damaged cornea is removed (lamellar keratoplasty or LK), or a disc the full thickness of the cornea (penetrating keratoplasty or PK) is excised. Removing the outer layer is much more difficult, but results in quicker healing. Once the damaged part of the cornea is removed, a corresponding section of the donor cornea is inserted in its place and sutured into position using extremely fine sutures.

Expected Outcome and Follow-Up

Restoration of normal vision. Eye protection is required for several days following surgery, and antibiotic plus corticosteroid eye drops are given to prevent infection and promote healing. Sutures are not removed until several months after surgery to ensure adequate healing.

Complications

Graft rejection.

Coronary Artery Bypass Graft

Alternate names—Heart bypass, coronary bypass.

Purpose

To restore adequate blood supply to cardiac muscles.

Indications

Reduction in blood flow to the muscles of the heart, as shown by myocardial infarction, ECG measurement or, more definitively, by diagnostic imaging such as angiography.

Anatomy

The heart itself is supplied with blood by the coronary arteries, branches of which supply different areas (Figure 77).

Some blood vessels in the body have very good surrounding circulation and can be removed to be used for grafts without causing serious problems. The internal mammary artery, the radial artery in the arm, and the saphenous vein in the thigh or calf are those most commonly used for grafts (Figure 78).

The internal mammary artery comes off the aorta and is therefore very close to the heart and can simply be connected to the coronary artery past the blockage point, meaning that only one junction is needed. Other vessels must be joined to the coronary arteries above and below the blockage. Of course, the mammary artery can only bypass a single coronary artery, so if multiple bypasses are required, other vessels will have to be used for the grafts.

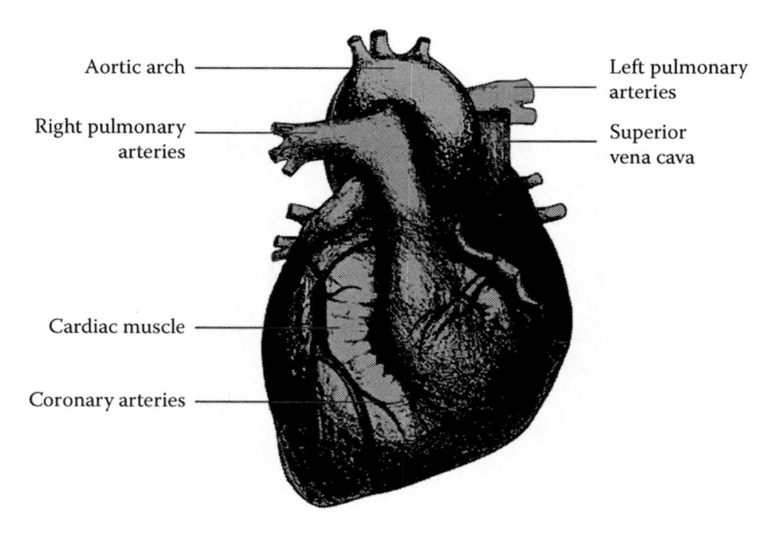

FIGURE 77
Human heart and associated blood vessels.

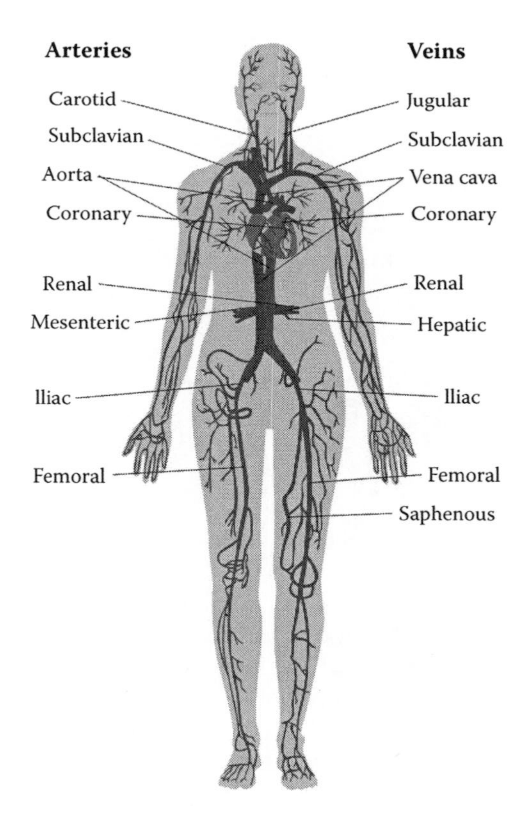

FIGURE 78
A color version of this figure follows page **176.** Circulatory system, showing the saphenous vein lower right.

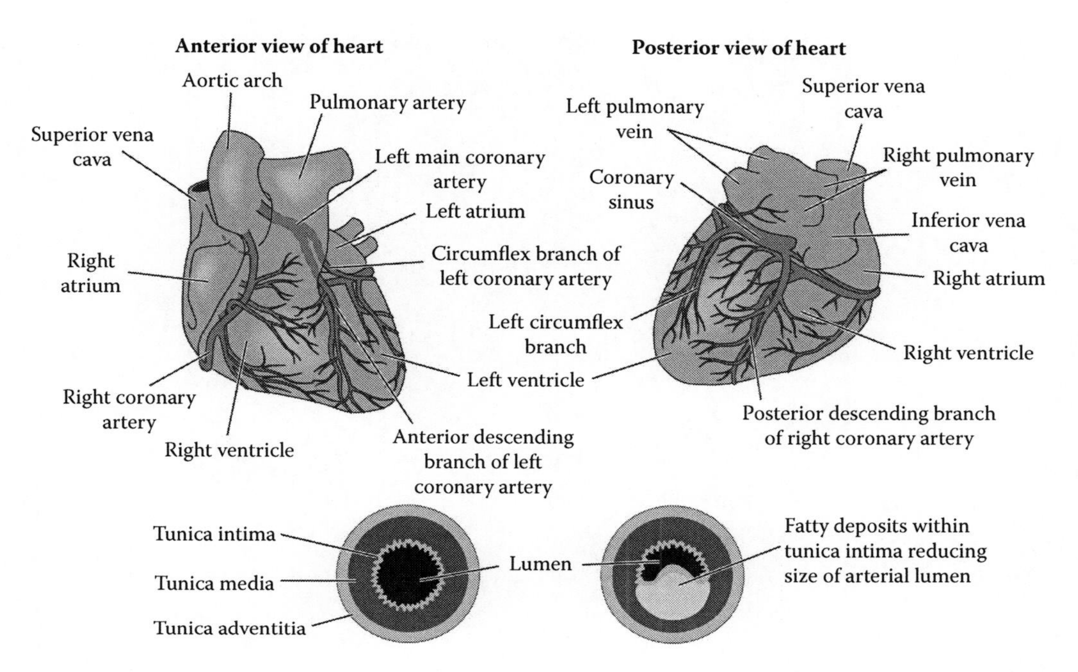

FIGURE 79
Heart anatomy and comparison between normal (left) and atherosclerotic (right) arteries.

Pathology

Arteriosclerosis (Figure 79) can cause narrowing of the coronary arteries to such an extent that muscle function, and therefore heart pumping effectiveness, is impaired (cardiac ischemia). If the blockage is severe enough, damage to or death of some of the cardiac muscle can result (myocardial infarction).

Physiology

The heart, being mostly muscle that is active all the time, requires an excellent blood supply to provide oxygen and nutrients and to remove wastes.

Signals from a center in the heart produce electrical signals that are carried throughout the heart to produce contractions. Various components provide delays to the signal so that contractions are coordinated and allow effective pumping.

Staffing

Major surgical team; the lead surgeon is a cardiologist. A perfusionist operates the heart–lung machine if used.

Equipment and Supplies

Major surgical, heart–lung machine or cardiac muscle stabilization device, graft vessel, chest retractors.

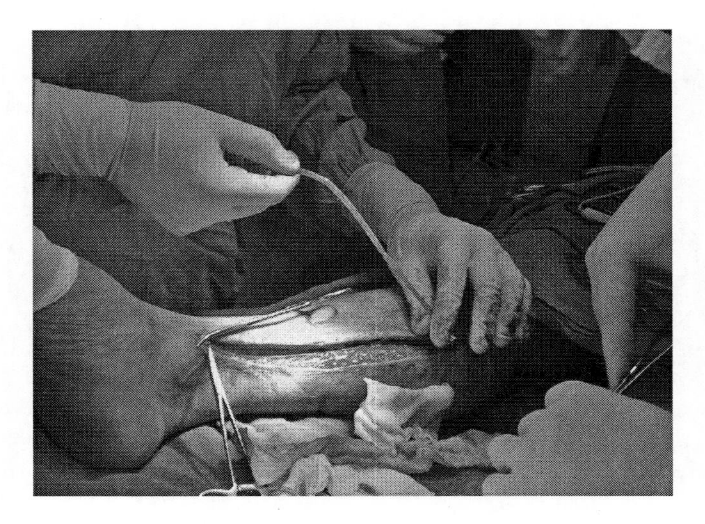

FIGURE 80
A color version of this figure follows page **176.** Stripping out a section of saphenous vein.

Preparation

Normal surgical. An anticlotting agent such as heparin is given.

Procedure

There are two general types of bypass surgery. The original form uses a heart–lung machine, bypassing and stopping the heart while surgery is performed. A newer method, sometimes called beating heart surgery, uses special devices to stabilize parts of the heart while it continues to beat. The traditional method is better for larger or multiple grafts, while the beating heart method has fewer side effects and needs smaller incisions. Both procedures require open access to the thoracic cavity, and both are major surgeries.

General anesthesia is induced.

The surgeon will have previously decided which vessel to use for the grafts, and this vessel is exposed by making an incision over the site. A section of the vessel is removed and prepared for grafting (Figure 80).

The chest is then opened.

For traditional surgery, a vertical incision is made over the sternum, and then the sternum is sawn through vertically. A chest retractor is placed in the cut and is mechanically adjusted to hold the chest open. The heart–lung machine is connected to bypass the heart, providing oxygenation and CO_2 removal, as well as temperature regulation during surgery. The heart is usually stopped using a potassium ion solution, which interrupts the chemical–electrical function of the heart. Newer methods use other drugs that have fewer harmful side effects than potassium. Once the heart is stopped and stable, the graft or grafts are connected (Figure 81, Figure 82).

One end is attached to the aorta or subclavian artery, and the other to a point just past the coronary artery blockage (Figure 83, Figure 84). If the mammary artery is used, only the end past the blockage needs to be attached. When all the connections are complete, the heart–lung machine is removed, the potassium solution is flushed out, and an electrical stimulus is applied to the heart to start it beating again.

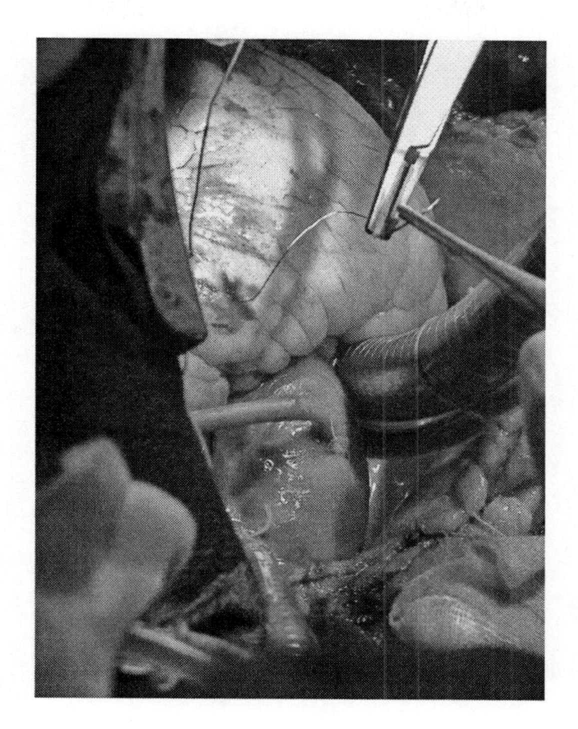

FIGURE 81
A color version of this figure follows page **176.** Grafting vein onto aorta.

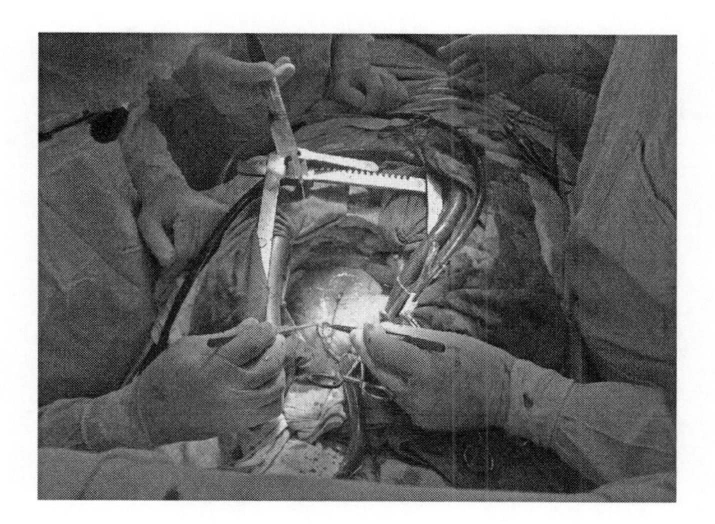

FIGURE 82
A color version of this figure follows page **176.** Open heart surgery.

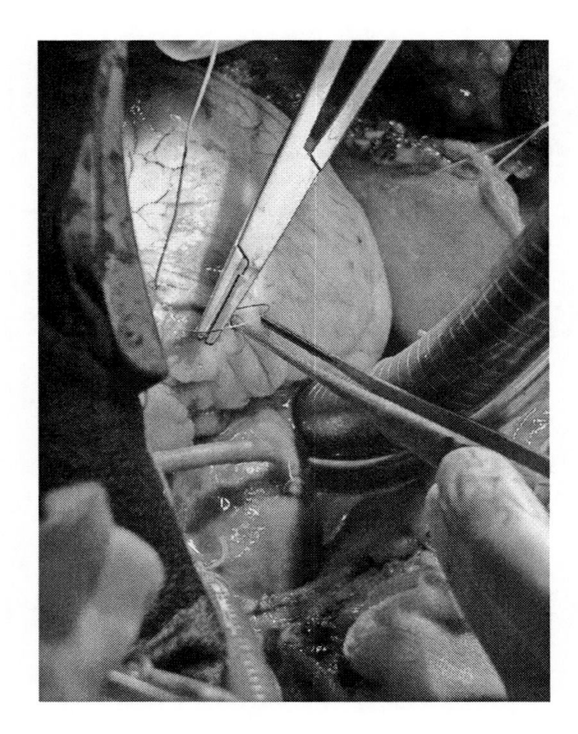

FIGURE 83
A color version of this figure follows page 176. Connecting graft vein to coronary artery.

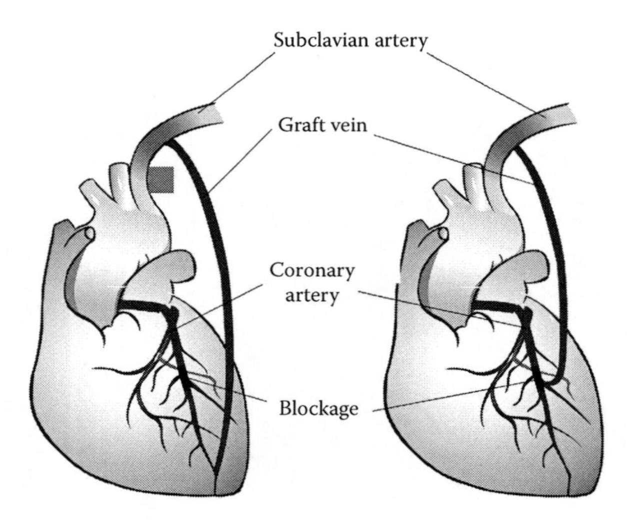

FIGURE 84
Two examples of bypasses from the subclavian artery to different locations on coronary arteries.

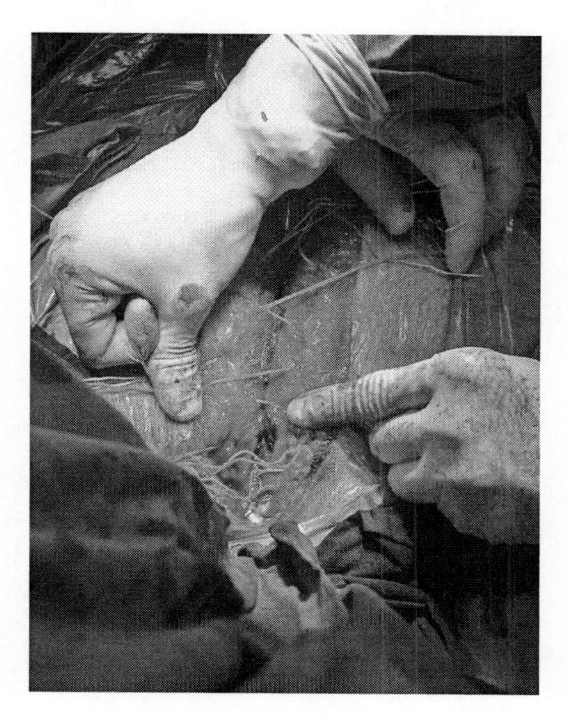

FIGURE 85
A color version of this figure follows page **176.** Closing the chest following cardiac surgery.

With beating heart surgery, smaller incisions are made in the chest wall, and the heart is accessed. A special stabilizing device is applied to the heart in the area of the graft so that there is very little movement in that portion. The heart continues to beat almost normally, allowing adequate blood flow. The grafts are then put in place in the same manner as with traditional surgery, and the stabilizing device is removed.

A temporary pacemaker may be connected if there are concerns about the patient's condition.

In both types of surgery, the grafts are checked for good blood flow and leaks or other problems, and if all is well, the incisions are closed, with drainage tubes in place (Figure 85).

The patient is often put on a ventilator immediately following surgery to reduce his physical load.

Expected Outcome and Follow-Up

Resumption of adequate blood supply to cardiac muscles and alleviation of previous symptoms. The patient is very closely monitored in an intensive care unit. The ventilator can usually be removed after a few hours. Physiotherapy can help avoid some complications. Normal activities are slowly resumed, with careful monitoring of each stage.

Complications

Failure of grafts, arrhythmias, blood clots, or kidney failure.

Craniotomy

Alternate names—n/a.

Purpose

To access the inside of the skull in order to remove a lesion or perform a repair of a damaged area.

Indications

The presence and location of the lesion is determined by EEG, physical examination, and diagnostic imaging.

Anatomy

The skull, or cranium, is a dome-shaped bone that protects the brain. It is lined with a three-layer membrane that further encloses the brain, holding a quantity of cerebrospinal fluid that provides shock-absorbing action.

Pathology

Various processes can produce lesions inside the skull. Benign or malignant tumors may develop, blood clots (hematomas) can form following a blow to the head, aneurysms may form in a weak area of a cerebral artery, infection can produce an abscess, and certain conditions can cause cerebrospinal fluid pressure to increase. Any of these conditions put extra pressure on the brain, causing symptoms that vary depending on the location. Headaches, sensory illusions or impairment, lack of coordination, communication difficulties, and intellectual impairment may occur, with stroke or even death a potential result. It is critical that the pressure from such lesions be released before permanent damage is done to the brain.

Physiology

n/a

Staffing

Normal surgical team.

Equipment and Supplies

Standard surgical, bone saws and drills, bone closures.

Preparation

Standard surgical. The skull is shaved.

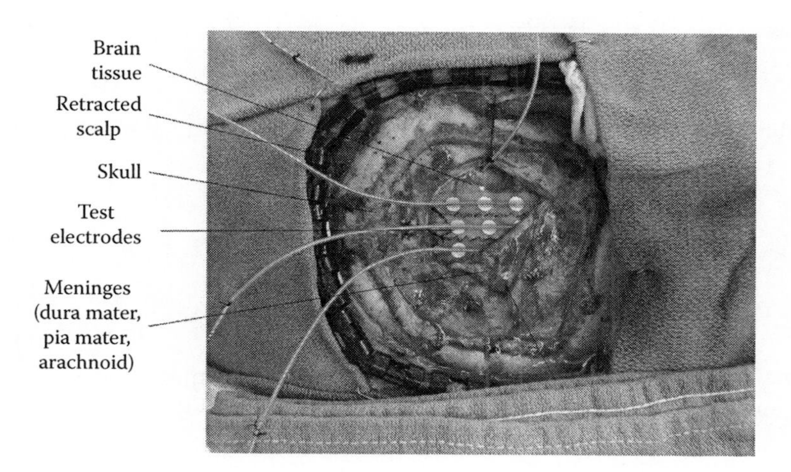

FIGURE 86
A color version of this figure follows page **176.** The brain exposed for testing and possible surgery.

Procedure

Craniotomy may be performed under general anesthesia, but is often done with local because the surgeon may need to communicate with the patient regarding the effects of specific actions on the brain.

An incision is made in the scalp above the lesion, and the scalp is pulled back to expose the skull. The scalp is well vascularized, so care must be taken to cauterize or tie off any blood vessels that are severed during this process. When the scalp is out of the way, a surgical drill or saw is used to cut through the skull. After suitable access is made, the linings of the brain are incised and pulled back to open the area of the lesion (Figure 86).

If possible, the lesion is removed and/or a repair is made; at the least a sample may be taken for lab analysis, or the internal pressure is simply relived if the underlying cause has resolved itself. Any blood vessels in the brain that were cut are sealed off, the brain linings are sutured, and the bone section that was removed is put back in place, held in position by special closures. The scalp incision is closed, and the patient brought out of anesthesia.

Expected Outcome and Follow-Up

Relief of symptoms, and/or procurement of a tissue sample for lab analysis. The results of the surgeon's observations and the lab findings may be used to help determine a course of treatment. Neurological evaluations are performed regularly following surgery to try to detect any problems before they become too severe. Medications are given to help avoid swelling of the brain that can occur due to the surgical intrusion. Normal activities are slowly resumed.

Complications

Bleeding is serious if it is inside the skull, because it can produce further pressure buildup. Some collateral damage to brain tissue may occur as an unavoidable part of the surgical process; this damage may have neurological consequences.

Cystectomy

Alternate names—Urinary cystectomy, bladder removal, urinary bladder removal.

Purpose

To remove a diseased portion of the urinary bladder.

Indications

Bladder cancer; possibly intractable bladder infections, endometriosis, or physical trauma to the bladder.

Anatomy

The urinary bladder is a muscular sac that receives urine from the kidneys via two ureters, with a sphincter muscle that holds urine in the bladder until urination. The urethra leads from the bladder to the outside (Figure 87).

Pathology

Cancer can develop in the wall of the bladder (Figure 88). As well, some infections can cause damage to the bladder wall and interfere with normal function. Accidents may damage the bladder.

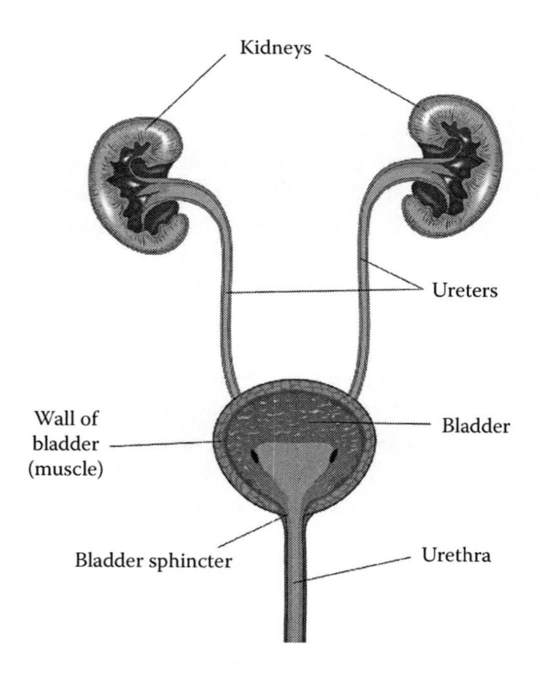

FIGURE 87
Urinary system.

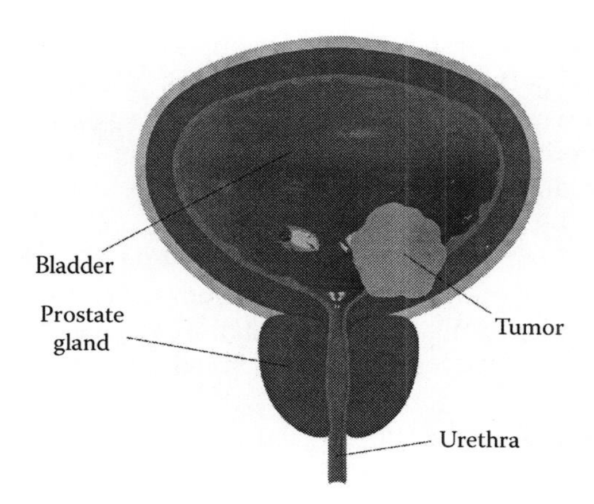

FIGURE 88
A color version of this figure follows page **176.** Bladder cancer.

Physiology

The bladder holds urine until it can be eliminated.

Staffing

Normal surgical team.

Equipment and Supplies

Standard surgical.

Preparation

Standard surgical.

Procedure

Somewhat different surgeries are performed depending on the extent of the cancer and the final goal of the process. All are performed under general anesthesia, and an incision is made in the lower abdomen to provide access to the bladder.

If the tumor is small and well defined, in an open part of the bladder, only a section of the bladder wall containing the tumor is removed, with a margin to try to ensure that all of the tumor is taken. The bladder wall is then sutured back together and the abdominal incision closed.

For larger tumors or other severe conditions, the entire bladder must be removed. Vessels supplying blood to the bladder are closed off and severed, muscles and ligaments supporting it are separated, and the ureters and urethra are cut away. The bladder is then removed from the abdomen.

After removal of the bladder, an alternate means of handling urine must be provided. The ureters are too small to be able to form a stoma through the abdominal wall, so a short section of the small intestine may be used. One end is closed off, the ureters are attached, and the other end is passed through an incision in the abdomen to form a stoma. Alternatively, a pouch may be formed using portions of intestine with a stoma leading through the abdominal wall. This approach means that urine is not continuously passed out through the stoma but is emptied by the insertion of a catheter through the stoma. Finally, a similar pouch may be connected to the urethra instead of a stoma, especially if the sphincter muscle in the ureter could be saved.

Bladder cancer has the potential of spreading to other nearby organs. If this is suspected, a radical cystectomy may be performed in which some of these organs are also removed. In females this may include the ovaries, uterus, fallopian tubes, and part of the vagina. In males, the prostate and seminal vesicles are removed. In both sexes, lymph nodes in the area are taken out.

After the basic procedure is complete, the abdominal incision is closed.

Expected Outcome and Follow-Up

Freedom from the cancer or other problem that necessitated surgery, with manageable handling of urine output. The patient is educated regarding care of the stoma or artificial bladder. For stomas, an exterior pouch is worn to collect urine. Since all procedures involve surgery on part of the intestine, food and water intake is not permitted at first; foods are reintroduced gradually after a couple of days.

Complications

Males may become impotent, and females will be infertile if a radical cystectomy was performed. There may be urine leakage from the site.

Cystoscopy

Alternate names—n/a.

Purpose

To examine the interior of the lower urinary system, including the bladder, the bladder sphincter, the urethra, and in males, the prostate.

Indications

Known or suspected problems in the lower urinary system. These may be indicated by blood in the urine, incontinence, urinary tract infections, painful or difficult urination, and injuries to the urinary tract.

Anatomy

The bladder collects urine from the kidneys and holds it until it can be released via urination. Sphincter muscles are at the exit of the bladder, and the urethra leads from the bladder to the outside. The bladder and urethra are held in place by a set of muscles and ligaments (Figure 89).

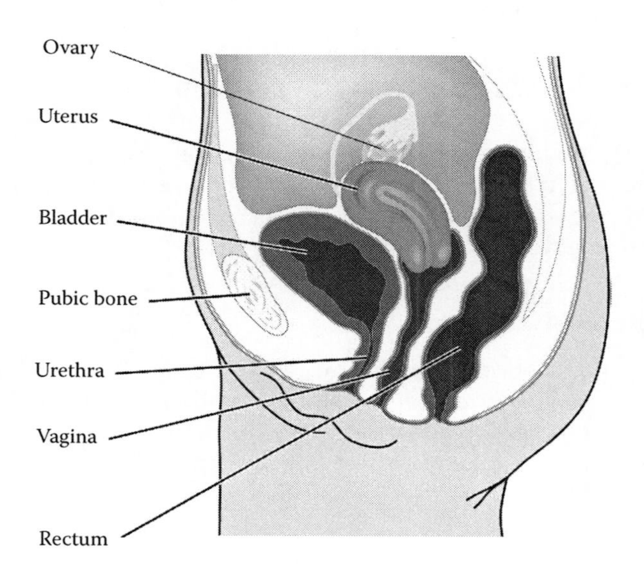

FIGURE 89
Bladder and adjacent organs, female.

The prostate is a gland that surrounds the urethra just below the bladder in males.

Pathology

If the bladder sphincter muscles are weakened, or if the muscles and ligaments supporting the bladder and urethra are weakened, involuntary urine flow (incontinence) may occur, especially at times of stress or pressure such as sneezing, coughing, exercising, or laughing.

Normally the prostate is small, but disease processes can cause it to become enlarged, which can cause problems with urination and sexual function. Prostate cancer is potentially fatal and must be detected and treated early for the best chance of survival.

Physiology

The bladder sphincter holds urine in the bladder until urination is voluntarily begun.

The prostate contributes components of seminal fluid.

Staffing

Physician, assistant.

Equipment and Supplies

Minor surgical, cystoscope.

Preparation

The patient is sedated.

Procedure

The cystoscope, a thin tube that may be rigid or flexible, is lubricated and inserted into the urethral opening. Fluid is pumped in through the tip of the scope to increase visibility, and light for video viewing is carried in both directions by fiber-optic channels in the scope. Another channel in the scope can be used to remove tissue samples or small stones. Video images are recorded for diagnostic purposes. When the desired images and samples have been obtained, the scope is removed.

Expected Outcome and Follow-Up

The acquisition of diagnostically useful images and other information; the removal of tissue samples for testing; the removal of stones or other objects or structures from the tract. The patient will probably experience painful urination for a day or two following the procedure; increased fluid intake may help alleviate this. Normal activities are resumed within a few days.

Complications

Perforated bladder.

Defibrillation

Alternate names—Synchronized cardioversion (in the case of atrial fibrillation correction).

Purpose

To stop cardiac fibrillation and restore normal cardiac contraction cycles.

Indications

Atrial or ventricular fibrillation of the heart. This is determined by physical symptoms and confirmed by electrocardiogram.

Anatomy

The heart is a muscular organ consisting of left and right atria, left and right ventricles, four valves, and associated blood vessels. These blood vessels carry the blood pumped by the heart out to various body parts and also carry blood to the heart itself. The atria are thinner walled than the ventricles. Nerves and other pathways in the heart carry electrical signals to control pumping contractions.

Pathology

Heart pumping effectiveness is only assured if the muscles of the heart contract in a coordinated fashion. The conductive pathways that help produce and coordinate contractions may be damaged, so that contractions are either weaker or less coordinated, or both. If the conductive pathways to the atria are compromised, the atria may begin to contract rapidly and not in coordination with ventricular contractions. This condition is called atrial fibrillation (A fib), and it results in reduced blood flow but still enough to sustain life. If the ventricular conduction pathways are disrupted, the ventricles may begin to contract in a completely uncoordinated way (V fib). Because the ventricles provide most of the pumping action of the heart, this results in cessation of blood flow and, if not corrected quickly, death.

The heart may reach a state where no muscle contractions occur at all, a condition called asystole. In this case, no blood flow takes place, and death will result quickly if contractions cannot be reinitiated.

When fibrillation occurs, an electrical shock delivered to the heart can restore normal contraction rhythm (defibrillation). This type of shock can restart contractions when asystole occurs, and may be referred to as defibrillation but technically is not, because there was no fibrillation to begin with.

Because muscle contractions caused by a defibrillation shock can cause limb flailing, and also because some of the electrical current may pass through other parts of the patient's body, other staff members must stay clear of the patient when shocks are delivered.

Physiology

Heart pumping effectiveness can be compromised by various problems within the heart. The muscles of the atria and especially the ventricles can be damaged by disease or lack of blood supply (cardiac infarction). The cardiac valves can be damaged by disease or age, thus allowing backflow of blood. Finally, the conductive pathways that help produce and coordinate contractions may be damaged, so that contractions are either weaker or less coordinated, or both.

Signals from a center in the heart produce electrical signals that are carried throughout the heart to produce contractions. Various components provide delays to the signal so that contractions are coordinated and allow effective pumping.

Staffing

Physician, nurses, possibly a respiratory therapist. Most hospitals have a team or teams that respond to cardiac emergencies, often called the code X team, where the X is a specific term for that institution, often blue.

Equipment and Supplies

Defibrillator, cardiac drugs, suction machine, endotracheal tube, hand ventilation bag (Figure 90, Figure 91).

Preparation

The preparations for atrial and ventricular defibrillation are quite different, because V fib is an extreme emergency situation, while A fib is relatively benign and can be handled in a more relaxed, scheduled manner.

With A fib, the patient is stabilized with medication as much as possible and is sedated before the procedure. Emergency equipment is on hand, and the patient is connected to a physiological monitor. The sites for defibrillator electrodes are shaved and cleaned carefully

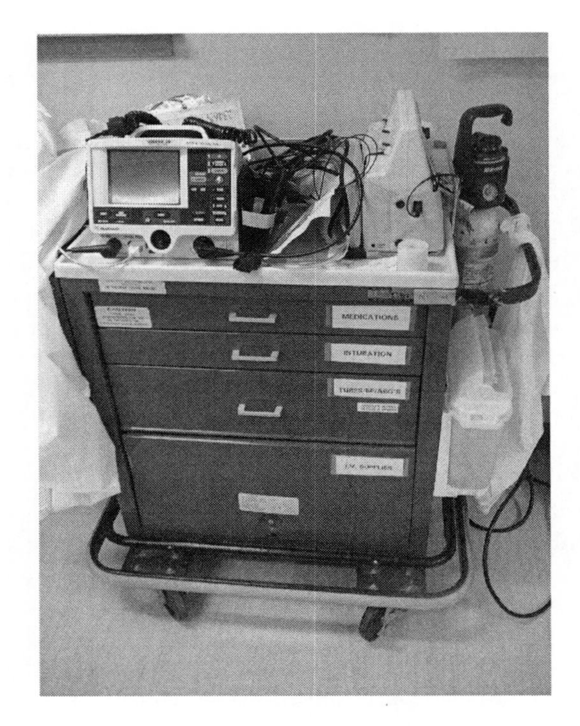

FIGURE 90
Crash cart with defibrillator. A suction pump and oxygen tank are also on the cart, with drugs and other supplies in the cart drawers.

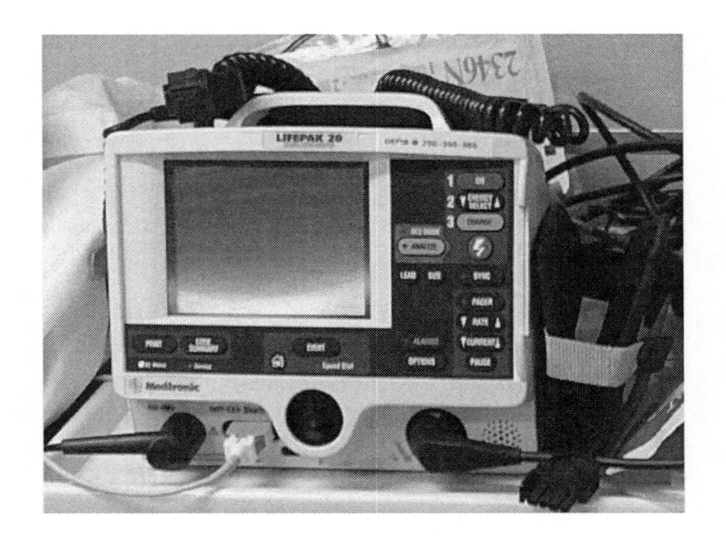

FIGURE 91
Defibrillator.

to give optimum electrical conductivity. The patient may be restrained physically, and a mouth guard may be provided because the defibrillation shock can cause significant contractions in other muscles of the body. Conductive paste or conductive adhesive pads may be placed on the chest in optimum locations.

V fib is performed in an extreme emergency mode. Cardiopulmonary resuscitation may be carried out until the team is ready to do the defibrillation, and some cardiac drugs may be administered, especially if the patient already has an IV line in place.

In either case, care must be taken to avoid conductive pathways between the paddles, such as trails of saline or blood, or bridges of conductive paste.

Procedure

For defibrillation to be successful in A fib, the shock must be delivered in coordination with the ongoing ventricular contractions, otherwise there is a risk of disrupting those contractions and causing V fib or asystole. The defibrillator must be capable of detecting the ventricular contractions and timing the discharge so that it occurs at the optimal point in the cycle to eliminate atrial fibrillation and reinitiate normal, coordinated atrial contractions. Electrodes on the patient's chest pick up ECG signals and transmit them to the defibrillator, where they are analyzed and the optimal shock point determined. The physician selects a level of shock that is enough to produce defibrillation without being too great. The defibrillator is charged, and the shock deliver paddles are placed on the patient's chest in optimal locations (Figure 92).

Staff is advised to keep clear of the patient, and the discharge buttons on the paddles are depressed. When the defibrillator determines that the optimal point in the ECG cycle is reached, the charge is delivered to the patient. The patient is monitored to check for successful defibrillation; if the procedure did not produce the desired effect it may be

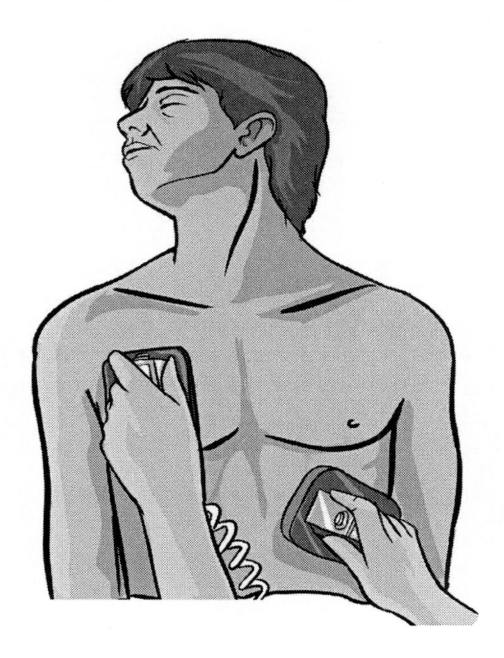

FIGURE 92
Defibrillation paddle placement.

repeated, sometimes after the administration of cardiac drugs, and often at a slightly higher energy setting.

The procedure for V fib is more hurried. The patient's chest is exposed, and paddles are placed on the chest. Sometimes ECG electrodes from the defibrillator are placed to provide an ECG signal, but units are designed to pick up an adequate signal from the paddles themselves. Conductive paste is smeared on the paddle faces to provide better conductivity. Cardiac drugs may be administered, often by injection directly into the heart. Once it is confirmed that the patient is in V fib or asystole, a relatively high energy level is selected, the defibrillator is charged, the operator advises other staff to keep clear, and the discharge buttons on the paddles are depressed, immediately delivering the shock. The ECG waveform is observed on the defibrillator, and if fibrillation is still occurring, another shock is delivered, usually at a somewhat higher energy level. Cardiac drugs may be administered between shocks. If normal ECG rhythm cannot be reestablished after several attempts at maximum energy level, action is usually discontinued and the patient pronounced dead.

Expected Outcome and Follow-Up

Defibrillation and resumption of normal cardiac contractions. The patient is monitored to ensure that normal cardiac function continues. Tests are performed to determine what and how much damage may have been done to the cardiac muscles during the fibrillation episode; this is more important following V fib.

Complications

Due to the high energy levels involved, the patient may have burns at the site where the paddles were placed.

Dilatation and Curettage

Alternate names—D&C.

Purpose

D&C has various purposes, depending on the situation: to obtain a tissue sample from the lining of the uterus in order to test for cancer, polyps, or fibroid tumors; to remove polyps or fibroids; to help reduce excessive menstrual bleeding; to perform early abortion; or to remove tissue from the uterine lining that remains after miscarriage, abortion, or birth.

Indications

Any of the above conditions.

Anatomy

The uterus is a pear-shaped smooth-muscle organ in which a fetus develops in pregnancy. The uterus opens into the vagina through the cervix (Figure 93).

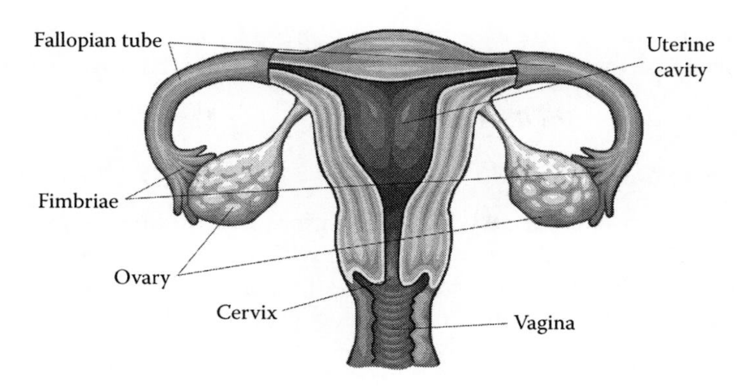

FIGURE 93
Uterus and associated structures.

Pathology

Excessive buildup of the uterine lining during the menstrual cycle can result in excessive bleeding and cramping. Small growths called polyps, or others called fibroid tumors (benign), may form in the lining; these structures may cause pain or blockages. Uterine cancer is the most common cancer of the female reproductive system and, as with other cancers, can be fatal if not treated or removed.

Physiology

During the menstrual cycle, the lining of the uterus thickens and increases in vascularization in preparation for a possible pregnancy. If no pregnancy occurs, the lining is shed as menstrual bleeding, and the cycle is repeated.

Staffing

Minor surgical, or physician (usually a gynecologist), assistant.

Equipment and Supplies

Vaginal speculum, cervical dilation instruments, curette.

Preparation

Local anesthetic or general anesthesia may be used.

Procedure

A speculum is inserted into the vagina and opened to provide access to the cervix. A series of tapered cervical dilation instruments, each larger than the last, are pressed into the cervix to slowly expand (dilate) the cervical opening. When the opening is sufficient, a curette is inserted into the uterus. The curette is spoon-shaped and has edges that can be used to scrape the lining of the uterus. The resulting material is aspirated, with samples taken for later lab analysis as required. When the scraping process is complete, the curette is removed, followed by the vaginal speculum. During the complete procedure, the physician checks tissues and structures that are visible for any abnormalities.

Expected Outcome and Follow-Up

Collection of tissue samples for analysis; alleviation of symptoms. Cramping is likely to occur following the procedure, but should abate within a day or two. Some bleeding may persist for a week or two. Sexual intercourse and tampon use should be avoided for 2 weeks, but other activities can be resumed almost immediately.

Complications

n/a

Discectomy

Alternate names—n/a.

Purpose

To remove an intervertebral disc in order to resolve problems that are associated with compression of spinal nerves in the area.

Indications

Back pain, associated pain and other symptoms resulting from spinal nerve compression including leg pain, weakness if the legs, and impaired bowel or bladder function. Physical examination and diagnostic imaging can determine that disc problems are the cause of these symptoms. Significant muscle weakness or bladder or bowel function impairment requires immediate intervention.

Anatomy

The spinal column is made up of a number of vertebral bones that surround the spinal cord. The vertebrae are connected with muscles and ligaments, and a fibrous disc between each pair serves as a bearing surface and shock absorber. Nerves branch out from the spinal cord between each pair of vertebrae to supply various areas of the body (Figure 94).

Pathology

Trauma, disease, or age can cause the intervertebral discs to thin out, break apart, or bulge (prolapse) either inward toward the spinal cord or outward into the area where spinal nerves are located. Any of these situations can result in pressure being placed on some of the spinal nerves, causing various symptoms depending on the location. This tends to

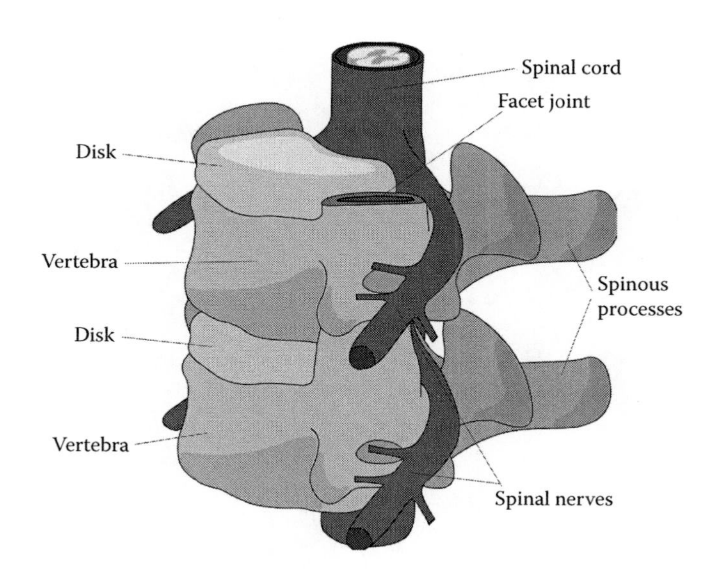

FIGURE 94
Spine anatomy.

happen most often in the lower back, in the lumbar or sacral areas. Symptoms of nerve compression in this area include back pain, pain along the course of the nerve, muscle weakness in the leg, or bladder or bowel dysfunction (Figure 95).

Physiology

Spinal nerves carry signals in and out of the spinal cord.

Staffing

Normal surgical team; usually a neurosurgeon.

Equipment and Supplies

Standard surgical, arthroscopy equipment if the procedure is done arthroscopically.

Preparation

Standard surgical.

Procedure

The procedure is usually done under general anesthesia. An incision is made along the spine to expose the area of the disc. Nerves and other structures in the area are gently moved aside to provide access to the disc. The disc material is removed. In some cases, the disc may be replaced by an artificial material. If necessary, the ends of the vertebral pair are abraded to encourage them to fuse together (spinal fusion); bone fragments from other parts of the patient's body may be added to increase fusion effectiveness. When the

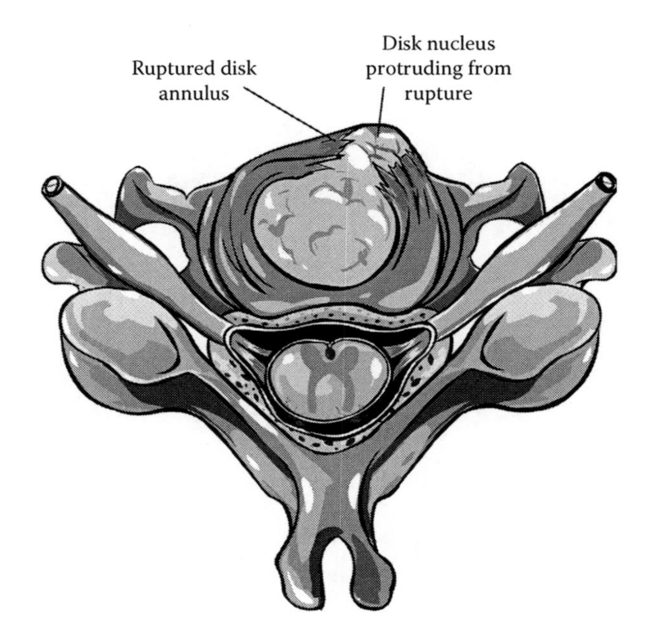

FIGURE 95
Ruptured disc from above.

activity involving the disc is completed, the surrounding tissues are moved back to their original locations and the incision closed.

Arthroscopic procedures are done in much the same manner but using smaller incisions and arthroscopy instruments.

Expected Outcome and Follow-Up

Alleviation of pain and other symptoms. The patient must avoid activities that might put stress on the surgical area for two to four weeks, though prescribed exercises are important to regain strength, mobility, and stability.

Complications

Nerve damage.

Electrocardiogram

Alternate names—ECG, EKG, heart tracing, cardiac monitoring, 12-lead.

Purpose

To obtain diagnostic or monitoring measurements of electrical activity in the heart.

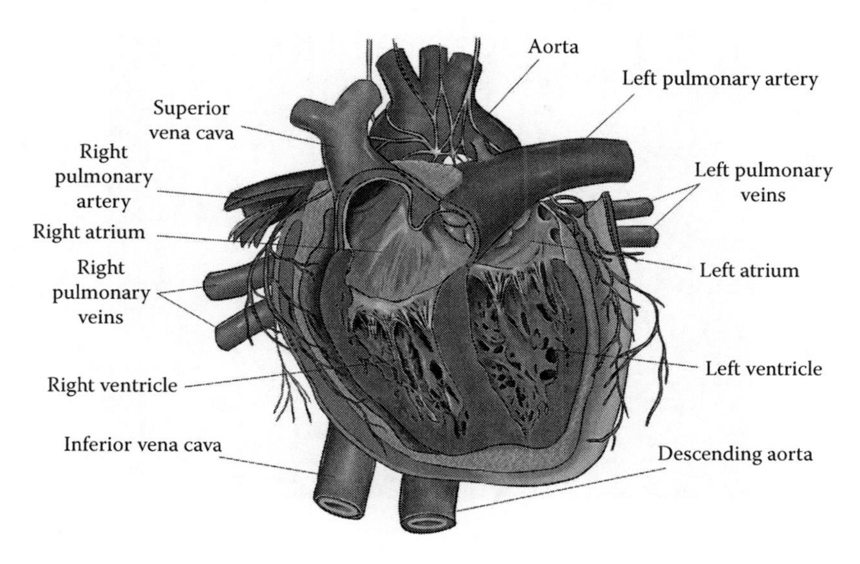

FIGURE 96
Cardiac anatomy.

Indications

Known or suspected heart problems.

Anatomy

The heart is a muscular organ consisting of left and right atria, left and right ventricles, four valves, and associated blood vessels (Figure 96). These blood vessels carry the blood pumped by the heart out to various body parts and also carry blood to the heart itself. The atria are thinner walled than the ventricles. Nerves and other pathways in the heart carry electrical signals to control pumping contractions.

Pathology

Heart pumping effectiveness can be compromised by various problems within the heart. The muscles of the atria and especially the ventricles can be damaged by disease or lack of blood supply (cardiac infarction). The cardiac valves can be damaged by disease or age, thus allowing backflow of blood. Finally, the conductive pathways that help produce and coordinate contractions may be damaged, so that contractions are either weaker or less coordinated, or both.

Physiology

Signals from a center in the heart produce electrical signals that are carried throughout the heart to produce contractions. Various components provide delays to the signal so that contractions are coordinated and allow effective pumping.

Staffing

For bedside ECG monitoring, nursing staff or physician; for 12-lead recordings, cardiology technologist; a cardiologist or internist will later interpret the results.

FIGURE 97
ECG machine with keyboard for patient data entry and LCD screen for previewing waveforms.

Equipment and Supplies

ECG monitor or machine. And ECG monitor is usually part of a physiological monitor that measures other vital signs such as blood pressure, respiration, blood oxygen saturation, and more; this information is displayed on a video screen at the bedside. An ECG machine performs only ECG measurements, in a greater variety and at higher accuracy than a monitor, and produces large format recording for analysis (Figure 97).

Preparation

The electrode sites are shaved and cleaned carefully to ensure optimal conductivity.

Procedure

Electrodes are placed on the skin in specific locations, and the signal is checked to ensure quality. For monitoring, alarm levels are set to provide a signal to staff if the heart rate becomes too high or too low. For a 12-lead, the recording is initiated, which stores the signals in the machine and allows them to be printed out on a large sheet of paper (Figure 98). For monitoring, the system begins receiving and analyzing the ECG signals.

Expected Outcome and Follow-Up

A useful recording of ECG activity is obtained for analysis and long-term monitoring. Depending on the results, a course of treatment for cardiac disease may be initiated.

Complications

Some patients may be sensitive to the adhesive used in the electrodes, and skin irritation may occur.

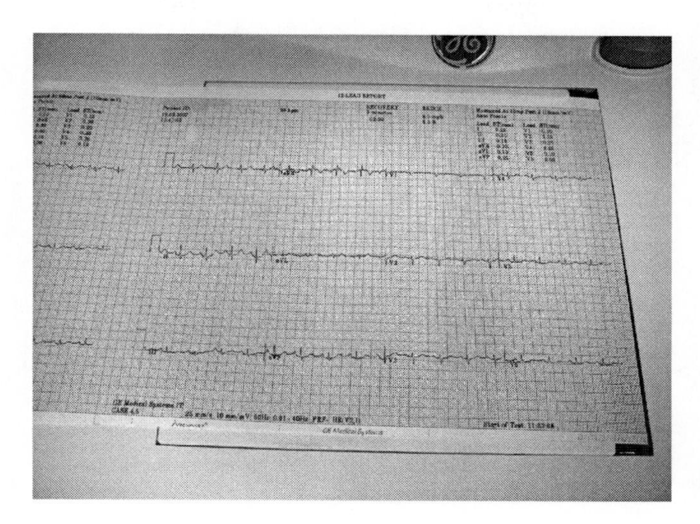

FIGURE 98
Output of waveforms from ECG machine.

Electroconvulsive Therapy

Alternate names—Shock treatment, shock therapy, electroshock therapy, ECT.

Purpose

To alleviate the symptoms of certain psychiatric conditions, especially intractable depression, mania, and schizophrenia.

Indications

Psychiatric conditions that do not respond to other forms of treatment such as medication or psychotherapy.

Anatomy

The brain is the center of the nervous system, providing cognition, sensation, motor control, and reflex activity (Figure 99).

Pathology

Abnormalities in the structure and/or function of the brain can produce various mental disorders.

Physiology

The brain functions as an electrochemical processor, handling input information from nerve receptors and providing output to muscle and other body systems control.

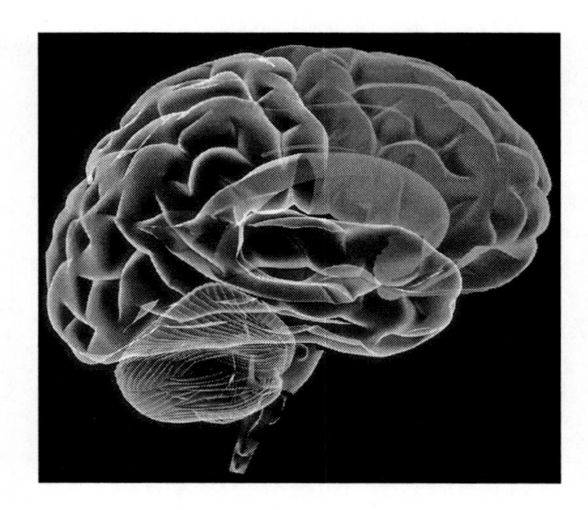

FIGURE 99
A color version of this figure follows page **176.** The brain.

Staffing

Psychiatrist, assistant.

Equipment and Supplies

Anesthetic supplies, ECT machine, muscle stimulator, vital signs monitor, EEG monitor. The latter two or three components may be integrated into the ECT machine, or they may be stand-alone. The ECT machine may also incorporate a sensor to detect movement of fingers or toes (Figure 100).

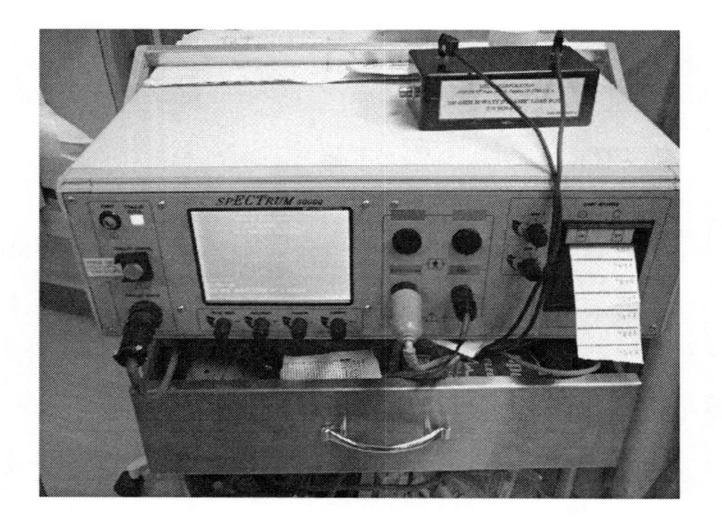

FIGURE 100
Electroconvulsive therapy machine, with controls for output waveforms, display screen, graphic recorder, and discharge button.

Preparation

The patient is sedated and an intravenous line started. The limbs are secured to the bed, and a mouth guard is put in place. ECG and EEG electrodes are applied and checked for contact. The ECT electrodes are placed, usually at or near the temples. Conductive paste is used to help ensure good conductivity. The amplitude and duration of the shock has been predetermined by the attending psychiatrist.

Procedure

An anesthetic agent is given to induce unconsciousness, and a muscle-paralyzing agent is given to prevent excess muscle contractions which could cause injuries or strains. The agent acts only on skeletal muscles, so heartbeat and respiration continue normally. A pressure cuff is placed on a limb, usually one leg, and inflated to prevent the paralyzing agent from reaching that area.

The operator ensures that all is ready and makes the final connections to the ECT electrodes. Shock buttons are depressed, causing the shock to be delivered. The shock induces a brain seizure. The effectiveness of the seizure is evaluated by observing the twitching of digits on the nonparalyzed limb or by use of a motion sensor on the digits. EEG and ECG waveforms are recorded. Occasionally the treatment is repeated.

After the treatment is determined to have been successful, a drug is given to counteract the paralyzing agent, and muscle control returns. The patient is returned to consciousness and the electrodes, restraint traps, and mouth guard removed.

Expected Outcome and Follow-Up

Alleviation of target symptoms. The patient's vital signs are monitored for a while following treatment to ensure all is well. Repeat treatments may be required, because they seem to have a cumulative effect.

Complications

Cognitive impairment, usually only short term; memory impairment in the form of amnesia and short-term memory loss; elevation of heart rate and blood pressure, sometimes to life-threatening levels; nausea and vomiting; muscle pain.

Electroencephalogram

Alternate names—EEG, brainwave monitoring.

Purpose

To obtain diagnostically useful recordings of brain electrical activity.

Indications

Known or suspected brain abnormalities, including stroke, brain tumors, certain mental illnesses, Alzheimer's, Parkinson's, and especially epilepsy. EEG may be used to determine brain death.

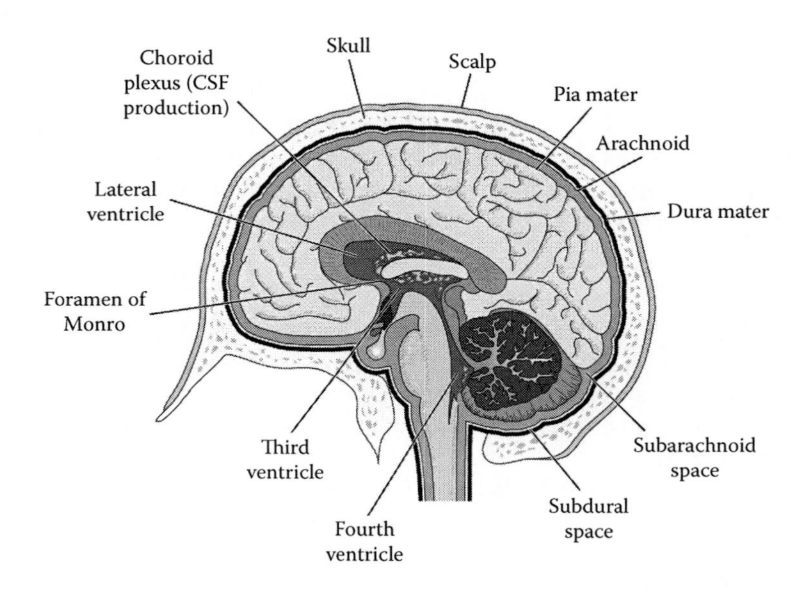

FIGURE 101
Cross section of brain.

Anatomy

The brain is the center of the nervous system, providing cognition, sensation, motor control, and reflex activity (Figure 101).

Pathology

Abnormalities in the structure and/or function of the brain can produce various disturbances of brain activity.

Physiology

The brain functions as an electrochemical processor, handling input information from nerve receptors and providing output to muscle and other body systems control.

Staffing

EEG technologist, psychiatrist.

Equipment and Supplies

EEG machine, stimulation apparatus (Figure 102).

Preparation

EEG electrodes are placed in specific positions in the patient's scalp and other areas of the head. Usually between 16 and 21 electrodes are used.

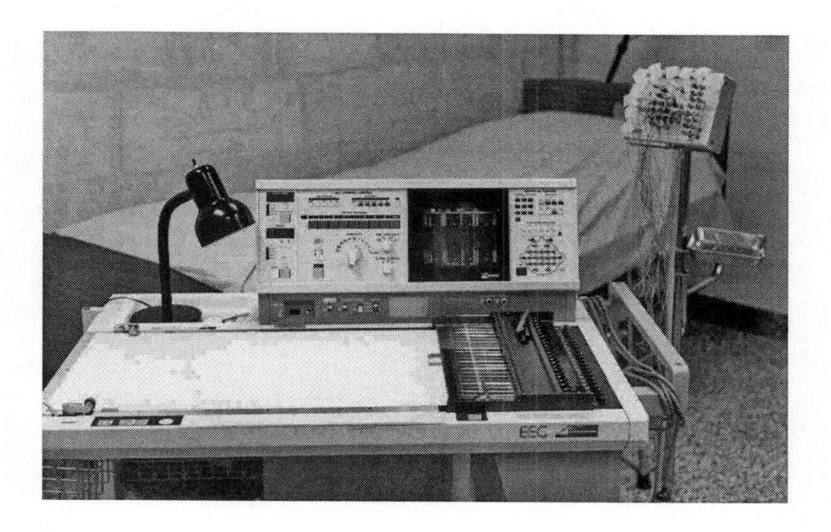

FIGURE 102
Electroencephalograph machine, with multichannel recorder, control panel, and patient electrode board (background right).

Procedure

The patient is connected to the EEG machine, and recording commences. The patient may be asked to perform certain activities, or various stimuli may be applied including flashing lights, viewing of certain pattern designs, and specific sounds.

Recordings may be made during sleep or, using an ambulatory recorder, during daily activities.

After a satisfactory recording is obtained, the electrodes are removed.

Expected Outcome and Follow-Up

Procurement of diagnostically useful recordings of brain electrical activity. The results of the test may help diagnose problems and guide a course of treatment.

Complications

Skin sensitivity to the electrodes may be seen. Seizures may be induced by the applied stimuli.

Episiotomy

Alternate names—Perineal incision, birth canal enlargement.

Purpose

To increase the diameter of the birth canal during labor in order to facilitate delivery.

Indications

Restricted size of the vaginal opening.

Anatomy

The birth canal consists of the exit to the uterus (the cervix) and the vagina, which are surrounded by the bones of the pelvis.

Pathology

Physical differences or deformities of the birth canal, or an unusually large fetus, can mean that the vaginal opening is not large enough for the fetus to pass through, at least not without traumatic tearing of the tissues in the area.

Physiology

When the fetus reaches full term, hormonal changes cause the uterus to begin contracting in cycles. The contractions gradually become stronger and more frequent, and the pressure of the contractions causes the cervix to dilate. When the cervix is dilated to the maximum extent, further uterine contractions and "pushing" by the mother move the fetus through the birth canal and into the outside world (Figure 103).

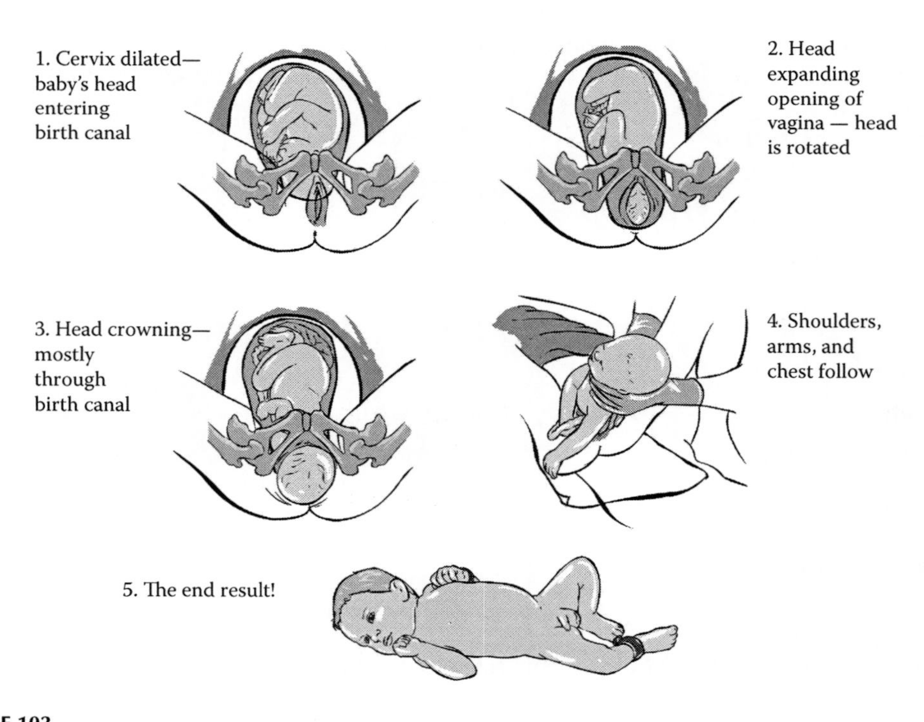

FIGURE 103
Birth stages. An episiotomy may be performed if the birth canal opening is restrictive.

Staffing

Physician, usually an obstetrician.

Equipment and Supplies

Minor surgical.

Preparation

Episiotomy is usually an urgent or emergency procedure, so preparation is not possible.

Procedure

A cut is made, usually from the edge of the vaginal opening closest to the anus, in the direction of the anus. The cut may be just through the adjacent vaginal and perineal tissue, or it may extend to the anus or even through the anal sphincter and possibly into the anal wall. This ideally opens the canal enough that delivery can take place. After delivery, the incision is sutured.

Expected Outcome and Follow-Up

Eased delivery of the fetus. The incision is checked to ensure it is holding together and to guard against infection, and the area is cleaned regularly and carefully.

Complications

Larger incisions result in greater chances of complications. Pain, sexual intercourse discomfort, and rectal damage may occur.

Fetal Monitoring

Alternate names—Fetal heart rate monitoring, nonstress test.

Purpose

To monitor the heart health of a fetus during pregnancy or labor.

Indications

Known or suspected fetal distress. Some facilities perform fetal monitoring in all pregnancies in order to catch problems at the earliest possible point should they occur. When monitoring is done as a baseline diagnostic test with the mother and fetus in a resting state, it is known as nonstress testing.

Anatomy

The fetus develops within the mother's uterus, a muscle-walled organ that is pear shaped before conception but expands as the fetus grows (Figure 104). The fetus is contained by a

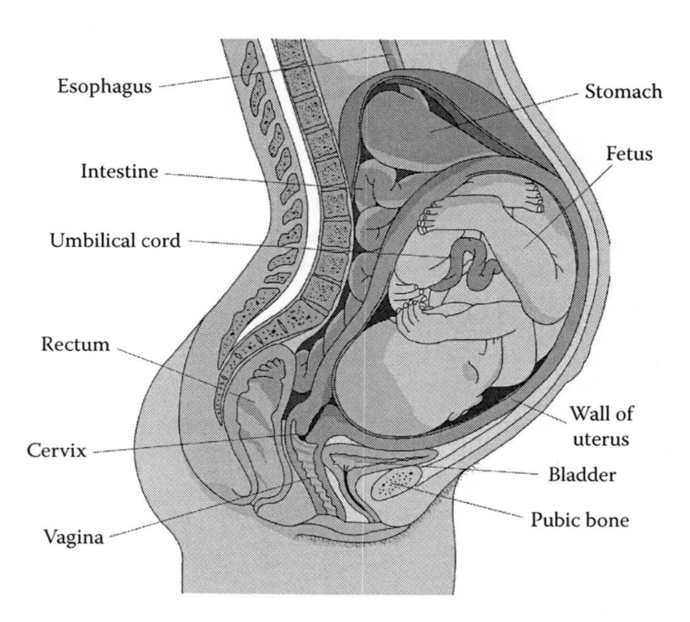

FIGURE 104
Pregnancy near full term.

thin-walled structure called the amniotic sac within the uterus; this sac is filled with fluid that helps protect the developing fetus. The fetus is connected to the mother via its umbilical cord, which carries blood in both directions between the fetus and the placenta. The placenta is a highly vascularized organ that is fetal tissue and has extensive connections to the wall of the uterus.

Pathology

Various problems can cause the fetus to become distressed. Such distress is often indicated by changes in the fetal heart rate. Problems may include development abnormalities such as heart defects, restriction of blood flow to the fetus because of umbilical cord or placental compromise, physical trauma due to accident, or harmful effects of drugs or alcohol ingested by the mother.

Physiology

During pregnancy, the fetus lives within the uterus and amniotic sac. When the fetus reaches full term, hormonal changes cause the uterus to begin contracting in cycles. The contractions gradually become stronger and more frequent, and the pressure of the contractions causes the cervix to dilate. When the cervix is dilated to the maximum extent, further uterine contractions and "pushing" by the mother move the fetus through the birth canal and into the outside world.

Staffing

Obstetrician and/or obstetrical nurse.

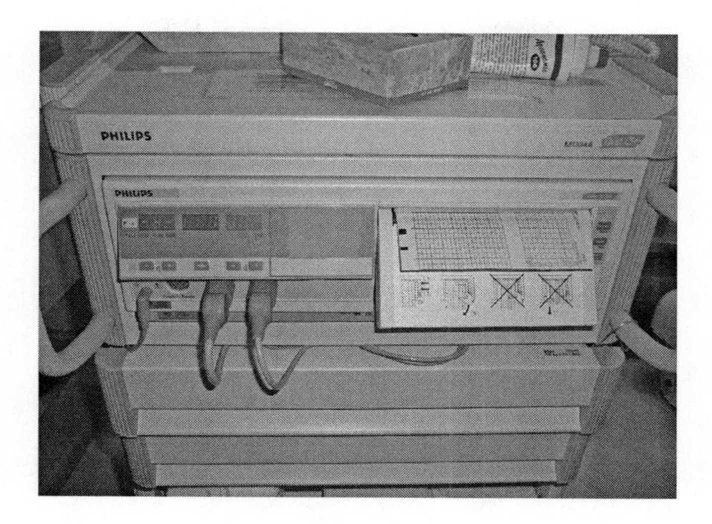

FIGURE 105
Fetal monitor with displays for fetal heart rate and relative uterine contraction strength (toco), chart recorder for these parameters, and connectors for transducers.

Equipment and Supplies

Fetal monitor and accessories. Monitors may be simple Doppler motion detectors that can provide an audible indication of the fetal heartbeat, or they may have multiple sensors that detect the fetal heart signals ultrasonically or electrically, uterine contractions, maternal ECG, and maternal pulse oximetry (Figure 105). Some monitors are able to monitor two fetuses at once. Certain models have radio transmitters that send measurement signals to a receiver, allowing the mother to move around while being monitored.

Preparation

The monitor is connected. Ultrasound transducers are strapped onto the mother's abdomen and ultrasound conducting gel is placed between the transducer and the skin to increase sound transmission (Figure 106). Transducer position is adjusted to give maximum heartbeat signal. If an electrical connection is used, the electrode cable is inserted through the vagina until it contacts the fetus' head. The cable is rotated so that the sharp spiral tip punctures the skin of the fetal scalp, providing good electrical contact.

Procedure

Fetal heart rate is observed and may be recorded. If uterine contractions are being monitored, they will be recorded on the same chart as the fetal heart rate so that the two factors can be correlated. If fetal distress is detected, appropriate action is initiated.

Expected Outcome and Follow-Up

Accurate indications of fetal health are obtained, and signs of fetal distress are noted and acted upon.

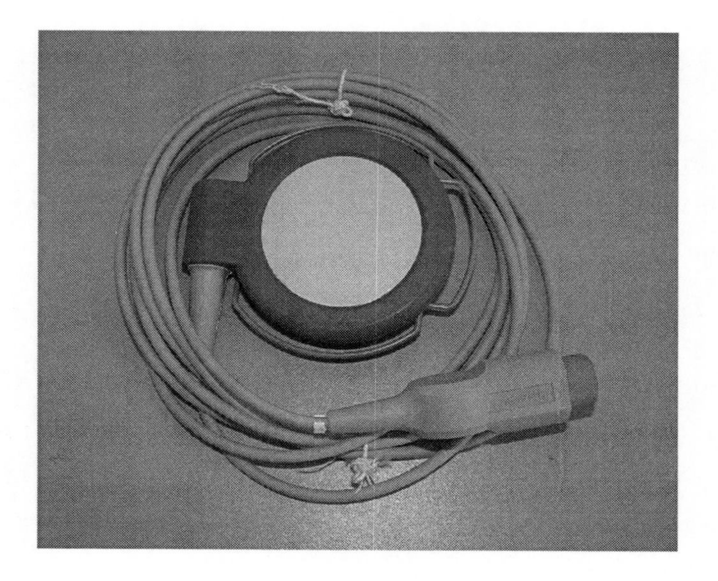

FIGURE 106
Fetal monitor ultrasound probe. Crystals in the face of the probe transmit a focused ultrasound beam to the fetal heart and also receive reflected signals from the heart in order for the monitor to analyze fetal heart rate.

Complications

Very few. Some mothers may be sensitive to the gel used. When fetal scalp electrodes are used, there is a possibility of damage to other areas of the fetus if placement is incorrect.

Gastrectomy

Alternate names—Stomach surgery, stomach removal, stomach resection, gastric resection.

Purpose

To remove all or part of the stomach, which is damaged by disease or trauma.

Indications

Disease or trauma of the stomach, including stomach cancer, intractable gastric ulcers, and large amounts of gastric polyps. Conditions are confirmed by diagnostic imaging.

Anatomy

The stomach is a pouch where the main digestion of food begins (Figure 107). Food enters the stomach from the mouth and esophagus via the esophageal sphincter, a ring of muscle that helps prevent food from being pushed back up the esophagus. Another sphincter, the pyloric, keeps food in the stomach until it is ready to be passed into the small intestine. The

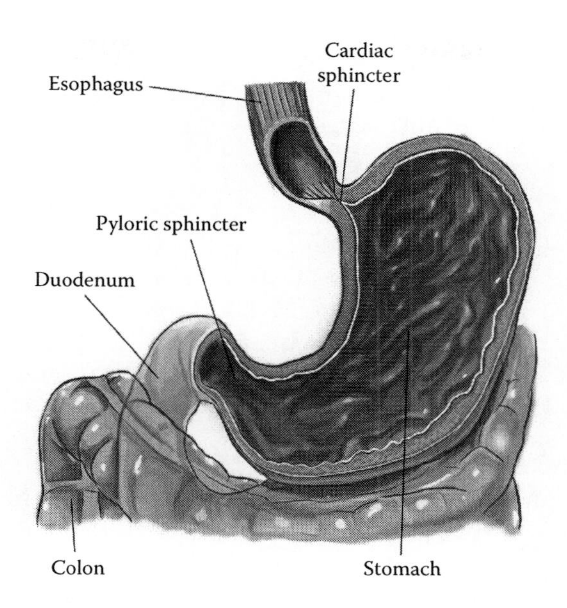

FIGURE 107
Stomach anatomy.

stomach walls are muscular to help move food around inside and promote digestion, but there is little absorption of nutrients in the stomach. From the stomach, food passes into the small intestine, which consists of three main sections: the duodenum, the jejunum (the longest section), and the ileum.

Pathology

Stomach cancer can form tumors in the wall of the stomach which, if untreated, can grow and cause direct problems or can spread to other parts of the body. Gastric ulcers can cause digestive upsets and internal bleeding; if they become perforated, peritonitis can result. Gastric polyps, lobe-shaped growths on the inner wall of the stomach, can interfere with digestion.

Physiology

Glands in the lining of the stomach produce digestive enzymes and hydrochloric acid to start digestion of food. They also produce mucous material to prevent the enzymes and acid from digesting the stomach itself. Further digestive enzymes are added in the small intestine and digestion continues. Nutrients from digested food are absorbed into the bloodstream from the small intestine.

Staffing

Normal surgical team.

Equipment and Supplies

Standard surgical.

Preparation

Standard surgical. A nasogastric tube is inserted to remove gastric fluids from the surgical area during and for a day or so after surgery.

Procedure

Gastrectomy is performed under general anesthesia. An incision is made from the sternum to the umbilicus, and the abdominal wall is retracted to provide access to the area. The diseased portion of the stomach is excised. If a complete gastrectomy is performed, the whole stomach is removed. In either case, associated blood vessels are isolated and cauterized or tied off. The remaining portions of the system are sutured together. In some situations, an artificial stomach may be constructed using parts of the small intestine to form a pouch.

Some forms of gastrectomy can be performed laparoscopically.

Following completion of work on the stomach, the abdominal incision is closed, with drainage tubes in place.

Expected Outcome and Follow-Up

Removal of the diseased or damaged part of the stomach, relief from symptoms. Oral intake is restricted for a couple of days, with hydration being provided intravenously. When the nasogastric tube is removed, clear fluids can be given, followed by soft foods as tolerated.

Complications

Nausea and vomiting.

Heart Transplant

Alternate names—n/a.

Purpose

To replace the patient's diseased or damaged heart with that of a donor, or possibly with an artificial heart.

Indications

Cardiac failure that cannot be alleviated by any other means.

Anatomy

The heart is a muscular organ consisting of left and right atria, left and right ventricles, four valves, and associated blood vessels (Figure 108). These blood vessels carry the blood pumped by the heart out to various body parts and also carry blood to the heart itself.

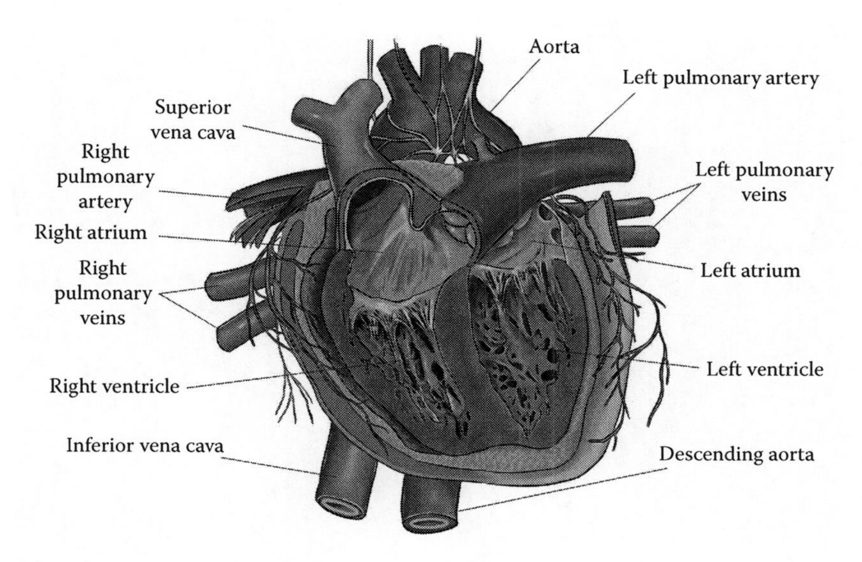

FIGURE 108
Heart and associated blood vessels.

The atria are thinner walled than the ventricles. Nerves and other pathways in the heart carry electrical signals to control pumping contractions. There are four heart valves: right atrioventricular (or tricuspid), pulmonary semilunar, left atrioventricular (or bicuspid), and aortic. The two valves between the atria and the ventricles, the bicuspid and tricuspid valves, have tendons called chordae tendineae that help hold them from being pushed back into the atria when the powerful ventricular contractions occur. The heart is surrounded by a membrane called the pericardium.

Pathology

Heart pumping effectiveness can be compromised by various problems within the heart. The muscles of the atria and especially the ventricles can be damaged by disease or lack of blood supply (cardiac infarction). The cardiac valves can be damaged by disease or age, thus allowing backflow of blood. Finally, the conductive pathways that help produce and coordinate contractions may be damaged, so that contractions are either weaker or less coordinated, or both.

Physiology

The heart pumps blood throughout the body to carry nutrients and wastes for use or disposal. Blood is pumped when the muscles of the heart wall contract, expelling blood from the interior chamber. Valves between the various parts of the heart keep blood flowing in the correct direction by preventing backflow. Flow proceeds from the venae cava (the main veins bringing blood back from the body to the heart) into the right atrium, through the right atrioventricular (or tricuspid) valve, into the right ventricle, through the pulmonary semilunar valve, through the pulmonary artery to the lungs, back via the pulmonary veins to the left atrium, through the left atrioventricular (or bicuspid) valve, into the left ventricle, out through the aortic valve, the aorta, and to the rest of the body.

FIGURE 109
Heart–lung machine monitored by a perfusionist during open heart surgery.

Signals from a center in the heart produce electrical signals that are carried throughout the heart to produce contractions. Various components provide delays to the signal so that contractions are coordinated and allow effective pumping.

Staffing

Complex surgical team, perfusionist.

Equipment and Supplies

Complex surgical, heart–lung machine (Figure 109), donor or artificial heart.

Preparation

The exact configuration and measurements of the patient's diseased heart are determined in advance and confirmed immediately before surgery using diagnostic imaging techniques. A donor heart must be available that is a suitable match in tissue type and size and is available soon enough after donation to be viable. The patient is sedated and the incision site cleaned. Antirejection (immunosuppressive) drugs are given for some time in advance of the surgery.

Procedure

A vertical incision is made over the sternum, and then the sternum is sawn through vertically. A chest retractor is placed in the cut and is mechanically adjusted to hold the chest open. The heart–lung machine is connected to bypass the heart, providing oxygenation and CO_2 removal, as well as temperature regulation during surgery. The heart is usually stopped

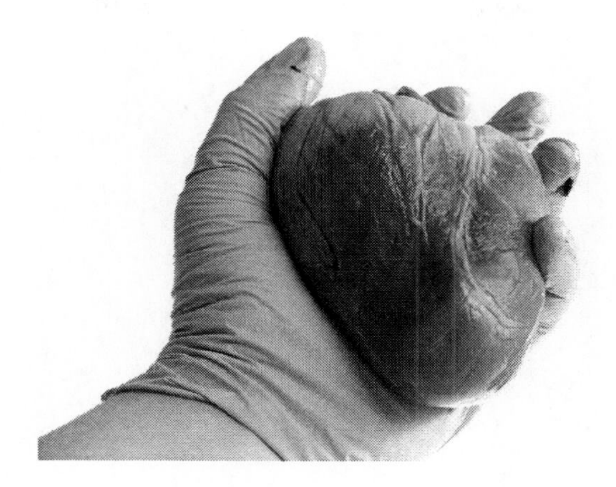

FIGURE 110
***A color version of this figure follows page* 176.** A human heart. This heart is not one to be used for transplant. More adjacent blood vessels would be included to allow for attachment to vessels in the recipient.

using a potassium ion solution, which interrupts the chemical–electrical function of the heart. Newer methods use other drugs that have fewer harmful side effects than potassium. Once the heart is stopped and stable, the pericardium is opened and the heart accessed. Blood vessels connecting to the heart are cut at appropriate locations to allow best attachment of those vessels from the donor heart. The heart is then removed (Figure 110).

The donor heart is placed and blood vessels reconnected. Any air present inside the heart must be removed before closure, or an air embolism can develop. The pericardium is closed. The heart–lung machine is removed, the potassium solution is flushed out, and an electrical stimulus is applied to the heart to start it beating again. Then the chest opening is closed and the skin incision sutured or stapled after thoracic drainage tubes are placed (Figure 111).

The patient will be put on a ventilator immediately following surgery to reduce his physical load.

Expected Outcome and Follow-Up

Normal functioning of the donor heart. Patients are given antirejection (immunosuppressive) drugs indefinitely, but are also regularly monitored for signs of rejection, more frequently immediately following surgery. Drugs and possibly pacemaker signals are used to help maintain normal cardiac rhythms for some time following surgery, until normal function resumes. Patients are monitored in a cardiac intensive care unit (CICU) for several days. Physical activities are slowly initiated in the CICU and increased when the patient can be moved to a step-down unit or general surgical ward. After discharge normal activities can be gradually resumed. Physiotherapy can aid greatly in restoring normal function.

Complications

Rejection of the donated heart is the most serious complication and may require a second transplant. Death can result if this is not possible and rejection cannot be controlled. Bleeding and infection are constant concerns. Other infections may occur as a result of the antirejection drugs used.

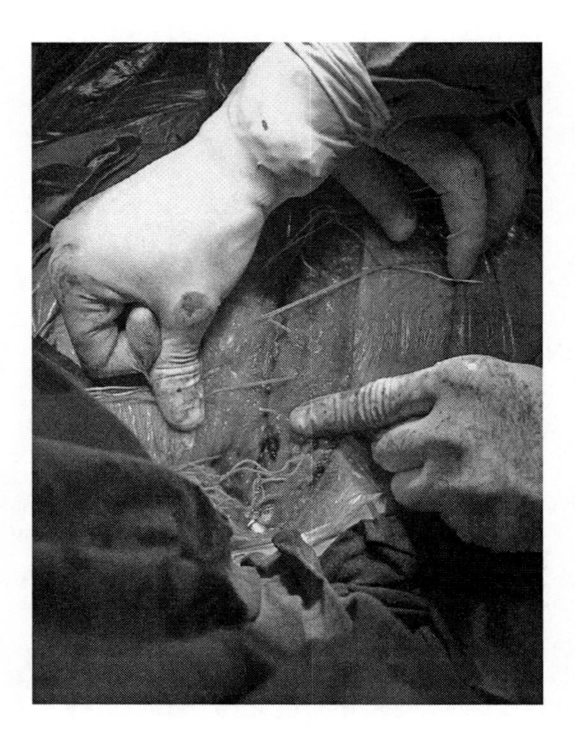

FIGURE 111
Closing the chest after surgery.

Heart Valve Surgery

Alternate names—Valve replacement, artificial heart valve surgery.

Purpose

To repair or replace a damaged heart valve.

Indications

Inadequate blood pumping by the heart; valve problems are confirmed by angiography and/or diagnostic imaging.

Anatomy

The heart is a muscular organ consisting of left and right atria, left and right ventricles, four valves, and associated blood vessels. These blood vessels carry the blood pumped by the heart out to various body parts and also carry blood to the heart itself. The atria are thinner walled than the ventricles. Nerves and other pathways in the heart carry electrical signals to control pumping contractions. There are four heart valves: right atrioventricular (or tricuspid), pulmonary semilunar, left atrioventricular (or bicuspid), and aortic (Figure 112). The two valves between the atria and the ventricles, the bicuspid and tricuspid valves,

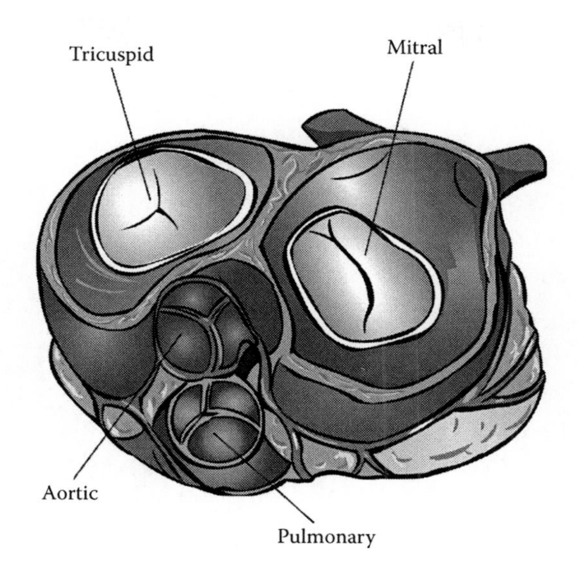

FIGURE 112
Heart valves.

have tendons called chordae tendineae that help hold them from being pushed back into the atria when the powerful ventricular contractions occur. The heart is surrounded by a membrane called the pericardium.

Pathology

Heart pumping effectiveness can be compromised by various problems within the heart. The muscles of the atria and especially the ventricles can be damaged by disease or lack of blood supply (cardiac infarction). The cardiac valves can be damaged by disease or age, thus allowing backflow of blood (valve prolapse). The valve itself may be weakened or damaged so that blood leaks past or through the valve, or the supporting chordae tendineae may be weakened or broken, similarly allowing backflow through the valve. Any backflow decreases the effectiveness of cardiac pumping, and when it is large enough, serious effects are evidenced such as weakness, blackouts, and shortness of breath. Death can eventually result. The mitral valve is most commonly affected by this.

Valves can also become narrowed; this stenosis prevents adequate blood flow and can have similar symptoms to valve prolapse.

Physiology

The heart pumps blood throughout the body to carry nutrients and wastes for use or disposal. Blood is pumped when the muscles of the heart wall contract, expelling blood from the interior chamber. Valves between the various parts of the heart keep blood flowing in the correct direction by preventing backflow. Flow proceeds from the venae cava (the main veins bringing blood back from the body to the heart) into the right atrium, through the right atrioventricular (or tricuspid) valve, into the right ventricle, through the pulmonary semilunar valve, through the pulmonary artery to the lungs, back via the pulmonary veins to the left atrium, through the left atrioventricular (or bicuspid) valve, into the left ventricle, out through the aortic valve, the aorta, and to the rest of the body.

Staffing

Major surgery team, perfusionist.

Equipment and Supplies

Standard surgery, heart–lung machine, replacement valve.

Preparation

The patient is sedated and the incision site cleaned. Details of the damaged valve have been previously determined by diagnostic imaging, including exact sizes and locations of damage. Anticoagulants are given to prevent clotting during and following the procedure.

Procedure

Adequate valve function can sometimes be restored by repair procedures, but usually replacement is required.

Replacement valves may be natural or artificial. Natural valves are taken from animals, usually pigs because their heart anatomy is very similar to that of humans. Artificial valves are made of metal and plastic and come in various configurations.

A vertical incision is made over the sternum, and then the sternum is sawn through vertically. A chest retractor is placed in the cut and is mechanically adjusted to hold the chest open. The heart–lung machine is connected to bypass the heart, providing oxygenation and CO_2 removal, as well as temperature regulation during surgery (Figure 113).

The heart is usually stopped using a potassium ion solution, which interrupts the chemical–electrical function of the heart. Newer methods use other drugs that have fewer

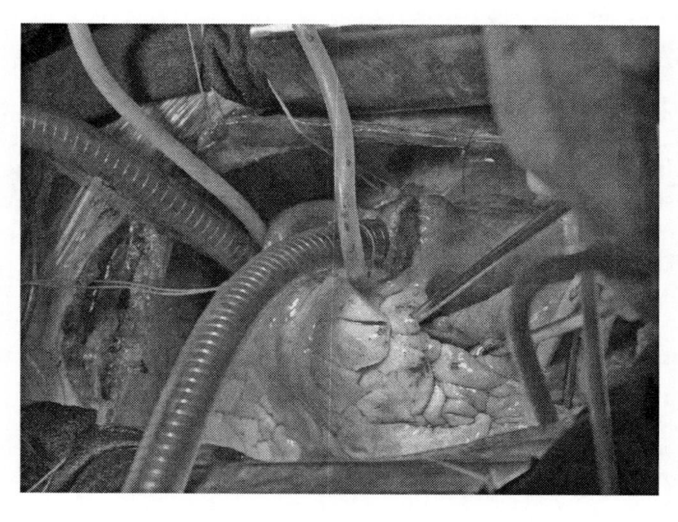

FIGURE 113
A color version of this figure follows page **176.** Heart–lung machine connections.

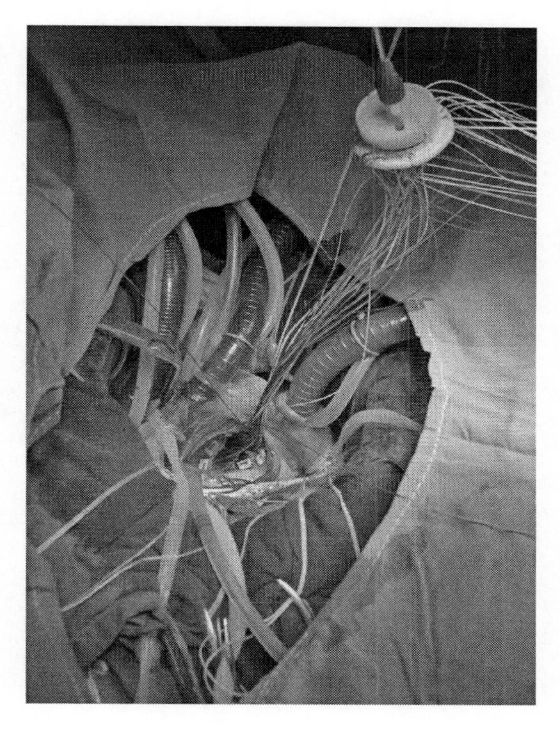

FIGURE 114
A color version of this figure follows page **176.** Artificial valve (at top) ready to be placed into heart. Note multiple sutures connecting valve and heart tissue.

harmful side effects than potassium. Once the heart is stopped and stable, the pericardium and the heart itself are opened to access the mitral valve. The diseased valve is removed by making an incision around the ring of tissue (annulus) that supports the valve. The resulting opening will match the previously selected replacement valve (Figure 114).

Sutures are preplaced around the annulus, the replacement valve is fitted, and the sutures are tightened (Figure 115).

Any air present inside the heart must be removed before closure, or an air embolism can develop. The heart wall is closed, followed by the pericardium. The heart–lung machine is removed, the potassium solution is flushed out, and an electrical stimulus is applied to the heart to start it beating again. Then the chest opening is closed and the skin incision sutured or stapled after thoracic drainage tubes are placed.

The patient may be put on a ventilator immediately following surgery to reduce his physical load.

Expected Outcome and Follow-Up

Resumption of adequate valve function and alleviation of previous symptoms. The patient is very closely monitored in an intensive care unit. If a ventilator was used, it can usually be removed after a few hours. Physiotherapy can help avoid some complications. Normal activities are slowly resumed, with careful monitoring of each stage.

Anticoagulants and antibiotics are given to all patients for some time following surgery.

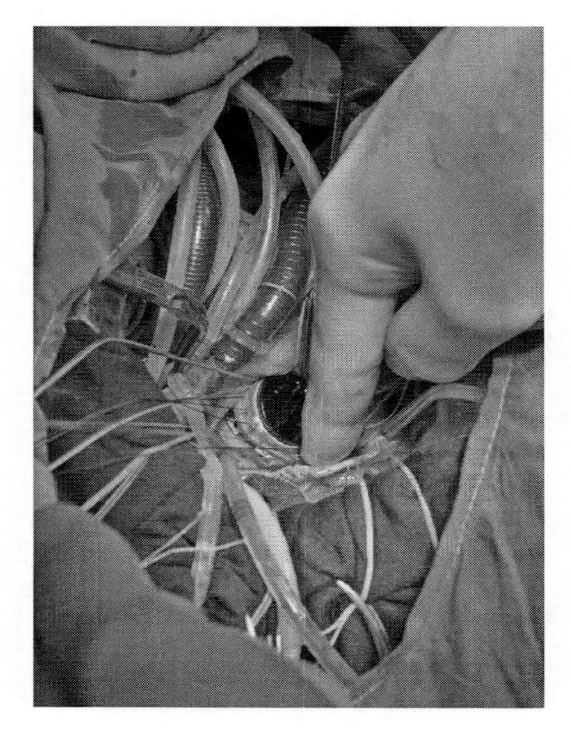

FIGURE 115
A color version of this figure follows page **176.** Valve inserted into heart.

Artificial valves usually last indefinitely, but anticoagulants always must be taken to prevent clots and embolisms. Animal valves do not require the long-term use of anticoagulants, but they may only last 10 to 15 years, at which time they must be replaced.

If the patient is a woman who may wish to become pregnant, organic valves must be used because the anticoagulant medication used with artificial valves can cause birth defects.

Complications

Blood clots leading to stroke or heart attack, valve rejection. Some mechanical valves can cause blood cell breakdown or hemolysis.

Hemodialysis

Alternate names—Renal dialysis.

Purpose

To remove toxins from the blood when disease or damage prevents the kidneys from doing so. Ideally the process allows the kidneys time to resume normal function, but this occurs only rarely. Dialysis may allow the patient to survive until a kidney transplant can be performed.

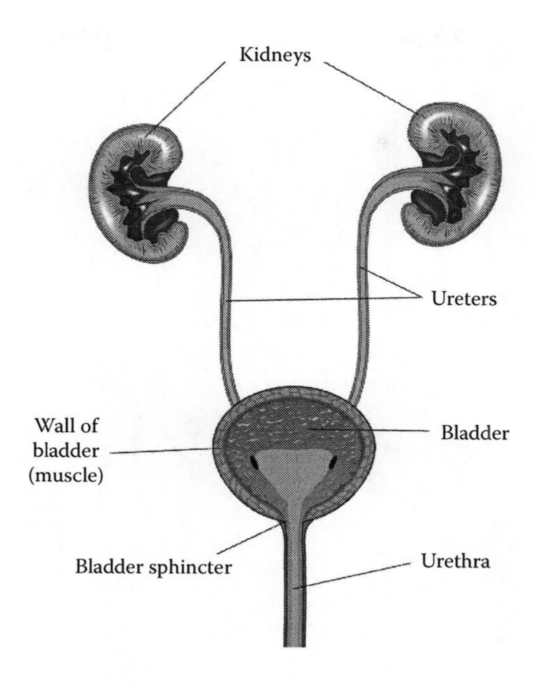

FIGURE 116
Urinary system.

Indications

Failure of kidney (renal) function as indicated by urine and blood tests.

Anatomy

The kidneys are a pair of organs located near the back wall of the abdominal cavity. They are well supplied with blood. The interior of the kidney is an open space called the pelvis, which collects urine and passes it to the ureters. The ureters connect to the bladder, which collects urine and holds it until it can be released via urination. Sphincter muscles are at the exit of the bladder, and the urethra leads from the bladder to the outside (Figure 116).

Pathology

Various disease processes or physical trauma can compromise renal function to the point where fluids and toxins build up in the body to intolerable levels.

Physiology

Kidneys take blood in from the body and filter out water and toxins, mainly urea. They also regulate electrolyte levels in the blood and produce hormones that are involved in blood pressure regulation.

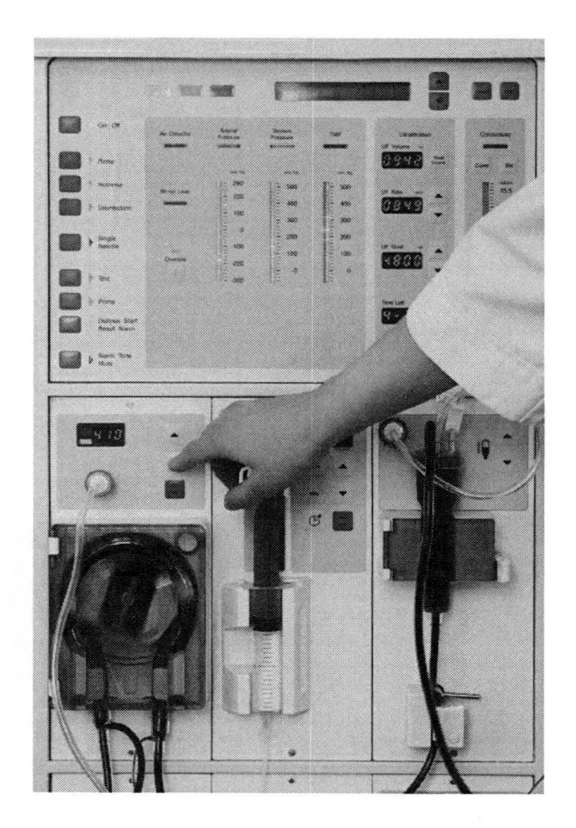

FIGURE 117
Hemodialysis machine.

Staffing

Nurse specialist, dialysis technologist.

Equipment and Supplies

Hemodialysis machine and associated supplies (Figure 117).

Preparation

An access point is placed, consisting of catheters or, for the normal longer-term use in hemodialysis, an arteriovenous (AV) fistula. The AV fistula is made by joining an artery and a vein together and locating the joint close to or at the skin surface. This allows needles to be inserted in opposite directions during the dialysis process. AV fistulas are much less susceptible to infection than are catheters. An anticoagulant is administered shortly before treatment begins. Fluid balance is critical in dialysis, so the patient is weighed before and after each session to ensure balance.

Procedure

Access needles are inserted into the AV fistula and used to divert part of the blood flow into the hemodialysis machine and then back to the patient. The machine pumps blood through a special assembly that removes toxins from the blood. Since relatively small amounts of blood are passed through the machine, effective reduction of toxins in the body takes a few hours and must be repeated a few times per week. After treatment is finished, the needles are removed from the AV fistula, and the patient is free until the next session.

Expected Outcome and Follow-Up

Removal of toxins from the blood. The AV fistula must be kept clean and free of infection.

Complications

Anemia, flu-like symptoms, low blood pressure, infection.

Hemorrhoidectomy

Alternate names—n/a.

Purpose

To remove problematic hemorrhoid tissue that does not respond to other treatments.

Indications

Hemorrhoids that cause bleeding, pain, or bowel restriction, or that protrude from the anus, and have not responded to nonsurgical treatments. Digital examination, anoscopy, or diagnostic imaging (usually using a barium enema) can provide definitive diagnoses.

Anatomy

The rectum is the final portion of the digestive tract (Figure 118). It is a pouch-like organ that is closed at the distal end by the anus, a sphincter muscle.

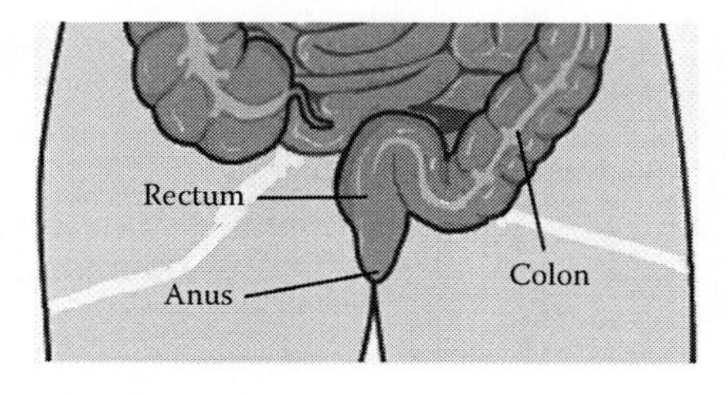

FIGURE 118
Lower digestive system.

Pathology

Hemorrhoids are masses of tissue originating with varicose veins that form in the rectum and anus, causing pain and itching and sometimes obstruction of the anal passage.

Physiology

The rectum receives fecal material from the colon and holds it until it can be evacuated. The anal sphincter keeps the rectum closed until evacuation.

Staffing

Physician, possibly an assistant.

Equipment and Supplies

Minor surgical, various special devices, depending on the technique to be used.

Preparation

The patient does not eat for at least 24 hours before the procedure, and is given laxatives and/or an enema to ensure that the rectum is empty. A sedative is given just before the procedure, and local anesthetic is administered.

Procedure

Various techniques are used.

The simplest method of removal consists of placing a small rubber band tightly around the base of the hemorrhoids. This cuts off the blood supply, and after a few days, the tissue dies and falls off. The base heals together and there is usually little or no bleeding. This method works best with hemorrhoids that protrude from the anus during evacuation straining.

A chemical can be injected into the base of the hemorrhoid, causing it to shrink.

An infrared probe can be used to gently cauterize or coagulate the hemorrhoidal tissue.

A combination of the three methods above can be used.

Finally, hemorrhoids can be excised surgically, with the base tied off to prevent bleeding.

Expected Outcome and Follow-Up

Removal of hemorrhoidal tissue and relief from symptoms. Stool softeners should be used for some time after the procedure to try to avoid straining during evacuation, which can cause extra bleeding.

Complications

Anal stenosis and incomplete healing.

Hip Arthroplasty

Alternate names—Hip replacement, hip surgery, total hip replacement, THR, hip prosthesis insertion.

Purpose

To replace a damaged or diseased hip joint that does not respond adequately to other forms of treatment.

Indications

Trauma to the hip joint; more commonly arthritic degradation of the joint. In either case, a range of medical and physical therapy treatments are tried thoroughly first, and if these methods do not provide sufficient pain relief and increased mobility, hip arthroplasty is considered.

Anatomy

The hip joint is a ball-and-socket type of joint, with the ball consisting of the head of the femur and the socket a depression (the acetabulum) in the side of the pelvis. There is a layer of cartilage on both surfaces. Muscles and tendons hold the joint in place.

Pathology

Trauma can damage the inner surfaces of the hip joint, but more commonly osteoarthritis (OA) produces rough spots on the mating surfaces of the bones in the joint, which can cause pain in itself (Figure 119). The OA can also cause the cartilage layers to wear thin or wear away completely, which produces even more pain, decreased range of motion, and decreased mobility.

Physiology

The cartilage in the joint allows smooth motion of the bones, and the structure of both the joint itself and the supporting tissues means that a wide range of motion is possible.

Staffing

Normal surgical team; the surgeon is an orthopaedic specialist.

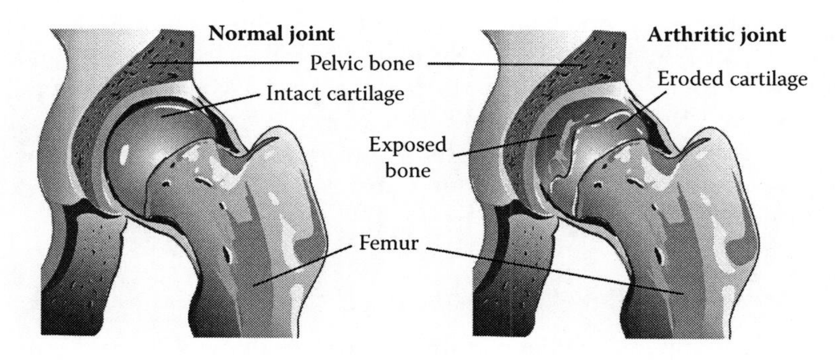

FIGURE 119
A color version of this figure follows page **176.** Normal and arthritic hip joints.

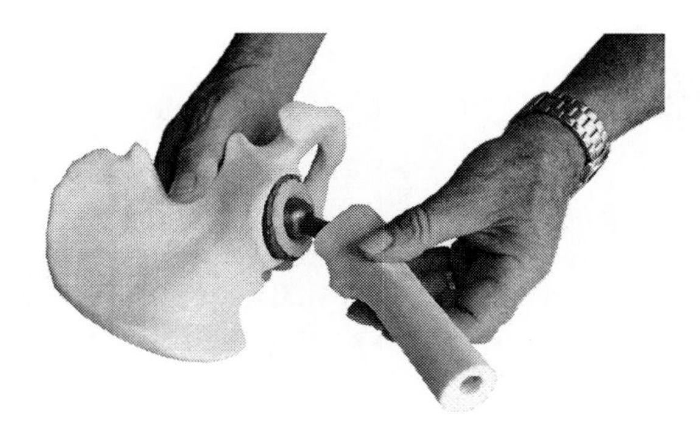

FIGURE 120
Model of an artificial hip joint.

Equipment and Supplies

Standard surgical, bone saws and drills, the artificial joint, special cements and hardware.

Preparation

The patient is evaluated for suitability for the surgery, with physical condition, other diseases, and mental fitness all considered. Obese patients may be required to lose weight before surgery can be performed. The patient's joint structure is evaluated and carefully measured, both directly and using diagnostic imaging, so that the correct size and shape of prosthesis can be chosen.

The type of prosthesis is chosen. It may be constructed of stainless steel, cobalt chrome alloy, or a ceramic material. The acetabular part may be lined with either plastic or ceramic (Figure 120).

The patient is sedated, and either general or regional anesthesia is initiated.

Procedure

An incision is made down the side of the upper thigh, and the ligaments and muscles in the area are moved to provide access to the joint itself. The head of the femur is removed from the acetabulum, and the upper part of the femur is cut away using a bone saw. Surgical power tools are used to remove the inner surface of the acetabulum and shape it to receive the artificial component. This component is then inserted into the hip bone and secured by either a tight fit, bone screws, or acrylic cement. The open end of the femur is then reamed out so the femoral part of the prosthesis will fit closely. This part is inserted into the femur and is held in place either by acrylic cement or a close, secure fit. The prosthetic ball is inserted into the prosthetic socket, and proper fit and motion are checked. Muscles and tendons are moved back into position, and the incision is closed.

Expected Outcome and Follow-Up

Installation of a new hip joint, relief of pain, and return of mobility and range of motion. The joint must be limited in movement for a few days after surgery to allow the supporting muscles and tendons time to tighten up and hold the joint securely in place. Antibiotic therapy is used to help prevent infection, and pain control medication is provided. Physiotherapy is used to gradually increase strength, mobility, and range of motion, but also to help avoid embolisms.

Complications

Dislocation of the prosthetic joint, since it does not fit quite as well as the natural joint; embolisms; infection; and nerve damage. The prosthetic joint may wear out prematurely or break, especially if the patient is very active or obese.

Hysterectomy

Alternate names—uterus removal.

Purpose

To remove the uterus and possibly some associated structures.

Indications

Fibroid tumors, endometriosis (inflammation of the lining of the uterus), excessive menstrual bleeding, cancer of the uterus or other associated structures.

Anatomy

The uterus is a pear-shaped smooth-muscle organ in which a fetus develops in pregnancy. The uterus opens into the vagina through the cervix. A pair of ovaries are on either side of the uterus and are loosely connected to the uterus via the fallopian tubes (Figure 121).

Pathology

Excessive buildup of the uterine lining during the menstrual cycle can result in excessive bleeding and cramping. Small growths called polyps, or others called fibroid tumors (benign) may form in the lining; these structures may cause pain or blockages (Figure 122).

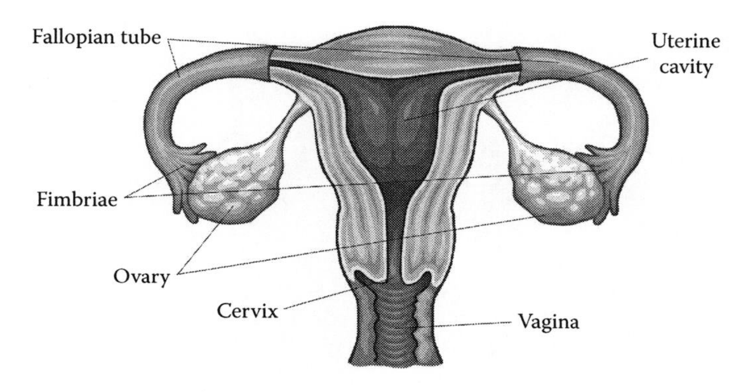

FIGURE 121
Uterus and associated structures.

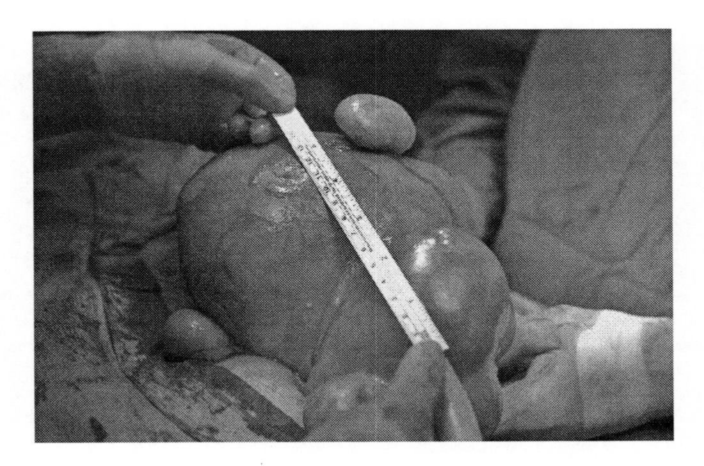

FIGURE 122
A color version of this figure follows page **176.** Uterine fibroid tumor.

Uterine cancer is the most common cancer of the female reproductive system and, as with other cancers, can be fatal if not treated or removed.

Physiology

During the menstrual cycle, the ovaries produce an ovum, which travels down the fallopian tubes to the uterus. The lining of the uterus thickens and increases in vascularization in preparation for implantation of a fertilized ovum and thus a possible pregnancy. If no pregnancy occurs, the lining is shed as menstrual bleeding, and the cycle is repeated.

Staffing

Normal surgical team.

Equipment and Supplies

Standard surgical.

Preparation

Standard surgical.

Procedure

Hysterectomy may be performed abdominally or vaginally, and either just the part of uterus (excluding the cervix), the whole uterus, or the uterus plus some associated structures, including the fallopian tubes, the ovaries, and nearby lymph nodes, may be removed. Ovary removal is called oophorectomy.

In an abdominal hysterectomy, an incision is made just above the pubic hairline, either horizontally or vertically. The horizontal incision extends almost to the pelvic bones on either side, while the vertical incision goes almost up to the umbilicus. Blood vessels supplying the tissues are isolated and cauterized or tied off, supporting structures are severed, and the target organ or organs are removed. The incision is closed.

For a vaginal hysterectomy, the vagina is held open with a speculum, and an incision is made near the top end of the vagina. This incision allows access to the uterus and other structures, and they are removed in the same manner as with an abdominal hysterectomy. The vaginal incision is closed.

Abdominal hysterectomies allow the surgeon more room to work and permit visualization of other abdominal structures and organs for possible diagnosis. Vaginal hysterectomies are less invasive and heal faster, and they leave no visible scar.

Expected Outcome and Follow-Up

Successful removal of the target organs and tissues, and cessation of symptoms. Hormone replacement therapy may be initiated if the ovaries have been removed.

Complications

Blood clots. Collateral damage to adjacent organs may occur.

Hysteroscopy

Alternate names—n/a.

Purpose

To visualize the inside of the uterus for diagnostic or treatment purposes.

Indications

Known or suspected problems with the cervix, the uterus, or its lining.

Anatomy

The uterus is a pear-shaped smooth-muscle organ in which a fetus develops in pregnancy. The uterus opens into the vagina through the cervix (Figure 123).

Pathology

Excessive buildup of the uterine lining during the menstrual cycle can result in excessive bleeding and cramping. Small growths called polyps, or others called fibroid tumors (benign) may form in the lining; these structures may cause pain or blockages. Uterine cancer is the most common cancer of the female reproductive system and, as with other cancers, can be fatal if not treated or removed. Cervical cancer is also relatively common.

Physiology

During the menstrual cycle, the lining of the uterus thickens and increases in vascularization in preparation for a possible pregnancy. If no pregnancy occurs, the lining is shed as menstrual bleeding, and the cycle is repeated.

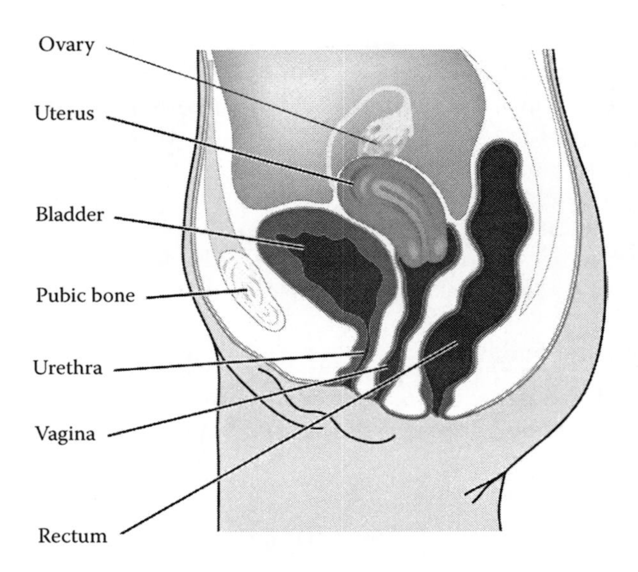

FIGURE 123
Female reproductive system.

Staffing

Physician (usually a gynecologist), assistant.

Equipment and Supplies

Minor surgical, hysteroscope.

Preparation

A local anesthetic is administered.

Procedure

A speculum may be inserted into the vagina and opened to provide access to the cervix. A series of tapered cervical dilation instruments, each larger than the last, are pressed into the cervix to slowly expand (dilate) the cervical opening. When the opening is sufficient, the hysteroscope is inserted. The uterus may be inflated with carbon dioxide to prove better visual access. The scope provides illumination and an optical channel that allows the physician to observe the inside of the uterus. There is also a lumen inside the scope that permits the insertion of devices to take tissue samples or remove small lesions such as polyps or fibroid tumors. When the examination is complete, the scope is removed.

Expected Outcome and Follow-Up

Adequate visualization of the interior of the uterus to provide diagnostic information; acquisition of tissue samples for lab analysis; removal of lesions and relief from symptoms they might have been causing. Some bleeding following the procedure is normal and should cease in a day or two. Cramping or pain may persist for a few hours.

Complications

Collateral damage to adjacent tissues or organs.

Ileostomy

Alternate names—n/a.

Purpose

To create an opening for the ileum thought the abdominal wall following permanent or temporary severing of the distal end of the ileum and part of the colon.

Indications

Diseases of the colon or trauma that require either removal of much of the colon including upper portions, or bypassing the colon to allow for healing. These conditions are diagnosed from physical examination, lab analysis of stool and blood, diagnostic imaging, and colonoscopy.

Anatomy

The ileum is the final section of the small intestine before it joins the large intestine, or colon (Figure 124). The colon empties into the rectum, which is closed off by the anus.

Pathology

A wide variety of problems can occur within the colon. Cancer, Crohn's disease, colitis and ulcerative colitis, polyps, diverticulitis/diverticulosis, Hirschsprung's disease, and

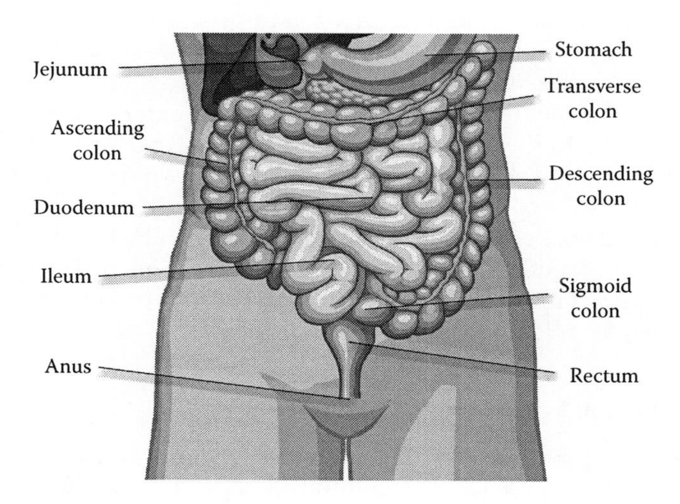

FIGURE 124
The lower digestive system.

irritable bowel syndrome can all cause bleeding and/or bowel malfunctions. If these conditions affect the colon right up to the ileum, there is not enough healthy colon tissue for a colostomy to be performed, and so an ileostomy must be done instead.

Physiology

The ileum helps absorb nutrients from digested food before it moved into the colon. The main functions of the colon are reabsorption of water and some electrolytes from this material and storage of waste products until they can be eliminated.

Staffing

Normal surgical.

Equipment and Supplies

Standard surgical.

Preparation

Extensive measures are taken before the procedure to ensure that the small intestine and colon are completely empty. A nasogastric tube is inserted. The patient is sedated, and general anesthesia is induced.

Procedure

A large incision is made vertically in the abdomen, usually just to one side of the navel. If the colon is to be removed, it is separated from supporting structures, and associated arteries and veins are closed off; the distal end of the portion to be removed is separated and closed off, the proximal end is separated, and the diseased or damaged section is removed from the abdomen. If the colon or part of it is to remain in place, the two parts are separated, and the section that will be allowed to rest is closed off.

A small opening is made in the abdominal wall on the side closest to the end of the ileum. This end of the ileum is drawn out through the opening and is everted, forming a "cuff" which has the inner lining of the ileum facing outward. The edges of the ileum are then sutured to the skin of the abdomen to form a stoma. Fecal material will now exit the body through the stoma, so a pouch is placed over the stoma to catch the material.

Alternatively, and if the ileostomy is expected to be permanent, parts of the ileum may be cut apart and sewn together again to form an internal pouch. This pouch is then connected to the abdomen with a stoma in the same way as above.

The abdominal incision is closed with drainage tubes in place.

Expected Outcome and Follow-Up

Elimination of the symptoms leading to the surgery, or provision of time for the bypassed section of the colon to rest and heal without the passage of fecal material.

Fluid and food intake is restricted for two or three days after surgery (hydration is provided intravenously), and then soft foods are gradually introduced.

The patient is educated concerning care of the ileostomy. An external pouch will need to be in place at all times, unless an internal pouch was created. These patients need just a dressing over the stoma, and will irrigate the stoma to allow for controlled release of fecal material.

Dietary modifications will be required to allow for the change in the digestive system.

If the ileostomy is temporary, after sufficient time has passed to allow healing of the damaged or diseased portion, the abdomen is reopened, and the ileum and the colon are reattached.

Complications

Blood clots. Some patients may suffer from psychological trauma.

In Vitro Fertilization

Alternate names—Test tube conception.

Purpose

To successfully initiate pregnancy when this cannot be accomplished naturally.

Indications

Inability to conceive as a result of dysfunctional fallopian tubes or other reasons, when otherwise the uterus could support a successful pregnancy.

Anatomy

Two ovaries are located on either side of the uterus and are loosely connected to the uterus via the fallopian tubes. In pregnancy, the fetus develops within the mother's uterus, a muscle-walled organ that is pear shaped before conception but expands as the fetus grows (Figure 125). The fetus is contained by a thin-walled structure called the amniotic sac within the uterus; this sac is filled with fluid that helps protect the developing fetus. The fetus is connected to the mother via its umbilical cord, which carries blood in both

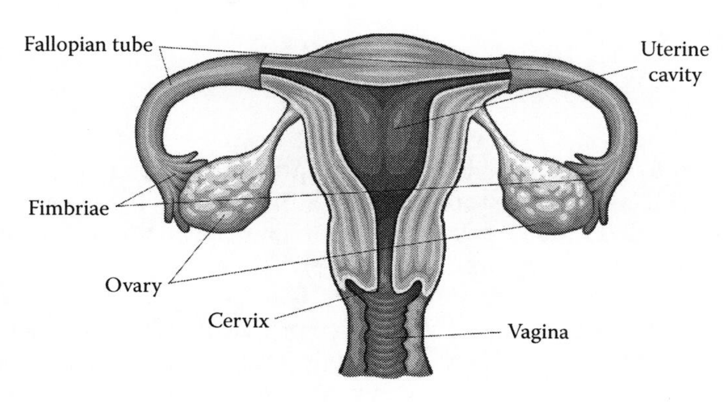

FIGURE 125
Female reproductive system.

directions between the fetus and the placenta. The placenta is a highly vascularized organ that is fetal tissue and has extensive connections to the wall of the uterus. The birth canal consists of the exit to the uterus (the cervix) and the vagina, which are surrounded by the bones of the pelvis.

Pathology

The fallopian tubes can be dysfunctional by reason of disease, trauma, or malformation that prevents the passage of ova from the ovaries to the uterus. This means that either an ovum cannot be fertilized or it cannot reach the uterus for implantation and the beginning of pregnancy.

Physiology

During the menstrual cycle, the ovaries produce an ovum, which travels down the fallopian tubes to the uterus. The lining of the uterus thickens and increases in vascularization in preparation for implantation of a fertilized ovum and thus a possible pregnancy. If no pregnancy occurs, the lining is shed as menstrual bleeding, and the cycle is repeated.

Staffing

Gynecologist, assistant, in vitro lab technology specialist.

Equipment and Supplies

Ovum removal needle, laparoscopic equipment or ultrasound machine for visualizing the ovaries and guiding the needle, lab apparatus for combining the ova and sperm and maintaining it in a suitable environment until it can be implanted in the mother, device for transferring the fertilized ovum or ova to the uterus.

Preparation

Fertility drugs may be given to the mother to stimulate production of multiple mature ova simultaneously. The presence of mature ova in the ovaries is confirmed by lab tests. Local anesthetic is administered.

Procedure

If the procedure is to be performed laparoscopically, incisions are made in the abdomen to allow access of the laparoscope tubes. A long, thin needle is inserted though the vagina or the laparoscopic tubes. The needle is guided to an ovary using ultrasound (for vaginal access) or visualization (for laparoscopic access). The needle penetrates the part of the ovary that contains the ovum, and the ovum is aspirated and collected. Several ova may be collected from different locations in the same procedure. The ova are then mixed with sperm and maintained in a suitable environment. They are observed to see if fertilization occurs, and if it does, the fertilized ova or zygotes are further monitored to ensure normal early development. Suitable zygotes are transferred to the mother's uterus where implantation may occur. If this part of the process is successful, pregnancy results.

Expected Outcome and Follow-Up

Successful pregnancy. Since multiple zygotes may be delivered to the uterus, a multiple pregnancy may result. Normal pregnancy care is provided.

Complications

Multiple pregnancies carry their own intrinsic risks.

Incisional Hernia Surgery

Alternate names—Ventral hernia surgery or repair, incisional or ventral herniorrhaphy.

Purpose

To repair an incisional hernia.

Indications

The presence of an incisional hernia.

Anatomy

The abdomen contains a number of organs including the stomach, the small and large intestines, the liver, the pancreas, the spleen, the kidneys, and the urinary bladder (Figure 126).

Pathology

Abdominal surgery produces scar tissue around the site of the incision. This may result in a weakened area which abdominal organs, especially intestines, can press against and eventually bulge out through.

Physiology

n/a

Staffing

Normal surgical team.

Equipment and Supplies

Standard surgical, possibly a mesh material to be used in repair.

Preparation

Standard surgical.

Procedure

Incisional hernia repair may be performed with open surgery or laparoscopically. If the herniated structure is large, open surgery may be required.

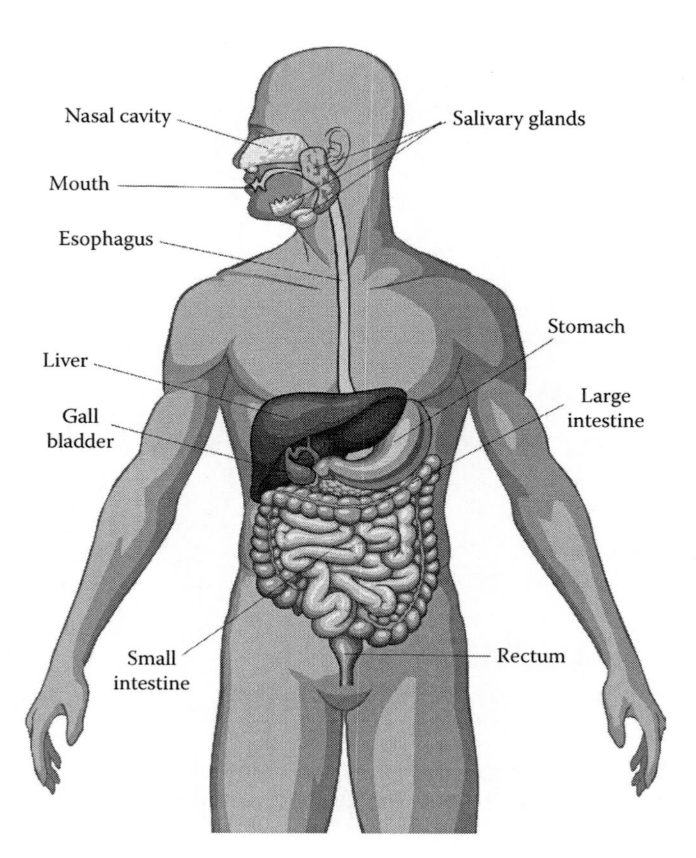

FIGURE 126
Human digestive system.

In either case, the herniated area is accessed by cutting through the tissue near the site. The protruding tissue is pushed back into place. It may have to be dissected from the edges of the hernia opening, especially if the condition has persisted for a long time. The opening may be sutured together, or preferably a mesh of plastic or stainless steel is placed under the weak or open area and sutured to the tissue well beyond the weak area.

The incision is closed.

Expected Outcome and Follow-Up

Permanent repair of the hernia. The incision in checked for integrity and bleeding. Activities are limited for several weeks to prevent rupture of the surgical site. Tissue grows into the mesh material, incorporating it into the natural structure and making it stronger.

Complications

Rupture of the surgical site.

Inguinal Hernia Repair

Alternate names—Herniorrhaphy.

Purpose

To repair an inguinal hernia.

Indications

The presence of an inguinal hernia.

Anatomy

The abdominal cavity in humans in enclosed by the ribs above, and to the upper sides and front, by abdominal muscles and the spinal column in the middle section, and by the pelvis in the lower section. The base of the cavity is formed by a set of muscles and other tissue and is penetrated by the digestive and urinary systems.

Pathology

The bottom of the abdominal cavity may be weak, especially in the areas where the intestines or urethra pass through. An upright posture means that physical pressure is applied to the base of the abdominal cavity by the abdominal organs. This can result in part of the abdominal organs, especially the intestines, protruding through the base structure. This is an inguinal hernia.

Physiology

n/a

Staffing

Normal surgical team.

Equipment and Supplies

Standard surgical, possibly a mesh material to be used in repair.

Preparation

Standard surgical.

Procedure

Inguinal hernia repair may be performed with open surgery or laparoscopically.

In either case, an incision is made in the abdomen, and the herniated area is accessed. The protruding tissue is pushed back into place. It may have to be dissected from the edges of the hernia opening, especially if the condition has persisted for a long time. The opening may be sutured together, or preferably a mesh of plastic or stainless steel is placed over the weak or open area and sutured to the tissue well beyond the weak area.

The incision is closed.

Expected Outcome and Follow-Up

Permanent repair of the hernia. The incision in checked for integrity and bleeding. Activities are limited for several weeks to prevent rupture of the surgical site. Tissue grows into the mesh material, incorporating it into the natural structure and making it stronger.

Complications

Rupture of the surgical site.

INTACS

Alternate names—Keratoconus repair.

Purpose

To repair a bulging cornea and improve vision.

Indications

A bulging cornea or keratoconus.

Anatomy and Physiology

The eye is the organ of visual sensation. It is spherical in shape, with a clear membrane (the cornea) covering the front; the cornea begins to focus light entering the eye. A thin liquid, the aqueous humor, is contained between the cornea and the colored part of the eye or iris. The iris can contract or relax to change the shape of an opening in its center, the pupil, controlling the amount of light entering the eye. A clear, fibrous lens behind the pupil focuses light and can change shape to alter the focal distance. The main body of the eye is filled with a semisolid fluid called the vitreous humor. At the back of the eye, the retina receives the focused light image and transforms it into electrical signals. These are carried from the eye to the brain by the optic nerve (Figure 127).

Pathology

Internal pressure can change the shape of the cornea.

Staffing

Ophthalmologist, assistant.

Equipment and Supplies

Operating microscope (Figure 128), INTACS devices, microsurgery equipment.

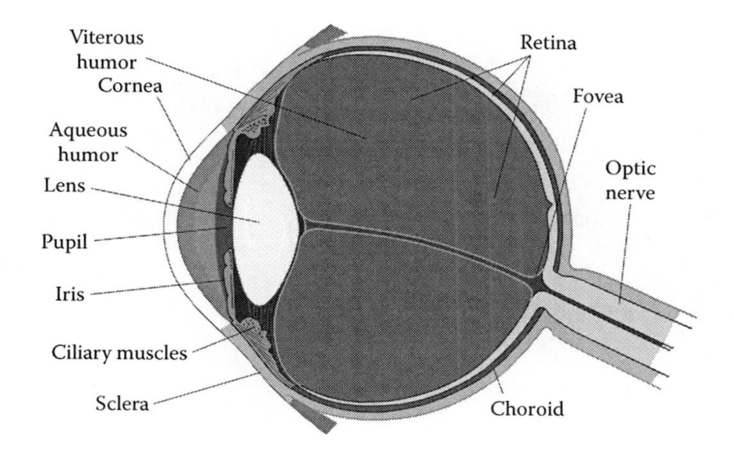

FIGURE 127
Eye anatomy.

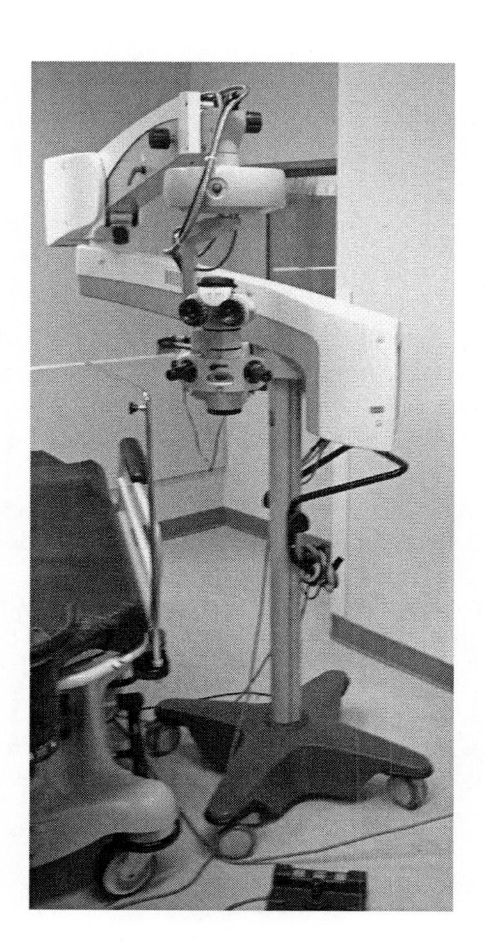

FIGURE 128
Operating microscope.

Preparation

A sedative may be given, and local anesthesia provided. The cornea will have been precisely measured and mapped to determine what correction is required to produce optimal vision.

Procedure

Specific layers of the cornea are separated microsurgically, and an INTACS ring is inserted. The ring adjusts the shape of the cornea to correct vision.

Expected Outcome and Follow-Up

Corrected vision.

Complications

Night vision disturbances, blurry vision.

Kidney Transplant

Alternate names—n/a.

Purpose

To replace a nonfunctional kidney with that of a donor.

Indications

Kidney failure, shown by lab tests of liver function; severe kidney damage by trauma shown by direct observation or diagnostic imaging. Patients are often on dialysis for years before they receive a kidney donation.

Anatomy

The kidneys are a pair of organs located near the back wall of the abdominal cavity (Figure 129, Figure 130). They are well supplied with blood. The interior of the kidney is an open space called the pelvis, which collects urine and passes it to the ureters. The ureters connect to the bladder, which collects urine and holds it until it can be released via urination. Sphincter muscles are at the exit of the bladder, and the urethra leads from the bladder to the outside.

Pathology

Various disease processes or physical trauma can compromise renal function to the point where fluids and toxins build up in the body to intolerable levels.

Physiology

Kidneys take blood in from the body and filter out water and toxins, mainly urea. They also regulate electrolyte levels in the blood and produce hormones that are involved in blood pressure regulation.

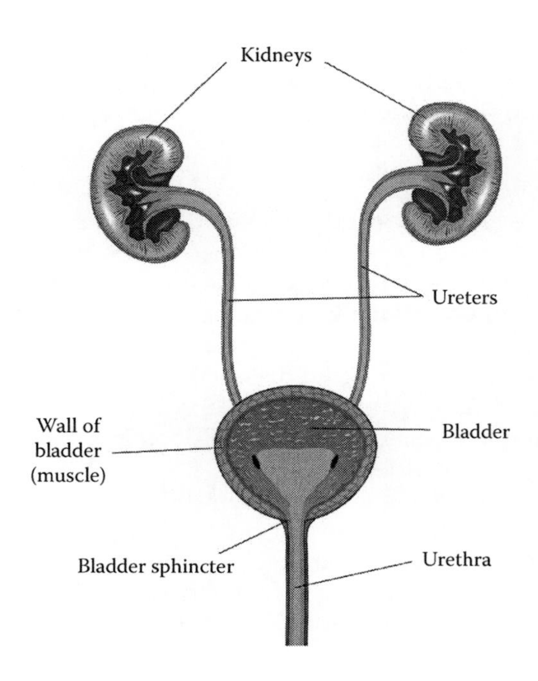

FIGURE 129
Urinary system.

Staffing

Complex surgical team.

Equipment and Supplies

Standard surgical, donated kidney.

Preparation

Antirejection (immunosuppressive) drugs are given for some time in advance of the surgery. Extensive tissue testing and matching is done to ensure the best possible match, to

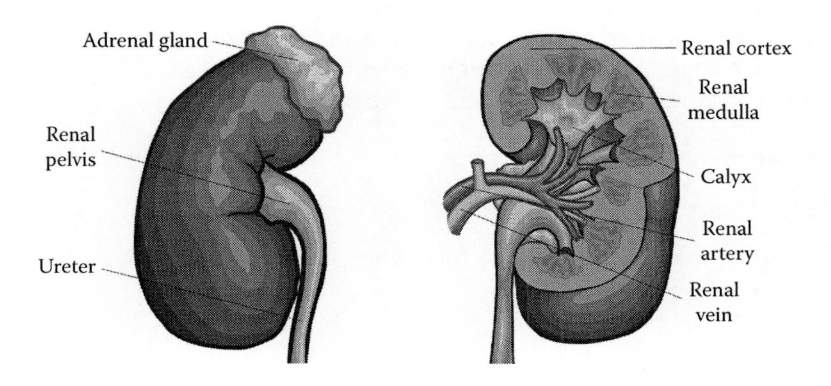

FIGURE 130
Kidneys.

reduce rejection issues. A donor kidney must be available that is a suitable match in tissue type and size and is available soon enough after donation to be viable. The patient is sedated and the incision site cleaned. A urinary catheter is placed.

Procedure

Though the body has two kidneys, it can usually function quite well with only one. For the patient, this means that only one kidney needs to be implanted. It also means that a kidney may be donated by a living donor. This allows more time for the procedure to be coordinated and may mean that a wider range of donors are available, increasing the degree of tissue match.

The donor kidney is removed along with associate blood vessels and the common bile duct; it is then placed in an environment that ensures optimal tissue health until it is implanted into the patient. In the case of a living donation, the donor kidney remains in the donor's body as long as possible.

An incision is made in the side of the abdomen and the tissues retracted for access. The donated kidney is placed in the abdominal cavity, usually low in the pelvis. The artery and vein supplying and draining the kidney are connected to the iliac artery and vein. The ureter of the donated kidney is attached directly to the patient's bladder. The diseased kidneys may be removed or left in place, depending on the type and degree of damage they have suffered.

After all connections and placements are checked, the incision is closed, with drainage tubes in place.

Expected Outcome and Follow-Up

Normal functioning of the donor kidney. Patients are given antirejection (immunosuppressive) drugs indefinitely, but are also regularly monitored for signs of rejection, more frequently immediately following surgery. Patients are monitored in an intensive care unit for several days. Physical activities are slowly initiated in the ICU and increased when the patient can be moved to a general surgical ward. After discharge normal activities can be gradually resumed. Kidney function tests are performed regularly to ensure adequate function.

Complications

Rejection of the donated kidney is the most serious complication and may require a second transplant. Death can result if this is not possible and rejection cannot be controlled, or the patient may have to go on dialysis. Bleeding and infection are constant concerns. Other infections may occur as a result of the antirejection drugs used.

Knee Arthroplasty

Alternate names—Knee replacement, total knee replacement, TKR, artificial knee joint surgery.

Purpose

To replace a damaged or diseased knee joint that does not respond adequately to other forms of treatment.

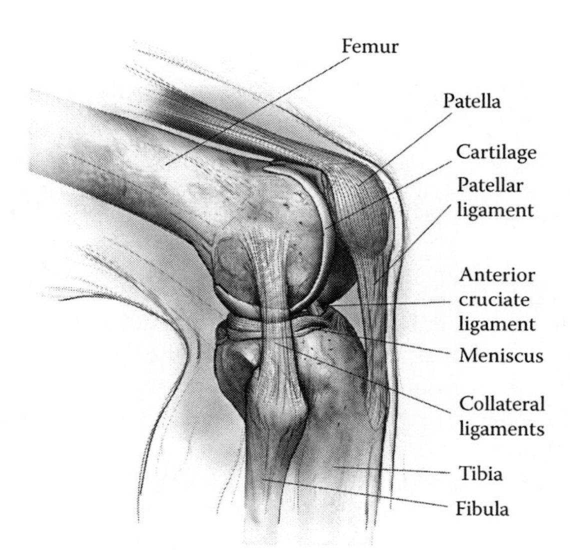

FIGURE 131
Knee joint anatomy.

Indications

Trauma to the hip joint; more commonly arthritic degradation of the joint. In either case, a range of medical and physical therapy treatments are tried thoroughly first, and if these methods do not provide sufficient pain relief and increased mobility, knee arthroplasty is considered. Knee condition is evaluated by physical examination, diagnostic imaging, and possibly arthroscopy.

Anatomy

The knee joint is complex, and consists of the lower end of the femur, the upper end of the tibia, the patella, and supporting muscles and tendons (Figure 131). There is a layer of cartilage on the mating bone surfaces.

Pathology

Trauma can damage the inner surfaces of the hip joint, and osteoarthritis (OA) produces rough spots on the mating surfaces of the bones in the joint, which can cause pain in itself. The OA can also cause the cartilage layers to wear thin or wear away completely, which produces even more pain, decreased range of motion, and decreased mobility.

Physiology

The cartilage in the joint allows smooth motion of the bones, and the structure of both the joint itself and the supporting tissues means that a smooth range of back and forth motion is possible, with very little side-to-side motion.

Staffing

Normal surgical team; the surgeon is an orthopaedic specialist.

Equipment and Supplies

Standard surgical, bone saws and drills, the artificial joint, special cements and hardware.

Preparation

The patient is evaluated for suitability for the surgery, with physical condition, other diseases, and mental fitness all considered. Obese patients may be required to lose weight before surgery can be performed. The patient's joint structure is evaluated and carefully measured, both directly and using diagnostic imaging, so that the correct size and shape of prosthesis can be chosen.

The type of prosthesis is chosen. It may be constructed of stainless steel, cobalt chrome alloy, or a ceramic material, plus a polyethylene component.

The patient is sedated, and either general or regional anesthesia is initiated.

Procedure

An incision is made over the knee and through the joint capsule. Because the knee joint depends greatly on the surrounding tendons and muscles to hold it in the correct position, these structures must be moved aside very carefully to allow access to the joint itself. When the joint is exposed, the end surfaces of both the femur and tibia are cut away and the remaining bone formed to receive the prosthetic components. The part attached to the femur is shell or cap shaped and is fitted over the newly shaped end of the femur. It is cemented in place (Figure 132).

The tibial portion of the new joint is then fitted; it may be attached with cement, or it may have a surface that encourages bone tissue to grow into it, thus providing a natural adhesion. A polyethylene plate is then attached to the prosthetic tibial surface to provide

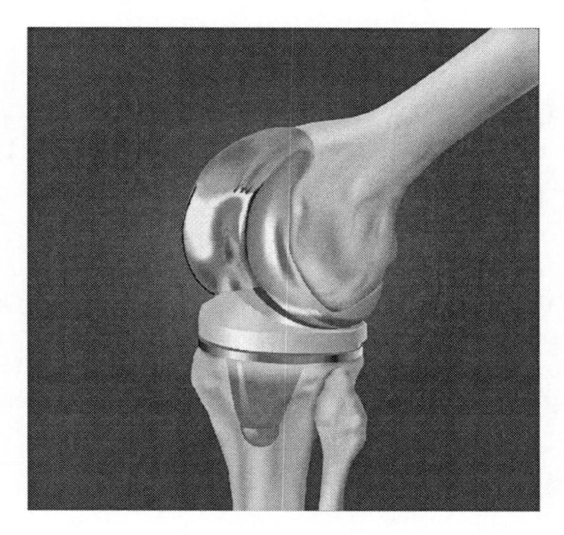

FIGURE 132
Artificial knee joint.

smooth motion and low friction. A second polyethylene structure is attached to the back of the patella to provide a low-friction surface as the patella moves over the metal or ceramic of the new joint. The components are all checked to be sure they are in the correct position, and the muscles and tendons are carefully returned to their normal positions. The incision is then closed.

Expected Outcome and Follow-Up

Installation of a new knee joint, relief of pain, and return of mobility and range of motion. The joint must be limited in movement for a few days after surgery to allow the supporting muscles and tendons time to tighten up and hold the joint securely in place. Antibiotic therapy is used to help prevent infection, and pain control medication is provided. Physiotherapy is used to gradually increase strength, mobility, and range of motion, but also to help avoid embolisms.

Knee prostheses generally last many years; failure is more common with very active or obese patients.

Complications

Dislocation of the new joint; embolism.

Labor and Delivery

Alternate names—Birth, childbirth, parturition.

Purpose

Successful delivery of a fetus or fetuses.

Indications

Full-term pregnancy.

Anatomy

The fetus develops within the mother's uterus, a muscle-walled organ that is pear shaped before conception but expands as the fetus grows (Figure 133). The fetus is contained by a thin-walled structure called the amniotic sac within the uterus; this sac is filled with fluid that helps protect the developing fetus. The fetus is connected to the mother via its umbilical cord, which carries blood in both directions between the fetus and the placenta. The placenta is a highly vascularized organ that is fetal tissue and has extensive connections to the wall of the uterus. The birth canal consists of the exit to the uterus (the cervix) and the vagina, which are surrounded by the bones of the pelvis.

Pathology

Following birth, the status of the infant may be evaluated at specific time intervals using the APGAR scoring method. Though named after its originator, Dr. Virginia Apgar, the

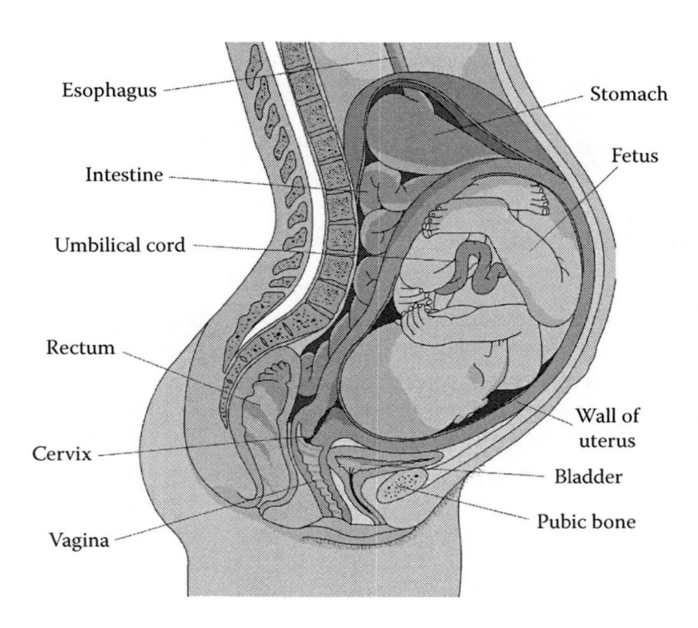

FIGURE 133
Pregnancy anatomy with near-full-term fetus.

term has been developed into an acronym for the five factors that are considered. Each factor is rated as either 0, 1, or 2, and the total gives a good indication of overall infant condition. The five factors are:

A—activity, or muscle tone
P—pulse presence and rate
G—grimace, or reflex response
A—appearance, mainly skin color
R—respiration quality

The total APGAR score can range from 0 to 10.

Physiology

When the fetus reaches full term, hormonal changes cause the uterus to begin contracting in cycles. The contractions gradually become stronger and more frequent, and the pressure of the contractions causes the cervix to dilate. When the cervix is dilated to the maximum extent, further uterine contractions and "pushing" by the mother move the fetus through the birth canal and into the outside world. Normally, the placenta separates from the wall of the uterus at this time and is pushed out of the uterus by residual contractions. The now-empty uterus contracts enough to prevent significant bleeding from the placental site.

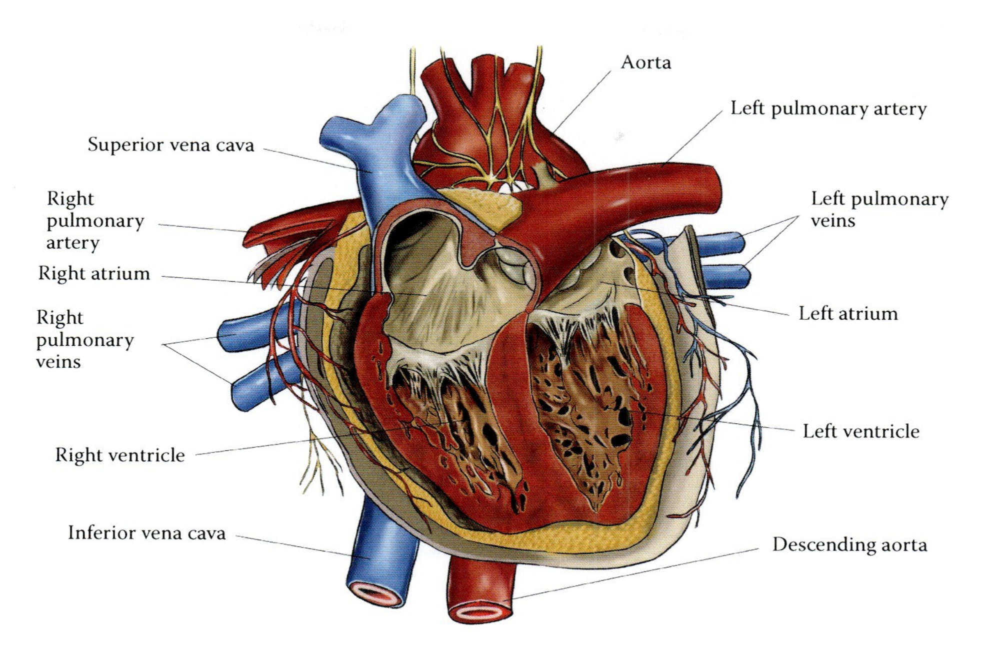

COLOR FIGURE 24
Cutaway view of heart and associated vessels.

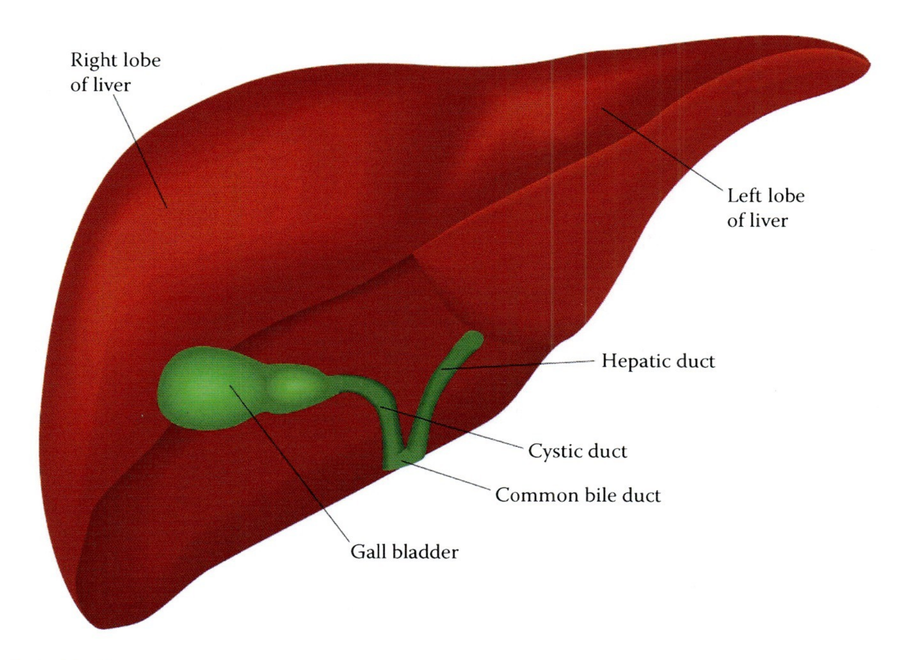

COLOR FIGURE 44
Liver anatomy.

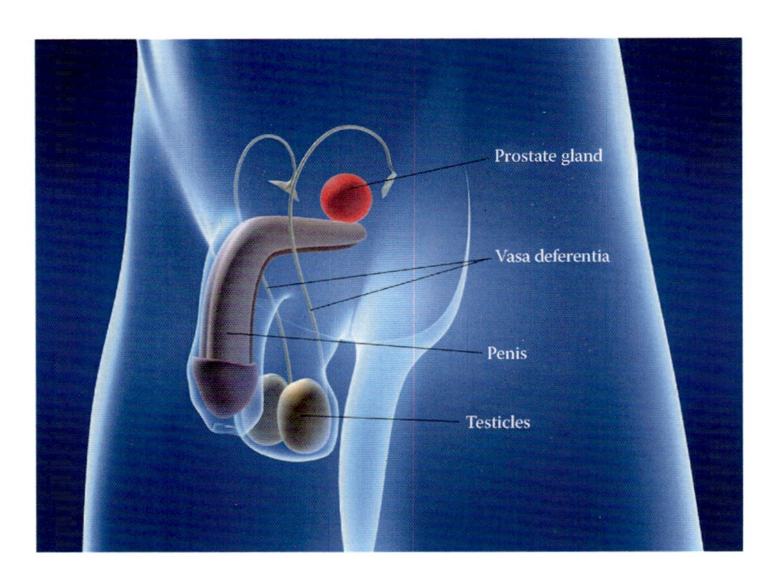

COLOR FIGURE 48
Male anatomy schematic showing prostate gland.

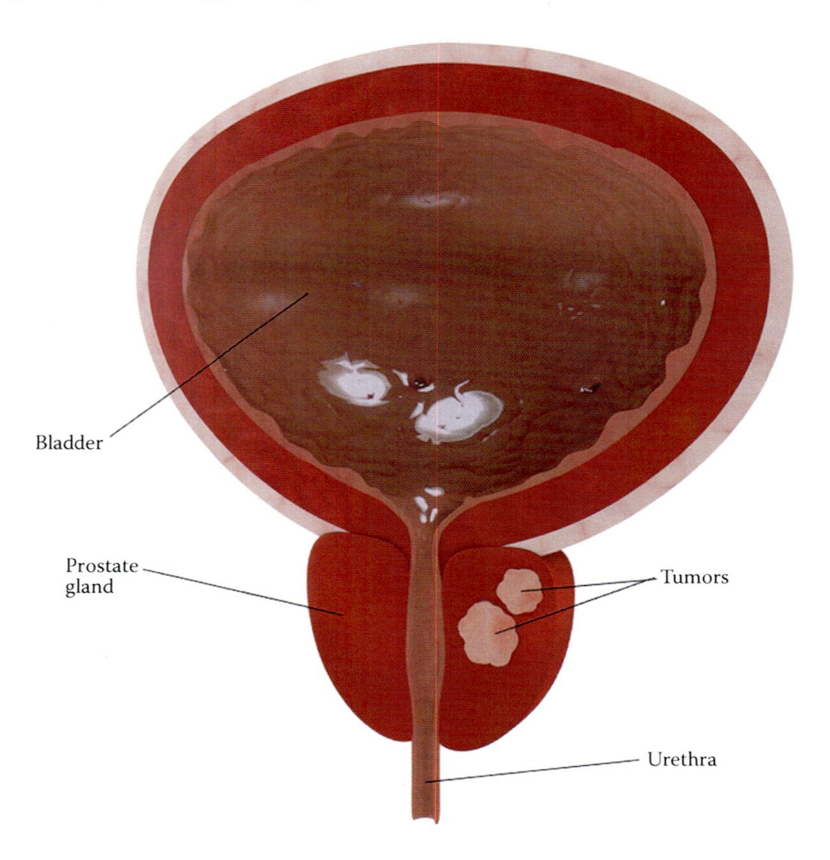

COLOR FIGURE 49
Prostate tumors.

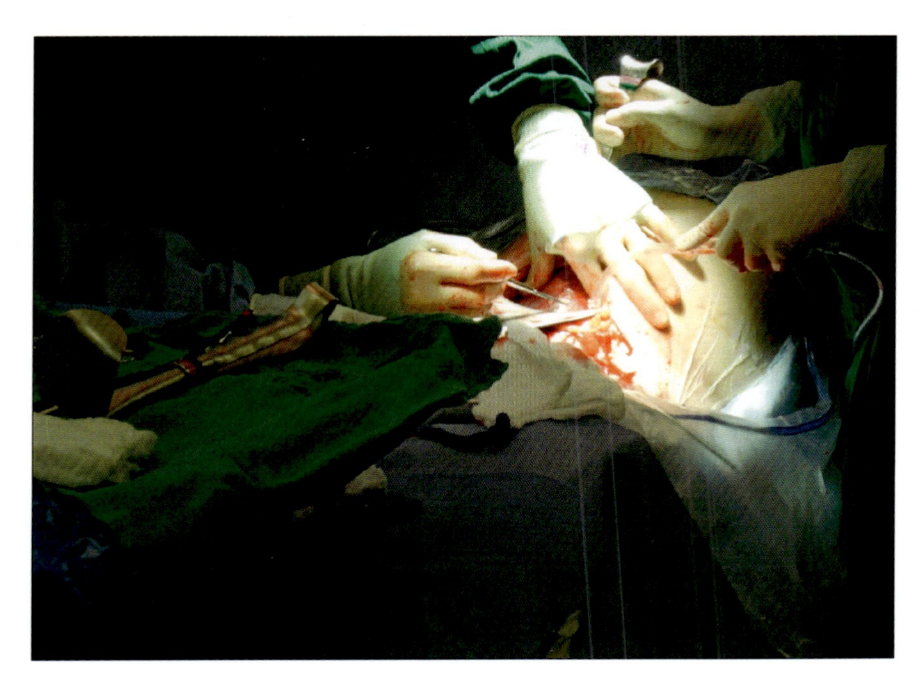

COLOR FIGURE 58
Caesarean section in progress.

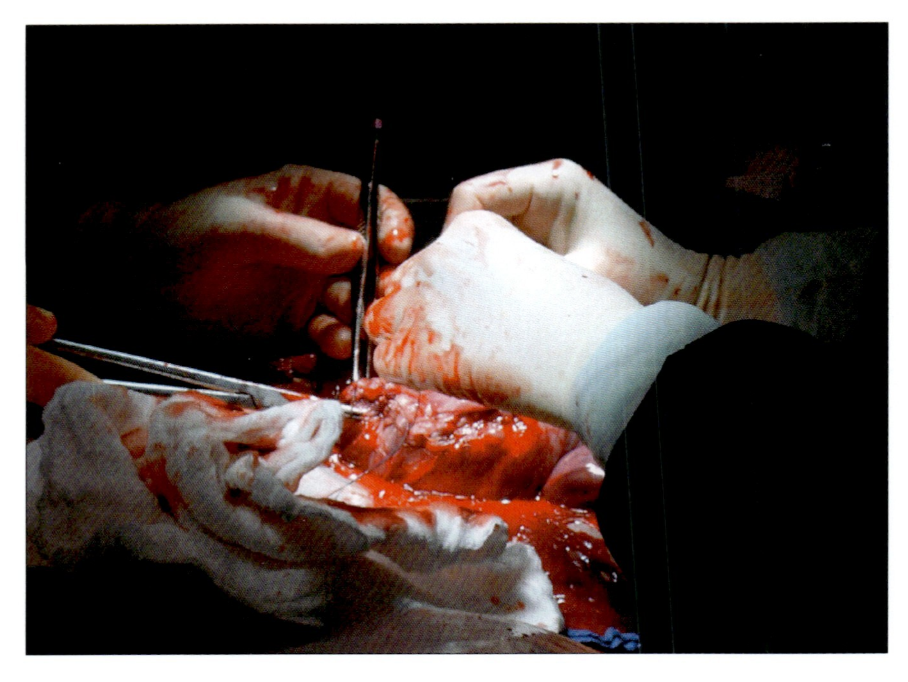

COLOR FIGURE 59
Suturing uterus following caesarean delivery.

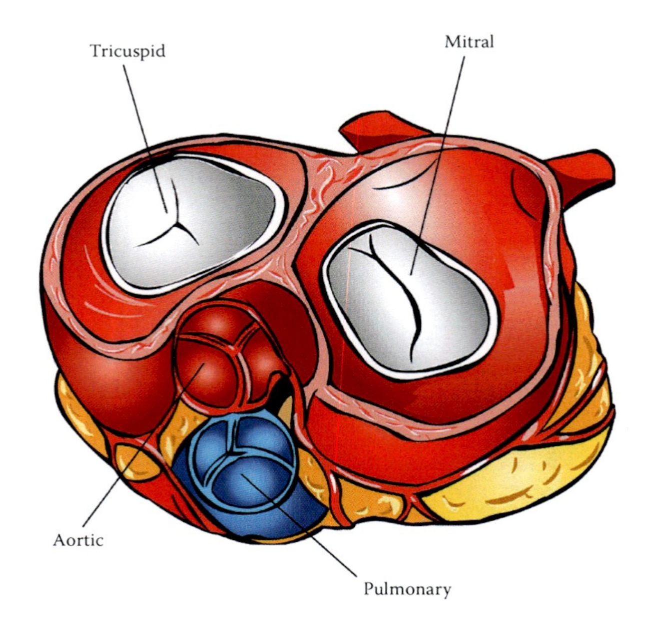

COLOR FIGURE 61
Heart valves.

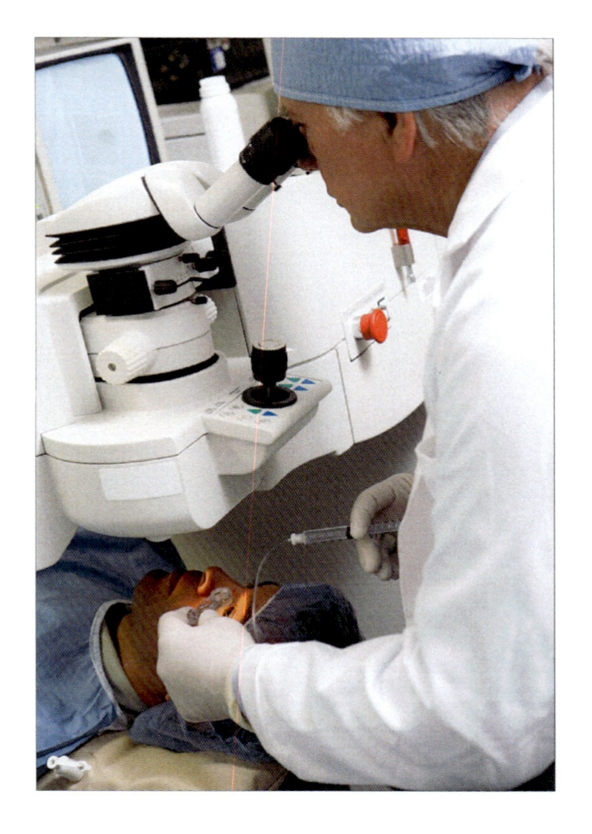

COLOR FIGURE 66
Surgeon performing eye surgery using an operating microscope.

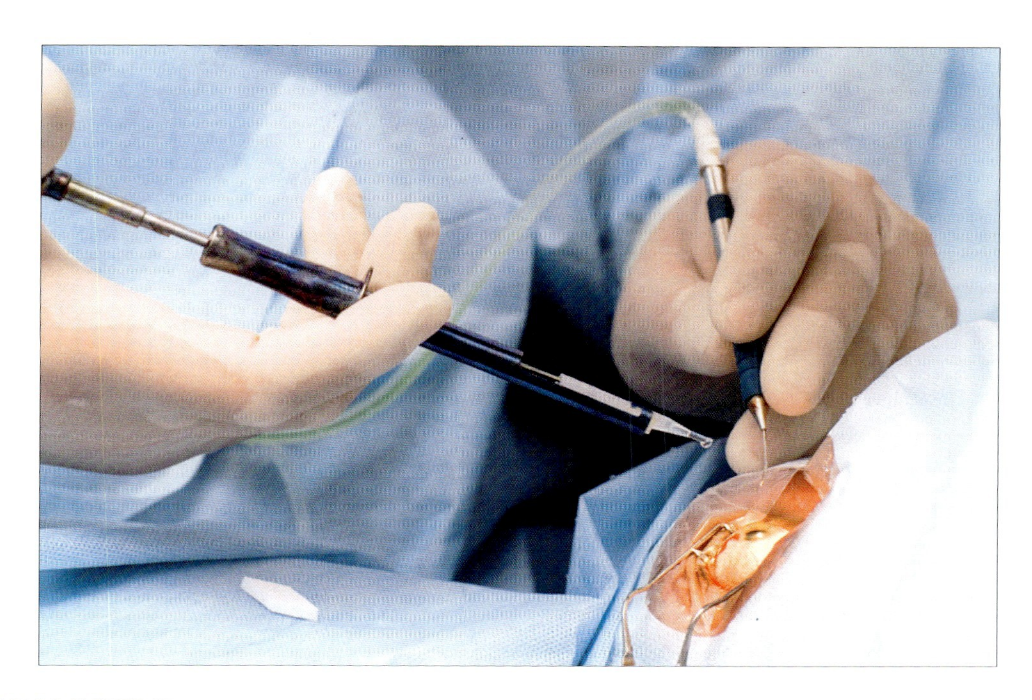

COLOR FIGURE 67
Cataract surgery.

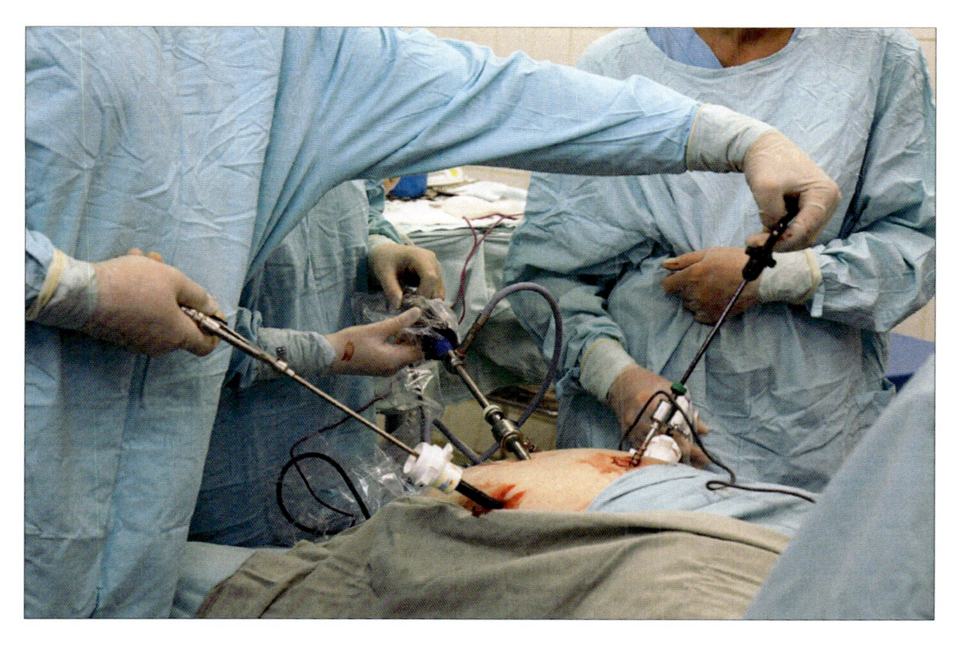

COLOR FIGURE 70
Laparoscopic surgery procedure.

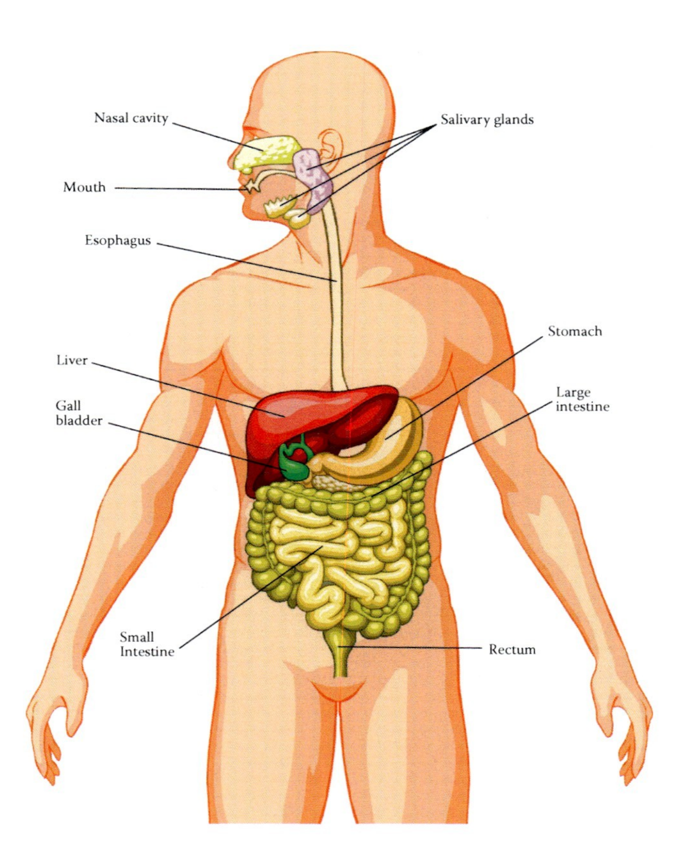

COLOR FIGURE 71
Digestive system.

COLOR FIGURE 72
Cutaway view of a colon tumor.

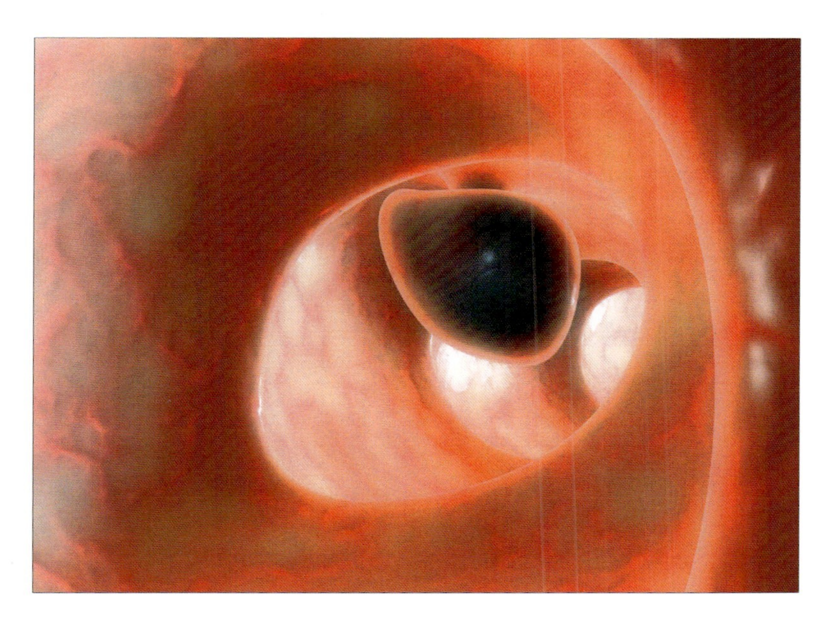

COLOR FIGURE 73
A large colon polyp.

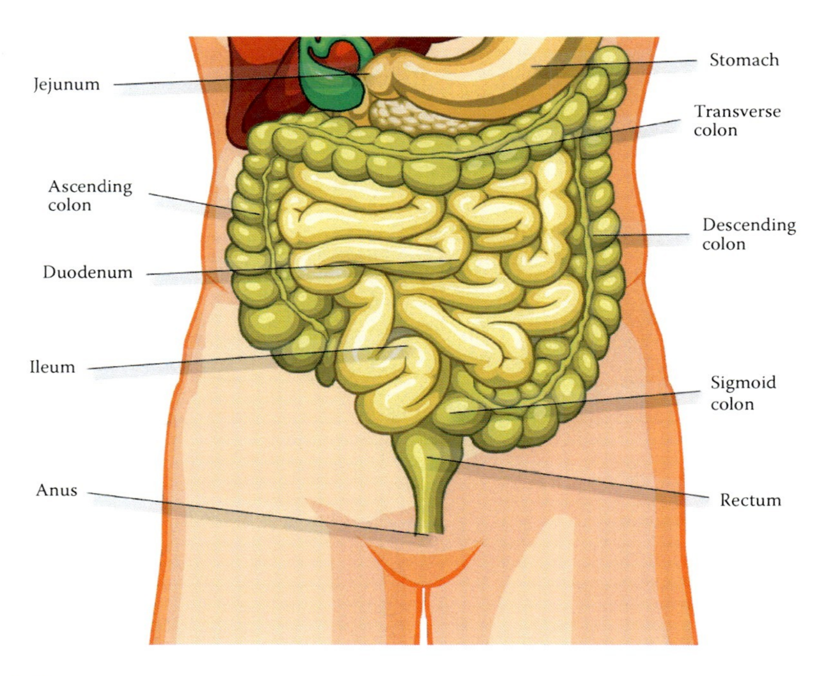

COLOR FIGURE 74
Intestinal organs.

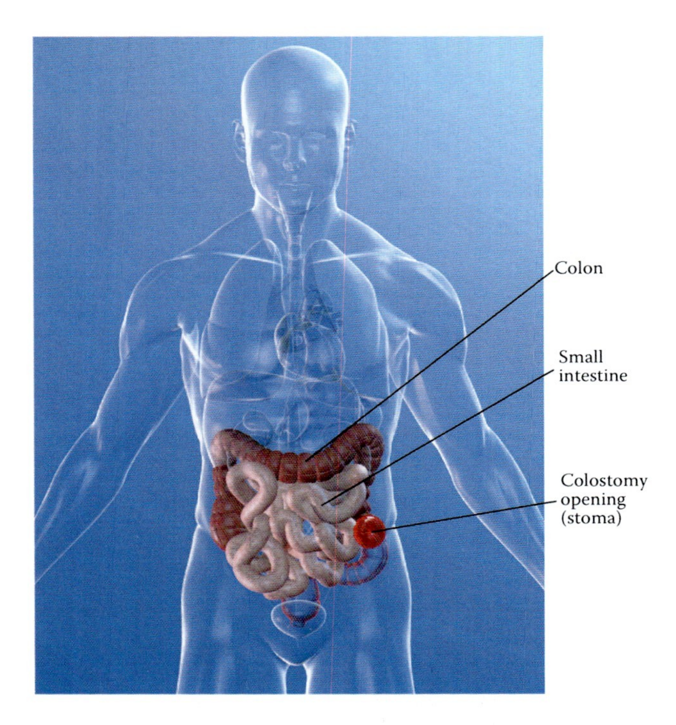

COLOR FIGURE 75
Colostomy formed at the lower end of the descending colon.

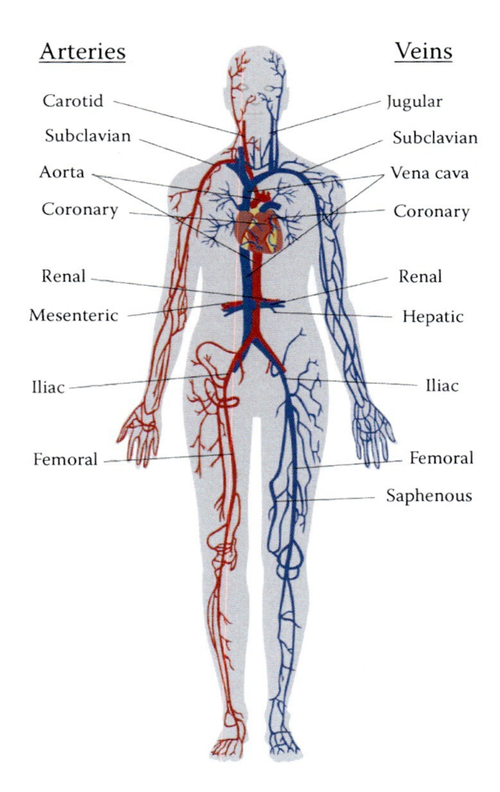

COLOR FIGURE 78
Circulatory system, showing the saphenous vein lower right.

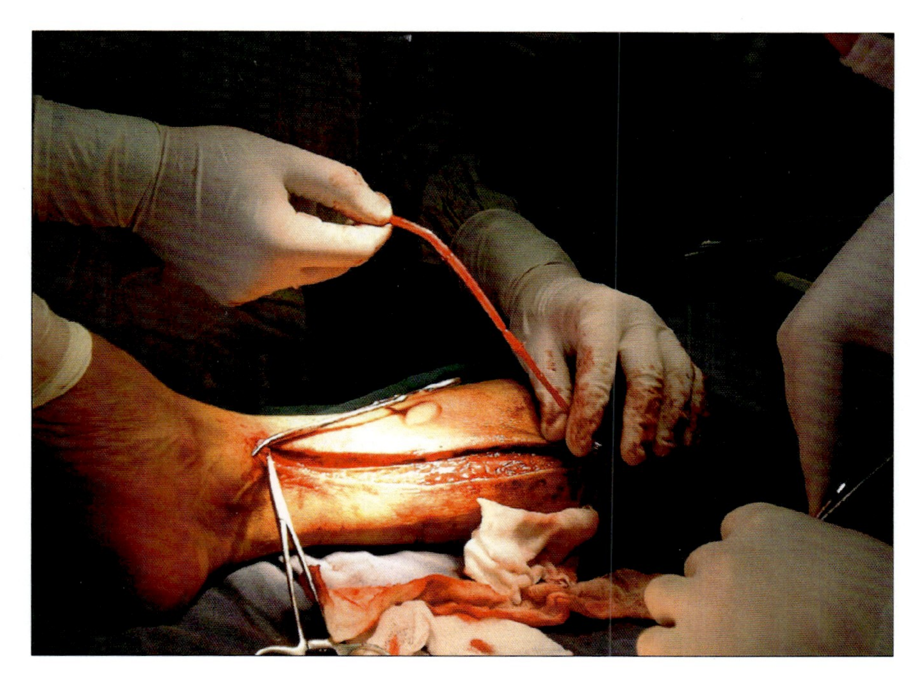

COLOR FIGURE 80
Stripping out a section of saphenous vein.

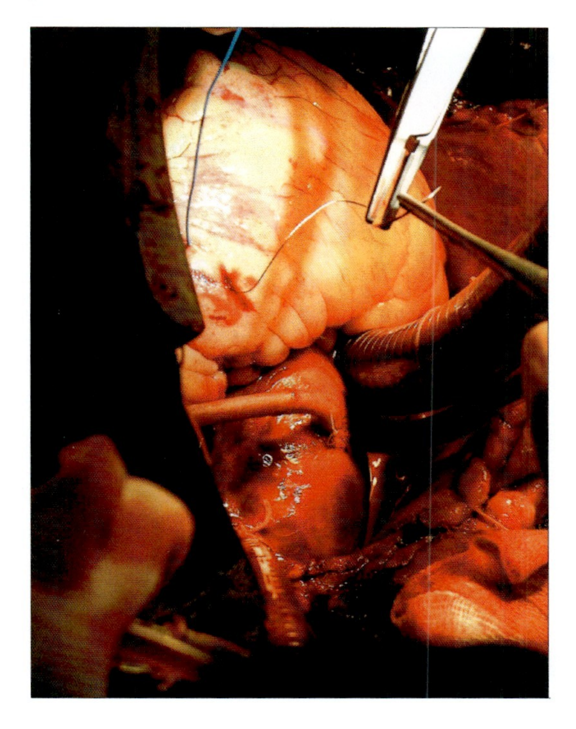

COLOR FIGURE 81
Grafting vein onto aorta.

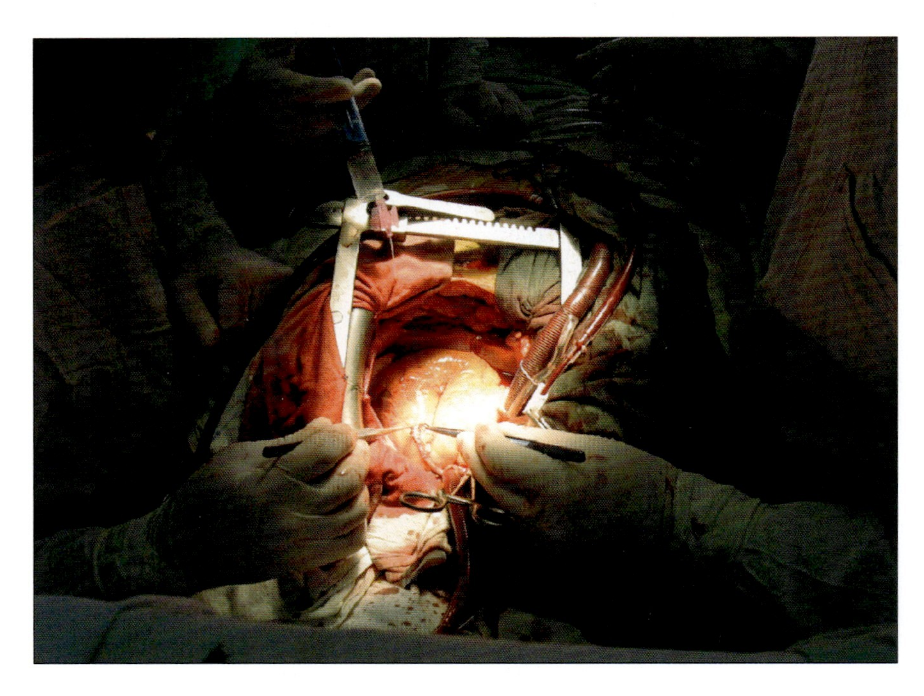

COLOR FIGURE 82
Open heart surgery.

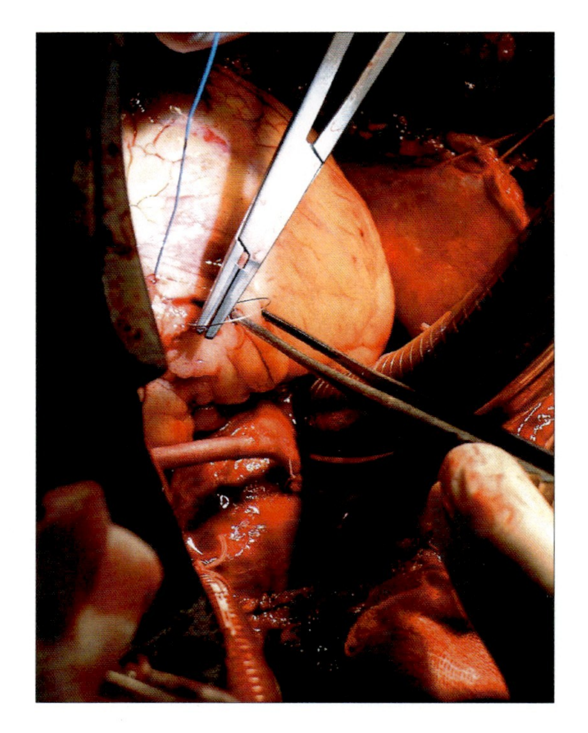

COLOR FIGURE 83
Connecting graft vein to coronary artery.

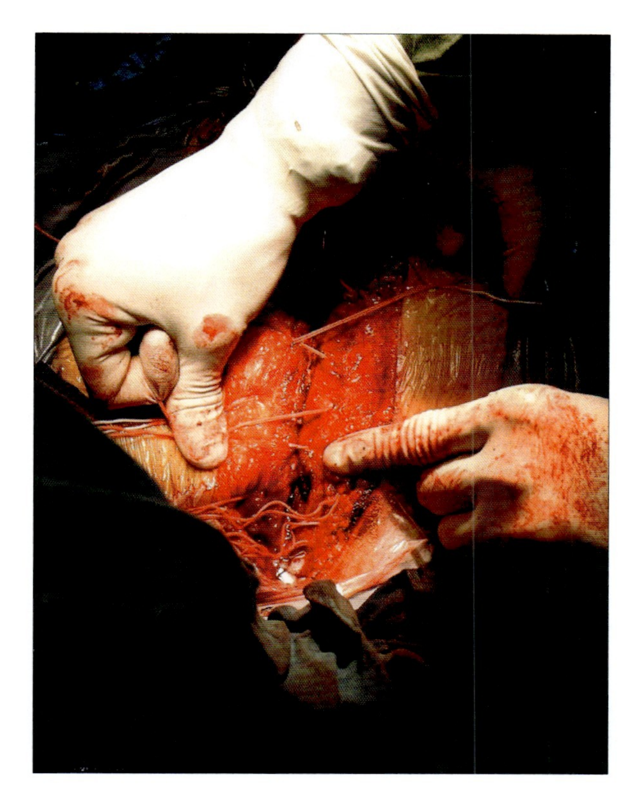

COLOR FIGURE 85
Closing the chest following cardiac surgery.

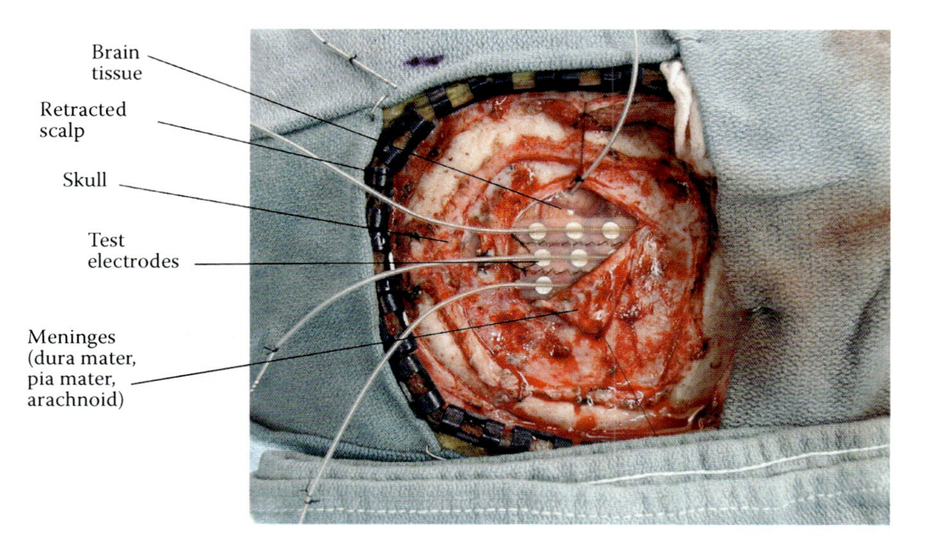

COLOR FIGURE 86
The brain exposed for testing and possible surgery.

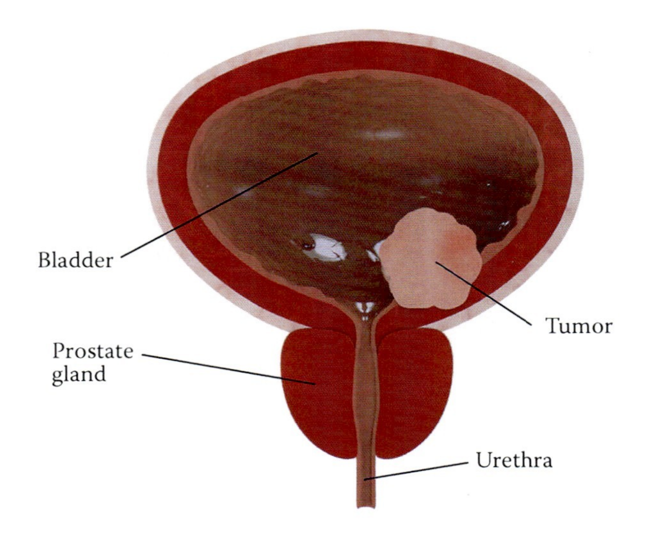

COLOR FIGURE 88
Bladder cancer.

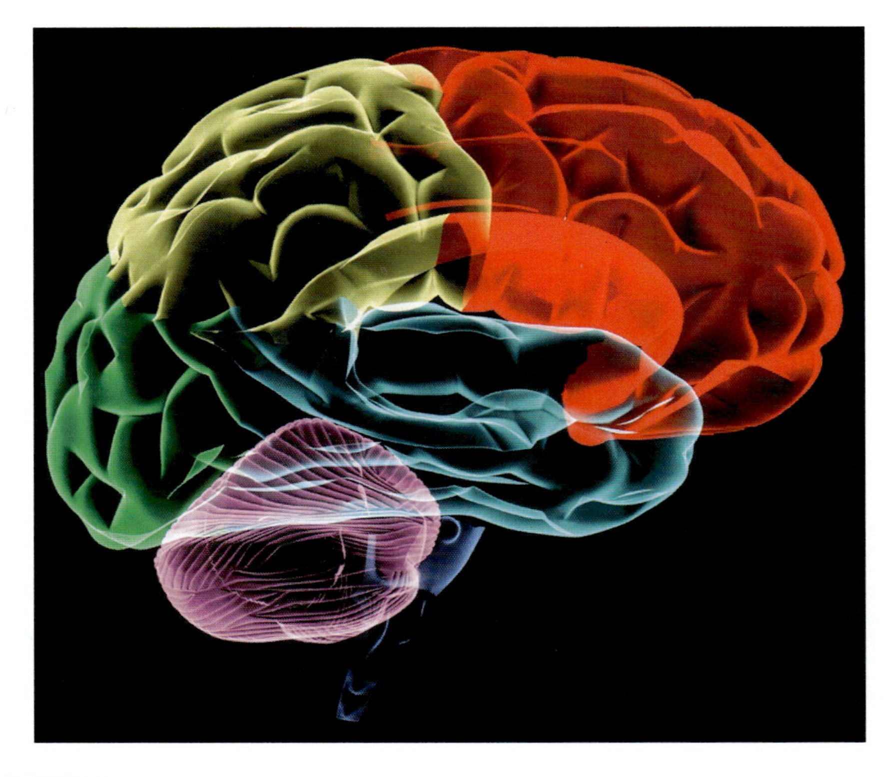

COLOR FIGURE 99
The brain.

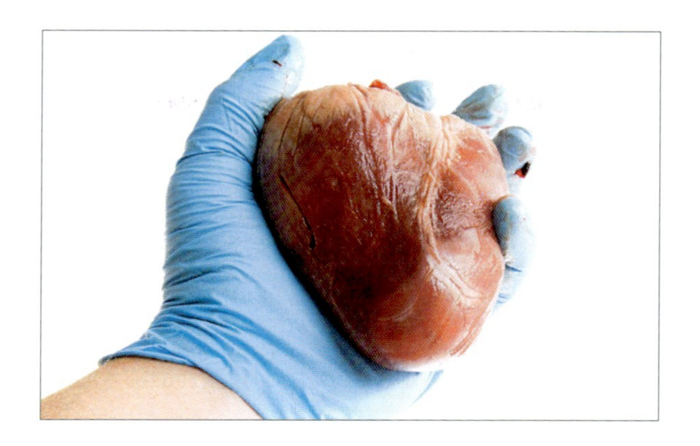

COLOR FIGURE 110
A human heart. This heart is not one to be used for transplant. More adjacent blood vessels would be included to allow for attachment to vessels in the recipient.

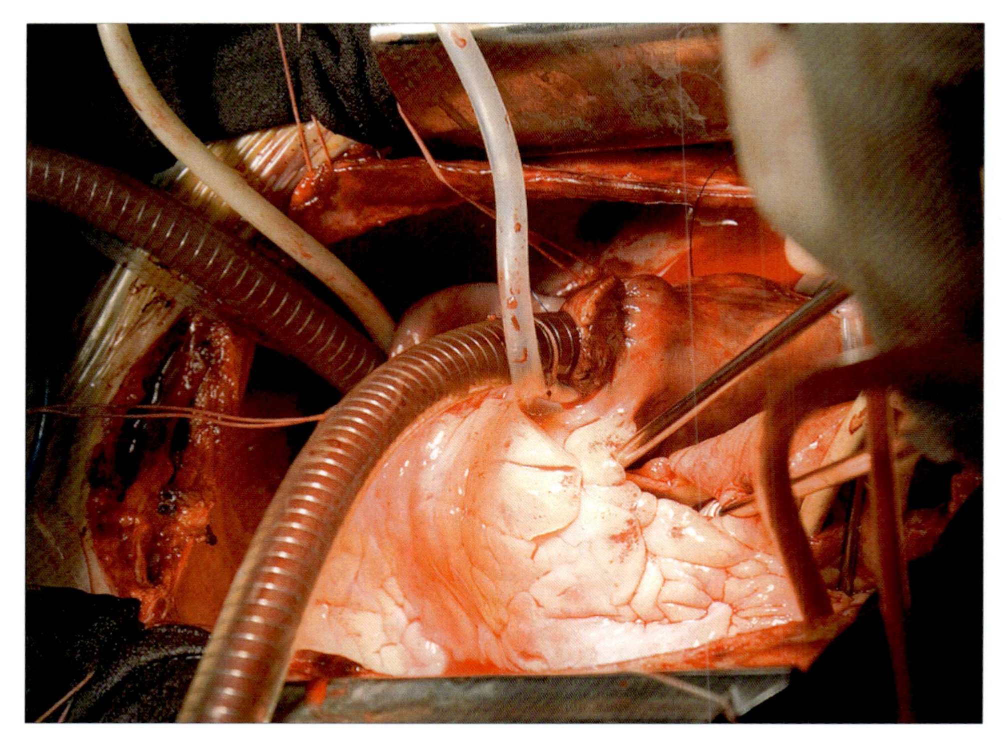

COLOR FIGURE 113
Heart–lung machine connections.

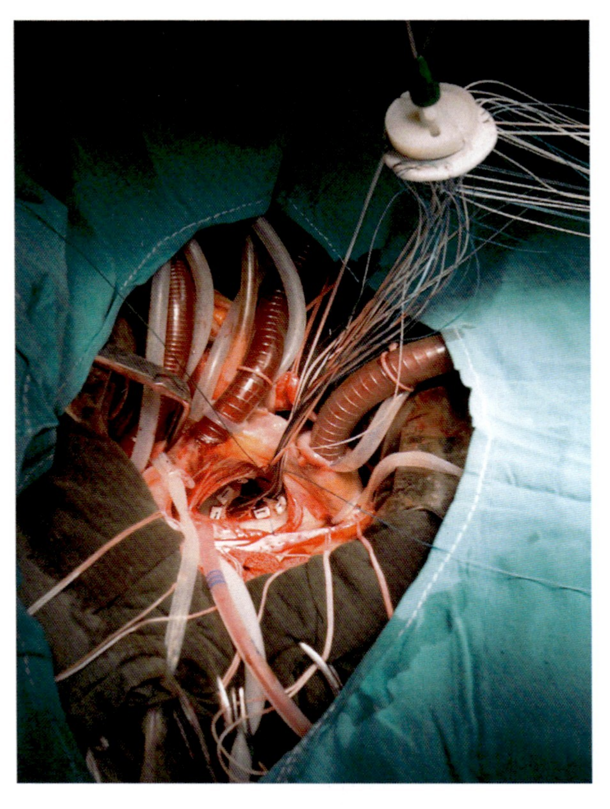

COLOR FIGURE 114
Artificial valve (at top) ready to be placed into heart. Note multiple sutures connecting valve and heart tissue.

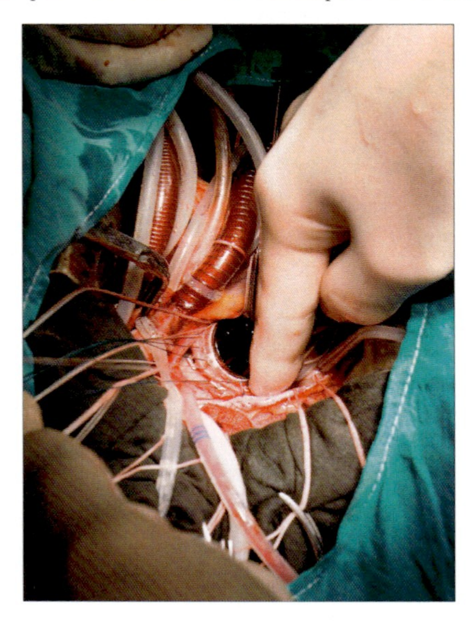

COLOR FIGURE 115
Valve inserted into heart.

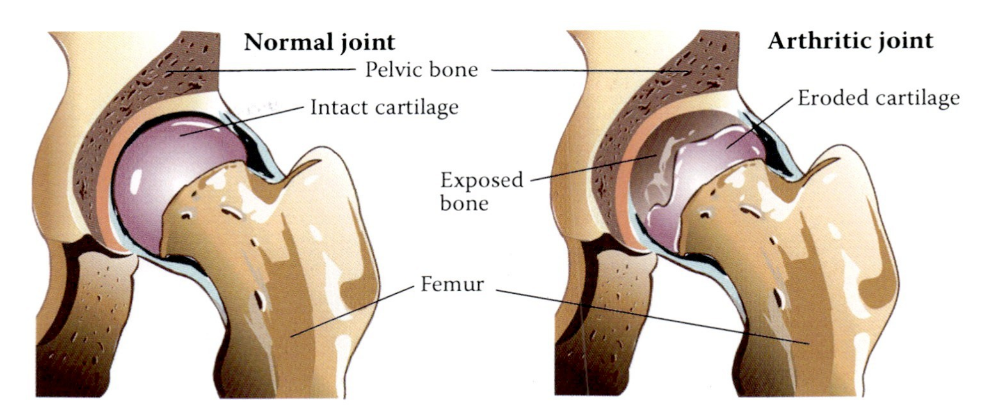

COLOR FIGURE 119
Normal and arthritic hip joints.

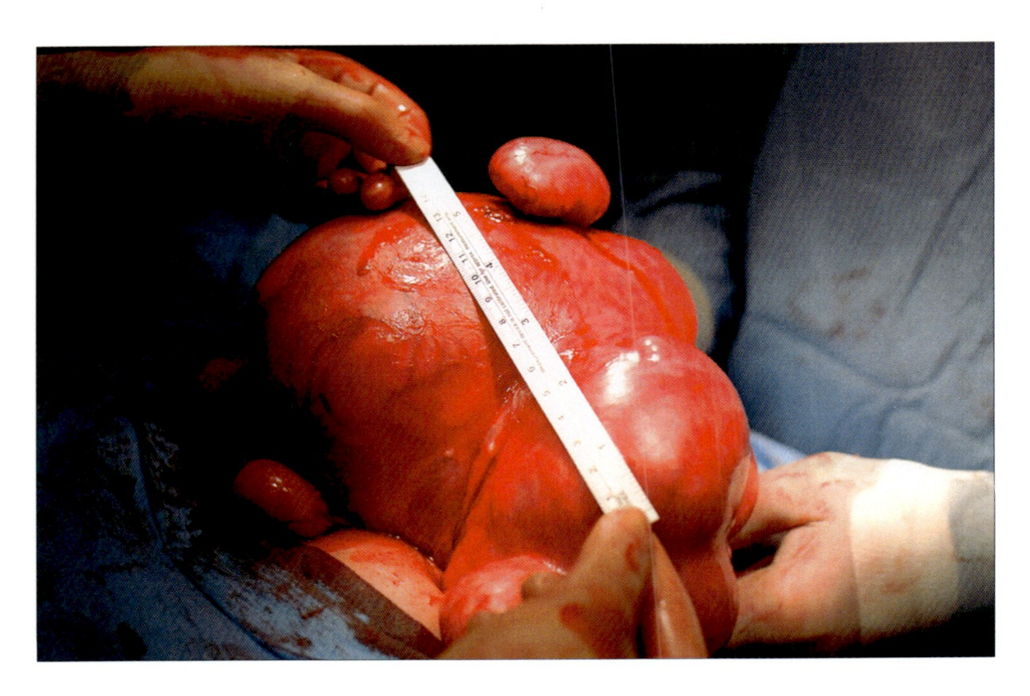

COLOR FIGURE 122
Uterine fibroid tumor.

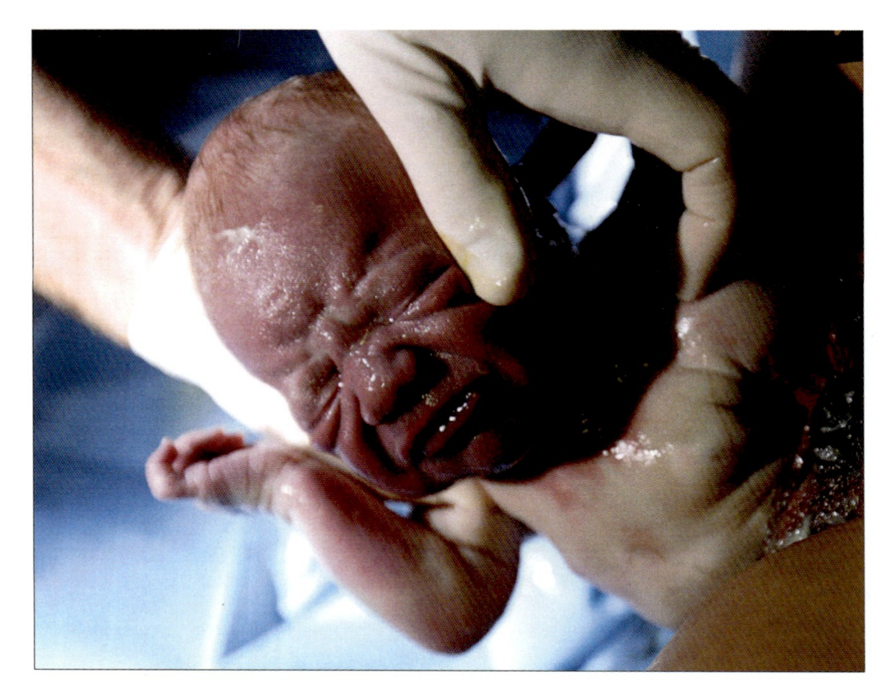

COLOR FIGURE 136
Newborn infant.

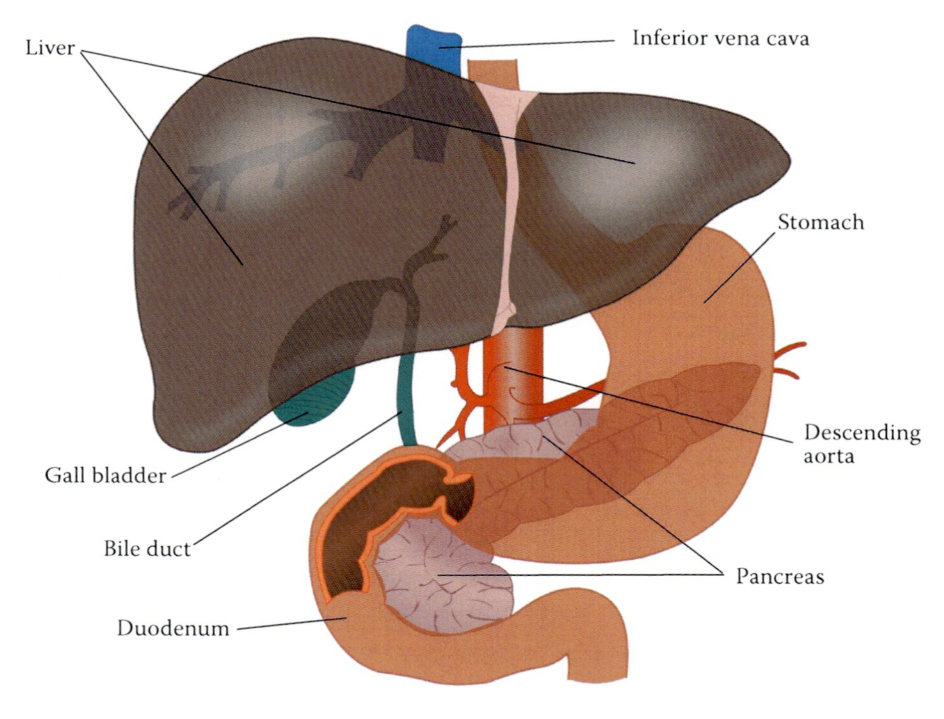

COLOR FIGURE 143
Liver and associated organs.

Staffing

Birthing coach or midwife, obstetrician, nurses. Staffing is extremely variable depending on the circumstances and the wishes of the parents-to-be. A woman can successfully deliver her baby without anyone else being present, though this is unusual.

Equipment and Supplies

Birthing bed or other birthing facilities such as birthing chair or water tank, fetal monitor, infant resuscitation equipment, devices to aid with difficulties.

Preparation

Prenatal classes can encourage exercises to help make labor and delivery easier, and provide techniques to cope with or reduce pain and anxiety. If it is necessary to expedite labor due to maternal or fetal considerations, oxytocin can be administered intravenously to induce contractions and the labor process.

Procedure

Labor and delivery often proceeds with little if any outside intervention (Figure 134, Figure 135).

If significant maternal or fetal distress in encountered and cannot likely be alleviated before birth, an emergency caesarean section may be performed. Fetal mispositions can sometimes be corrected through abdominal manipulations and/or maternal position changes. If the fetus is lodged partway through the birth canal and is not progressing, forceps or suction may be used to help extract the fetus. If the opening to the vagina is restrictive, an episiotomy may be performed to enlarge the passage (Figure 136).

1. Uterus contracting, baby's head presses on cervix

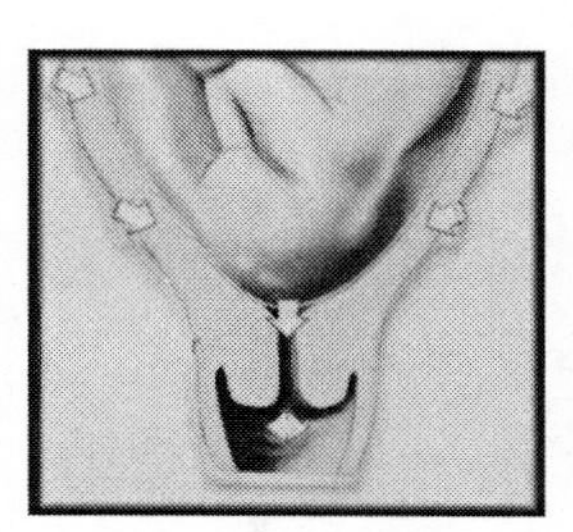

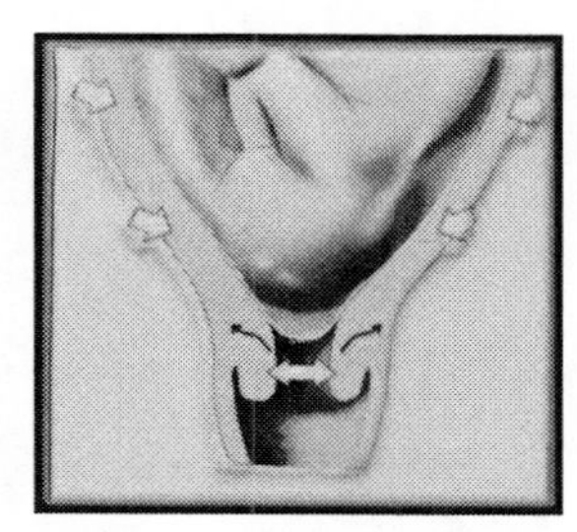

2. More contractions, cervix dilates

3. Baby moves through birth canal

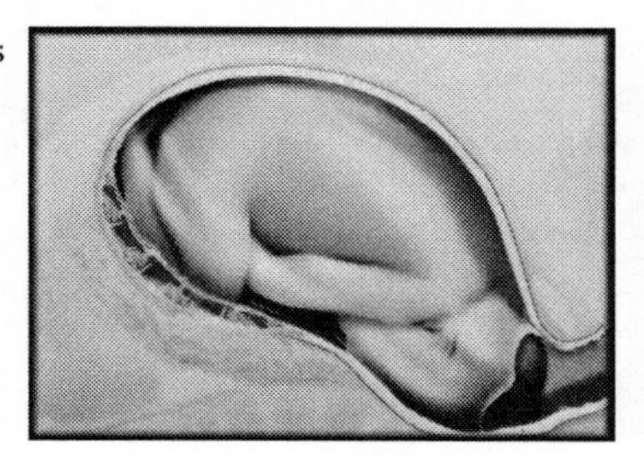

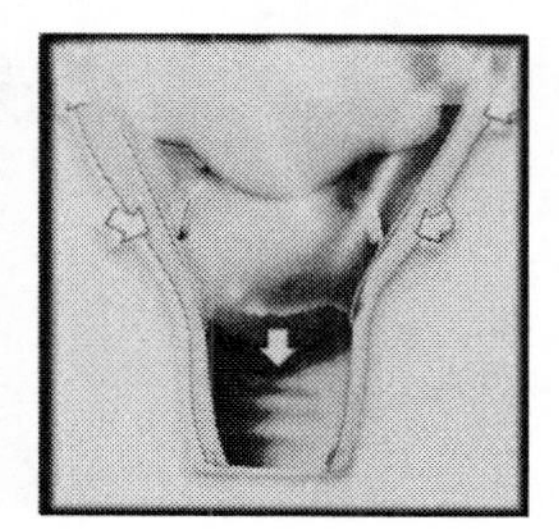

4. Placenta is expelled

FIGURE 134
Stages of labor and delivery.

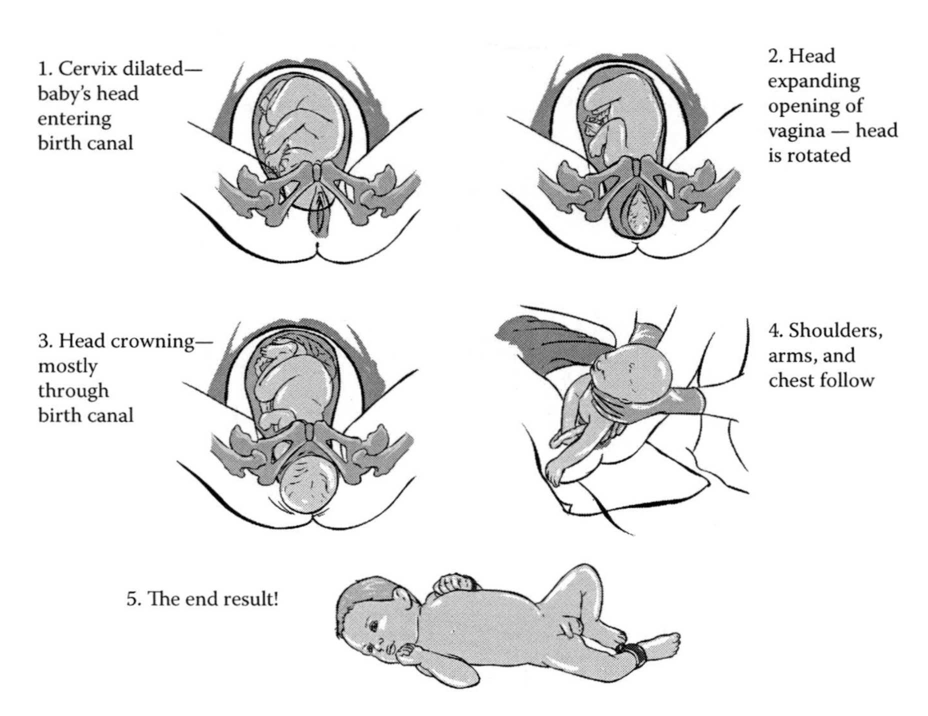

FIGURE 135
Labor and delivery.

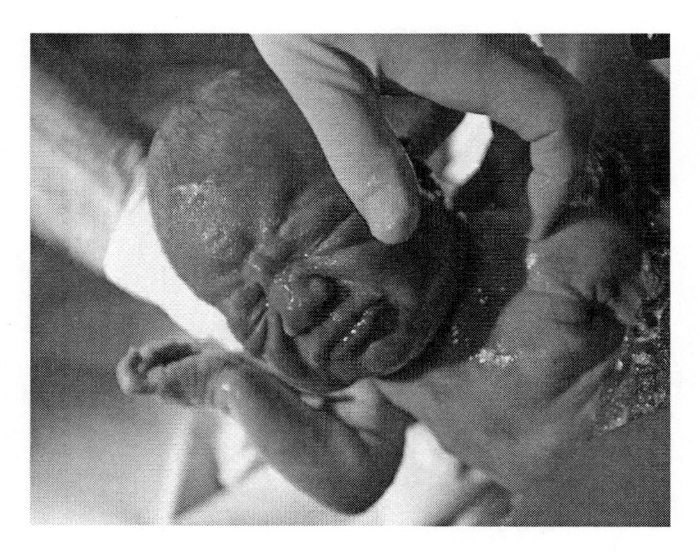

FIGURE 136
A color version of this figure follows page **176.** Newborn infant.

Expected Outcome and Follow-Up

Successful delivery of the fetus or fetuses. Following delivery, the infant is examined for any abnormalities, and its status is evaluated, usually using the APGAR score. It may require suctioning of airways and a warm environment. If the infant is significantly premature or has other physical problems, it may be taken to a neonatal intensive care unit. Less severe but still important concerns may mean that the infant is placed in an incubator for some time. Few or no problems mean that the infant can remain in a simple bassinette or with the mother.

The mother is monitored for excess bleeding, blood pressure, and infection. Breastfeeding is encouraged, partly because it encourages further uterine contractions, which can help reduce bleeding.

Complications

Failure of labor to progress, fetal or maternal distress, hemorrhage, severe drop in blood pressure, maternal heart failure or stroke.

Laminectomy

Alternate names—Spinal decompression.

Purpose

To remove part of the bony structure of a vertebra in order to relieve back pain.

Indications

Back pain that does not respond to other treatment methods.

Anatomy

The spinal column is made up of a number of vertebral bones that surround the spinal cord. The vertebrae are connected with muscles and ligaments, and a fibrous disk between each pair serves as a bearing surface and shock absorber. Nerves branch out from the spinal cord between each pair of vertebrae to supply various areas of the body (Figure 137, Figure 138).

Pathology

If a spinal nerve is compressed, various symptoms are produced, depending on the location. This tends to happen most often in the lower back, in the lumbar or sacral areas. Symptoms of nerve compression in this area include back pain, pain along the course of the nerve, muscle weakness in the leg, or bladder or bowel dysfunction. If the compression takes place inside the spinal canal, the exact cause is hard to determine without surgery. Compression can be caused by a bulging disk, bone spurs, or tumors (Figure 139).

Physiology

Spinal nerves carry signals in and out of the spinal cord.

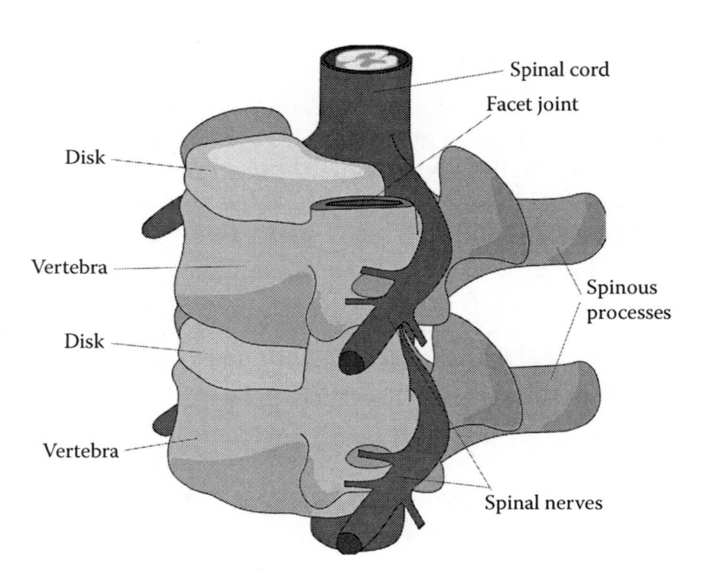

FIGURE 137
Vertebrae, disks, spinal cord, and spinal nerves.

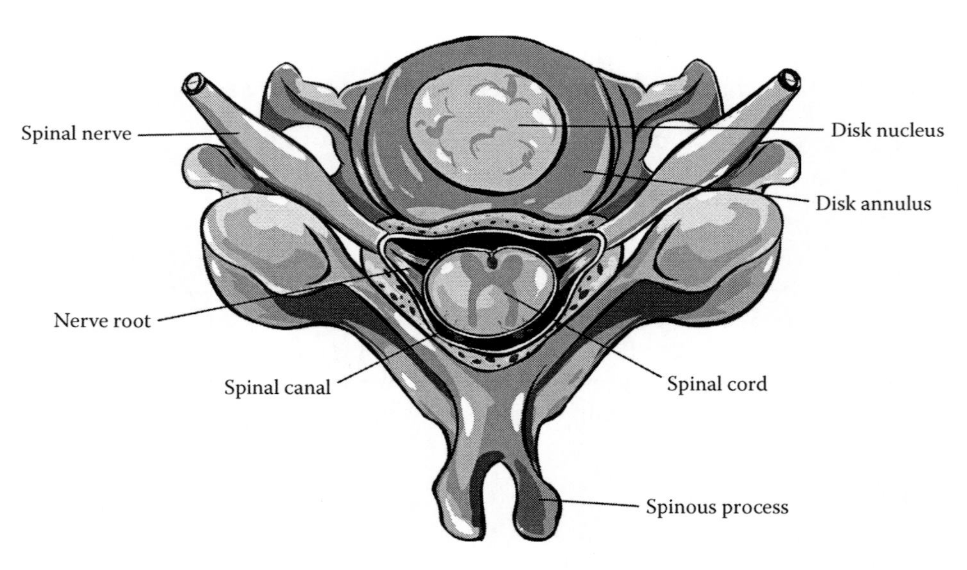

FIGURE 138
Cross section of spinal column.

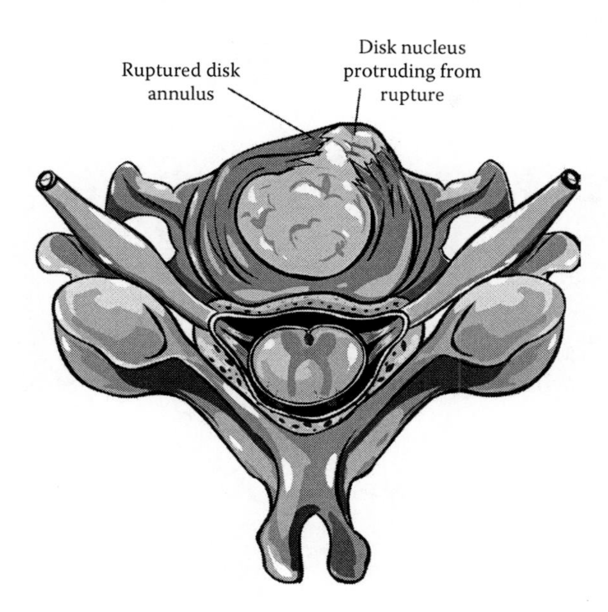

FIGURE 139
Cross section of spinal column showing ruptured disk.

Staffing

Normal surgical team; usually a neurosurgeon.

Equipment and Supplies

Standard surgical.

Preparation

Standard surgical.

Procedure

General anesthesia is induced. An incision is made through the skin and muscle over the target vertebra, and the tissues retracted. The bone above the area is cut away, and an opening is made in the ligament that holds adjacent vertebrae together. This provides access to the spinal canal and exposes the cause of nerve compression. The structure causing compression is removed or reduced if possible. If necessary, spinal fusion will be performed by exposing the faces of adjacent vertebrae and inserting a small piece of the patient's bone, which is fixed in place by screws. After these processes are completed, the incision layers are sutured and closed.

Expected Outcome and Follow-Up

Alleviation of pain and other symptoms. The patient must avoid activities that might put stress on the surgical area for 2 to 4 weeks, though prescribed exercises are important to regain strength, mobility, and stability.

Complications

Nerve damage.

Laryngectomy

Alternate names—Larynx removal, voice box surgery or removal.

Purpose

To remove the larynx when it is damaged beyond repair or if cancerous tumors are present.

Indications

Severe damage to the larynx; larynx cancer.

Anatomy

The larynx is a cartilaginous structure surrounding the trachea. It contains reed-like structures called vocal cords.

Pathology

Physical trauma can damage the larynx. Cancerous tumors can develop.

Physiology

The larynx can vibrate as air passes through it, producing certain speech sounds. It is also involved in breathing and swallowing.

Staffing

Normal surgical team.

Equipment and Supplies

Standard surgical.

Preparation

Standard surgical.

Procedure

A permanent tracheotomy is done to allow air to enter the lungs during and after the surgery. An incision is made in the neck to allow access to the larynx, and the section with the tumor or the whole larynx is removed. The incision is then closed, but the tracheostomy remains.

Expected Outcome and Follow-Up

Removal of cancerous tumors or damaged structures. The patient is instructed in the care of the tracheostomy. New methods of producing speech sounds that were formerly produced by the larynx are developed as much as possible. Swallowing food is not possible for 2 or 3 weeks following surgery, so feeding must be done via a stomach tube.

Complications

Psychological distress.

LASIK

Alternate names—Laser eye surgery. LASIK stands for laser in-situ keratomileusis.

Purpose

To improve vision.

Indications

Astigmatism, myopia.

Anatomy and Physiology

The eye is the organ of visual sensation. It is spherical in shape, with a clear membrane (the cornea) covering the front; the cornea begins to focus light entering the eye. A thin liquid, the aqueous humor, is contained between the cornea and the colored part of the eye or iris. The iris can contract or relax to change the shape of an opening in its center, the pupil, controlling the amount of light entering the eye. A clear, fibrous lens behind the pupil focuses light and can change shape to alter the focal distance. The main body of the eye is filled with a semisolid fluid called the vitreous humor. At the back of the eye, the retina receives the focused light image and transforms it into electrical signals. These are carried from the eye to the brain by the optic nerve (Figure 140).

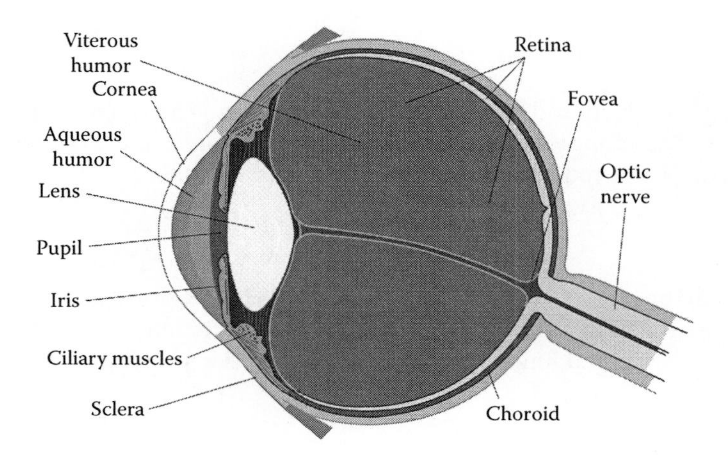

FIGURE 140
Vertical section of the eye.

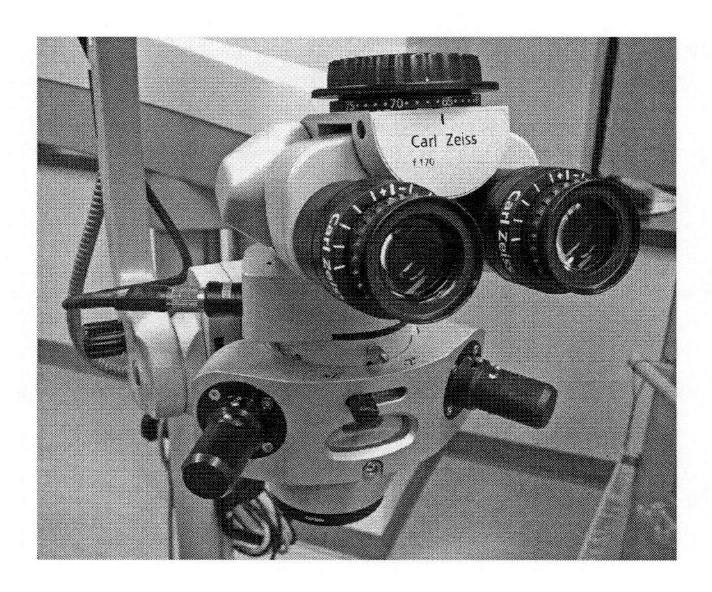

FIGURE 141
Operating microscope viewing head. Focus, magnification, and position can all be adjusted manually from the head, but many functions can also be adjusted using a foot switch system, thus freeing the surgeon's hands.

Pathology

Clear vision may be impaired by improper focusing of light.

Staffing

Ophthalmologist, assistant.

Equipment and Supplies

Operating microscope (Figure 141), excimer laser, microsurgery devices.

Preparation

A sedative may be given and local anesthesia provided. The cornea will have been precisely measured and mapped to determine what correction is required to produce optimal vision. A speculum is used to hold the eyelids open.

Procedure

The outer layer of the cornea is cut to create a flap, which is pulled to the side. The excimer laser beam is focused on the underlying layer of the cornea and is activated and moved around to ablate predefined areas and amounts of the cornea. The flap is replaced and smoothed out.

Expected Outcome and Follow-Up

Corrected vision. The corneal flap heals in place without stitches. Vision is tested in several follow-up visits to ensure continued correction. A shield is placed over the eye to provide protection and to block light while the cornea heals. Antibiotic drops are used for a few weeks to help avoid infections.

Complications

Corneal perforation, improper healing of the corneal flap, ongoing light sensitivity, loss of vision correction.

Lithotripsy

Alternate names—Extracorporeal shock wave lithotripsy, ESWL.

Purpose

To break apart kidney stones into small enough pieces that they can be passed out through the urinary tract.

Indications

Kidney stones that are causing serious difficulties for the patient and are not likely to pass on their own. Stones are confirmed by diagnostic imaging.

Anatomy

The kidneys are a pair of organs located near the back wall of the abdominal cavity (Figure 142). They are well supplied with blood. The interior of the kidney is an open space called the pelvis, which collects urine and passes it to the ureters. The ureters connect to the bladder, which collects urine and holds it until it can be released via urination. Sphincter muscles are at the exit of the bladder, and the urethra leads from the bladder to the outside.

Pathology

Kidney stones, or renal calculi, form in the renal pelvis. They may be composed of various substances, most commonly calcium oxalate. Uric acid may form stones, and more rarely they

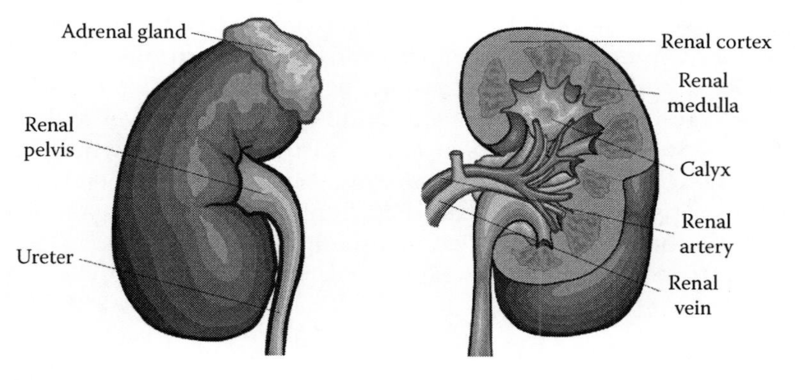

FIGURE 142
Kidney anatomy.

are made of calcium phosphate or cystine. Stones form under a number of circumstances, including a diet high in calcium oxalate or the precursors of uric acid, chronically inadequate hydration, recurrent urinary tract infections, certain metabolic disorders, or some disease processes. The stones grow gradually and are often passed out through the ureters, bladder, and urethra, sometimes with little or no discomfort but often accompanied by excruciating pain. The pain is mostly caused by the stone blocking the flow of urine, which produces backpressure and stress in the kidneys; this pain is called renal colic. Some stones may become too large to be passed, and alternate means of removing them are needed.

Physiology

Kidneys take blood in from the body and filter out water and toxins, mainly urea. They also regulate electrolyte levels in the blood and produce hormones that are involved in blood pressure regulation.

Staffing

Physician (usually a urologist), assistant, lithotripsy equipment operator, diagnostic imaging technologist.

Equipment and Supplies

Lithotripsy machine. This device produces high-intensity sound waves (shock waves) that can be focused on a very specific location within the body. The waves are produced by a high-intensity electrical spark or by the action of a powerful electromechanical crystal.

Preparation

The patient is sedated and then immersed in a water bath or placed on a lithotripsy table. Some restraints may be placed to help the patient remain motionless.

Procedure

The exact location of the target stone is determined by fluoroscopy, and the shock wave focusing system aims the equipment at the stone. Shock wave production is begun, usually about one or two per second. The source of the shock waves is outside the body (extracorporeal), and the waves may be transmitted into the patient through an immersion water bath or via a rubber bladder that is pressed against the patient's side or back. Fluoroscopic images help to continuously guide the focus point of the shock waves to that they are concentrated on the kidney stone target. The images also may indicate when the stone begins to break apart, though often it stays in a concentrated mass even when it is shattered into smaller particles. Application of the treatment continues until the team decides that the stone is likely to be broken sufficiently; this usually takes an hour or so. The shock wave generator is then turned off, and the patient is able to move about.

Expected Outcome and Follow-Up

Reduction of the stone to small enough pieces that they can be passed through the urinary tract. Though the stone particles are smaller, they may still cause considerable discomfort when they are passed; this may take several days. Greatly increased fluid consumption may help flush out the stone particles. There may be some blood in the urine, because the shock waves can cause some minor collateral damage to kidney tissue adjacent to the stone being broken. Patients may be asked to strain their urine when particles of stone are being passed so that they can be analyzed. The composition of the stones may indicate a course of treatment to help avoid future stones, such as dietary modifications, increased hydration, or medication.

Complications

Persistent abdominal pain, incomplete breakage of the stone, minor damage to adjacent kidney tissue.

Liver Transplant

Alternate names—n/a.

Purpose

To replace a nonfunctional liver with that of a donor.

Indications

Liver failure, shown by lab tests of liver function; severe liver damage by trauma shown by direct observation or diagnostic imaging.

Anatomy

The liver is a large abdominal organ, most of which is on the right side of the body just under the diaphragm. It is very well vascularized, and is connected to the small intestine via the common bile duct, which also connects to the gall bladder (Figure 143).

Pathology

Disease and chemical or physical trauma can all affect liver function and eventually alter its structure. Alcohol abuse, poisoning by various chemicals and drugs, and metabolic disorders such as hemochromatosis can all affect liver structure and function. Diseases such as jaundice (a buildup of bilirubin in the body), cirrhosis (the development of scar tissue and nodules in the liver), hepatitis, various cancers of the liver, and tuberculosis may also compromise the liver. If the common bile duct is damaged or missing (in children this may be a birth defect known as biliary atresia), bile and toxins may back up in the liver, causing damage.

Physiology

The liver performs a variety of functions, including the production of bile, which is used in the digestion process, the breakdown and excretion of toxic substances from the blood, and production of certain amino acids. The liver can store excess glucose from digestion in the form of glycogen, and it also produces cholesterol, which is an important metabolic compound.

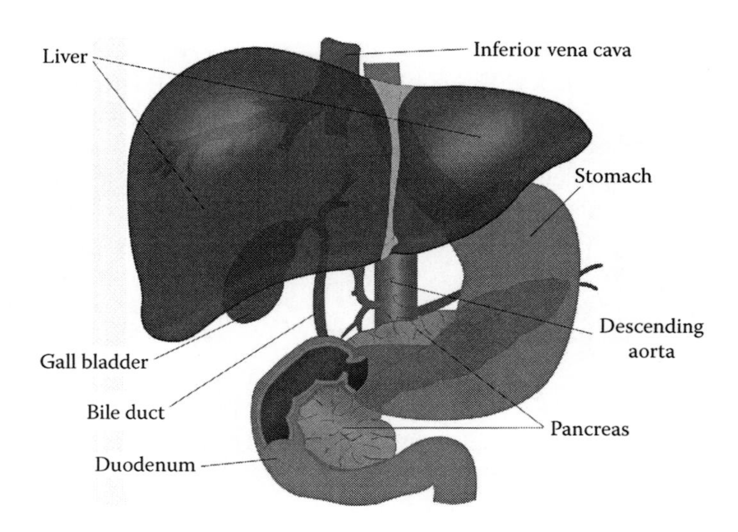

FIGURE 143
A color version of this figure follows page **176.** Liver and associated organs.

Staffing

Complex surgical team.

Equipment and Supplies

Standard surgical, donated liver.

Preparation

Extensive tissue testing and matching is done to ensure the best possible match, to reduce rejection issues. A donor liver must be available that is a suitable match in tissue type and size and is available soon enough after donation to be viable. The patient is sedated and the incision site cleaned. Antirejection (immunosuppressive) drugs are given for some time in advance of the surgery.

Procedure

The liver is a resilient organ, which means that the donor organ can possibly be reduced in size to match the needs of the patient, without compromising function. This is a benefit for transplanting a large donor liver into a child; only part of the liver may be used.

The donor liver is removed along with associate blood vessels and the common bile duct; it is then placed in an environment that ensures optimal tissue health until it is implanted into the patient.

A large incision is made in the abdomen and the abdominal wall layers retracted to allow access to the liver. The major blood vessels supplying the liver are clamped and severed, leaving enough to allow connection to the donor organ's vessels. The donated liver or portion is put in place and the blood vessels connected. The common bile duct is connected as well. It may be necessary to construct an artificial bile duct using a small part of the patient's intestine. After all connections and placements are checked, the abdominal incision is closed, with drainage tubes in place.

Expected Outcome and Follow-Up

Normal functioning of the donor liver. Patients are given antirejection (immunosuppressive) drugs indefinitely, but are also regularly monitored for signs of rejection, more frequently immediately following surgery. Patients are monitored in an intensive care unit for several days. Physical activities are slowly initiated in the ICU, and increased when the patient can be moved to a general surgical ward. After discharge normal activities can be gradually resumed. Liver function tests are performed regularly to ensure adequate function.

Complications

Rejection of the donated liver is the most serious complication and may require a second transplant. Death can result if this is not possible and rejection cannot be controlled. Bleeding and infection are constant concerns. Other infections may occur as a result of the antirejection drugs used.

Mastectomy, Lumpectomy

Alternate names—Breast conservation surgery, partial mastectomy.

Purpose

To remove known or suspected cancerous tissue from the breast.

Indications

Abnormalities in breast tissue detected by palpation, mammography, ultrasound, or other imaging techniques. These may include lumps or other masses, or changes in tissue when compared to earlier, baseline images.

Anatomy

The breast consists mainly of fat cells and milk-producing glands (Figure 144). Breast tissue can have a wide range of textures.

Pathology

Breast cancer and other diseases of the breast can produce abnormal masses within the breast. These masses may be completely self-contained, or they may extend to various degrees into the breast tissue and possible surrounding organs. Breast cancer can metastasize or spread to distant locations in the body.

Physiology

Breasts produce milk following pregnancy; when not lactating, they are essentially dormant.

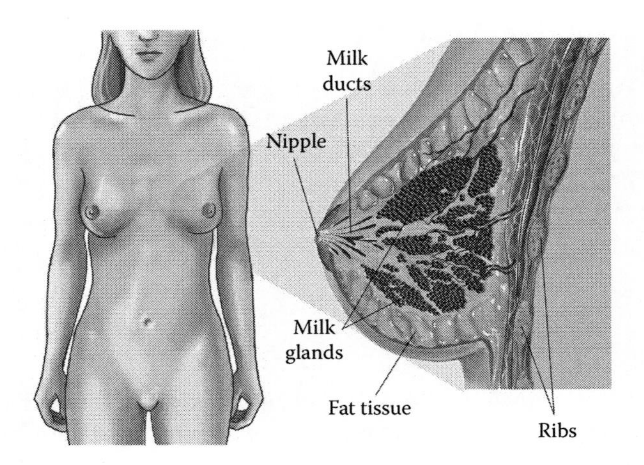

FIGURE 144
Breast anatomy.

Staffing

Minor surgical team.

Equipment and Supplies

Standard surgical.

Preparation

Standard surgical. Chemotherapy may be given prior to surgery in order to reduce the size of the tumor and hopefully kill any small groups of cancer cells beyond the lump.

Procedure

If a diagnosis of breast cancer has been made or is suspected and the cancerous tissue is determined to be completely contained in a node or lump, a removal of the lump and possibly some surrounding tissue may be the only surgery required to eliminate the disease. Some associated lymph nodes may also be removed as a prophylactic measure.

The procedure is usually performed using general anesthesia.

An incision is made to provide optimal access to the target area of the breast. The tumor is located and excised, possibly along with some surrounding tissue. The tissue will be sent to the lab for examination and definitive diagnosis. The incision is then closed; depending on the size of the incision, sutures may be used to close, but adhesive strips may be sufficient.

Expected Outcome and Follow-Up

Removal of all cancerous tissue from the body. Breast reconstruction surgery (mammoplasty) may be performed at a later date to restore a more normal appearance. Activities, especially those involving lifting of any weight, are restricted for a few days. Radiation therapy is usually begun within a few days of surgery.

Complications

Scarring, loss of sensation in the breast.

Mastectomy, Radical

Alternate names—n/a.

Purpose

To remove the breast and all cancerous tissue or potentially cancerous tissue in the area.

Indications

Abnormalities in breast tissue detected by palpation, mammography, ultrasound, or other imaging techniques. These may include lumps or other masses, or changes in tissue when compared to earlier, baseline images.

Anatomy

The breast consists mainly of fat cells and milk-producing glands. Breast tissue can have a wide range of textures.

Pathology

Breast cancer and other diseases of the breast can produce abnormal masses within the breast. These masses may be completely self-contained, or they may extend to various degrees into the breast tissue and possible surrounding organs. Breast cancer can metastasize or spread to distant locations in the body.

Physiology

Breasts produce milk following pregnancy; when not lactating, they are essentially dormant.

Staffing

Normal surgical team.

Equipment and Supplies

Standard surgical.

Preparation

Standard surgical.

Procedure

If a diagnosis of breast cancer has been made and the cancerous tissue cannot be determined to be completely contained within the breast, removal of the breast and various associated structures including muscles and lymph nodes may be necessary.

The procedure is usually performed using general anesthesia.

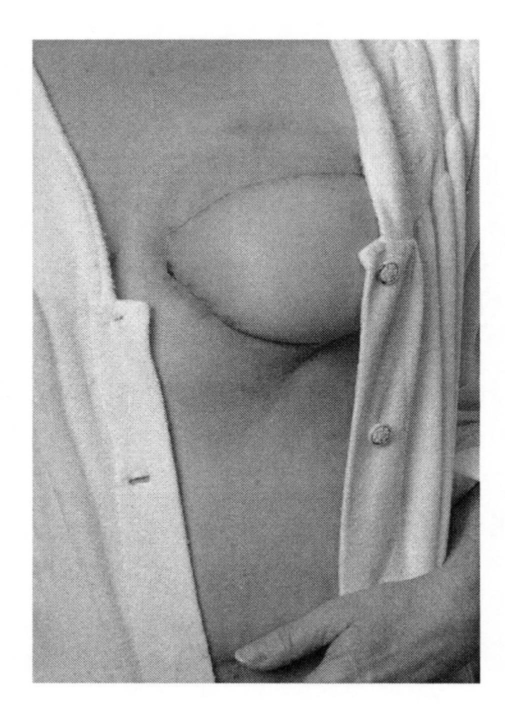

FIGURE 145
Reconstructed breast following mastectomy.

An incision in made in such as way as to preserve as much skin as possible, to aid in later reconstructive surgery. All breast tissue is removed, including the nipple and areola, as well as portions of several groups of underlying muscles, lymph nodes in the area, and any fat or connective tissue in the area. Care must be taken to avoid damage to nerves in the area. All tissues will be sent to the lab for examination and definitive diagnosis. The incision is then closed, with drain tubes in place.

Expected Outcome and Follow-Up

Removal of all cancerous tissue from the body. Breast reconstruction surgery (mammoplasty) may be performed at a later date to restore a more normal appearance (Figure 145).

Activities, especially those involving lifting of any weight, are restricted for several days. Radiation therapy is usually begun within a few days of surgery. Physiotherapy is important, because there is a significant loss of muscle tissue.

Complications

Edema of the arm, nerve damage, recurrence of cancer.

Mastectomy, Simple

Alternate names—n/a.

Purpose

To remove the breast and all cancerous tissue or potentially cancerous tissue.

Indications

Abnormalities in breast tissue detected by palpation, mammography, ultrasound, or other imaging techniques. These may include lumps or other masses, or changes in tissue when compared to earlier, baseline images.

Anatomy

The breast consists mainly of fat cells and milk-producing glands. Breast tissue can have a wide range of textures.

Pathology

Breast cancer and other diseases of the breast can produce abnormal masses within the breast. These masses may be completely self-contained, or they may extend to various degrees into the breast tissue and possible surrounding organs. Breast cancer can metastasize or spread to distant locations in the body.

Physiology

Breasts produce milk following pregnancy; when not lactating, they are essentially dormant.

Staffing

Normal surgical team.

Equipment and Supplies

Standard surgical.

Preparation

Standard surgical.

Procedure

If a diagnosis of breast cancer has been made and the cancerous tissue is determined to be completely contained within the breast, a removal of the breast itself may be required to eliminate the disease, without removing any associated structures. Lymph nodes in the area may be removed as a prophylactic measure. If there is known or strongly suspected involvement of the other breast, both may be removed in the same surgery.

The procedure is usually performed using general anesthesia.

An incision is made along the side of the breast, and all breast tissue is removed, sometimes including the areola and nipple. Lymph nodes may be dissected out and removed at this point. All tissues will be sent to the lab for examination and definitive diagnosis. The incision is then closed.

Expected Outcome and Follow-Up

Removal of all cancerous tissue from the body. Breast reconstruction surgery (mammoplasty) may be performed at a later date to restore a more normal appearance. Activities, especially those involving lifting of any weight, are restricted for several days. Radiation therapy is usually begun within a few days of surgery.

Complications

Scarring, recurrence of cancer.

Myringotomy

Alternate names—Ear tube insertion, myringocentesis, tympanotomy, tympanostomy.

Purpose

To relieve pressure in the inner ear caused by fluid buildup.

Indications

Recurrent inflammation of the inner ear, including fluid buildup that does not respond to other forms of treatment.

Anatomy

The ear consists of the outer portion or pinna, the auditory canal, the outer tympanic membrane or eardrum, the inner ear containing three tiny bones (the malleus or hammer, the incus or anvil, and the stapes or stirrup), the inner ear membrane, the semicircular canals that are involved in balance, and the cochlea, a spiral structure with nerve receptor hairs inside that connect via the auditory nerve to the brain (Figure 146). The eustachian tube connects the inner ear to the back of the throat; the tube is normally collapsed.

Pathology

Infections of the inner ear can cause fluid to build up. This can result in severe pain and, if untreated, can cause permanent damage to the ear and subsequent hearing loss. Inflammation of the eustachian tube can prevent normal drainage of fluids from the inner ear.

Physiology

Sound waves are carried into the outer tympanic membrane, where they cause vibration. This vibration is carried and amplified by the small bones within the inner ear, and then to the inner membrane. This sets up vibrations within the cochlea, and specific receptor hairs vibrate harmonically with different sound frequencies. Nerve signals sent to the brain allow interpretation of sounds. Structures within the semicircular canals provide information to the brain about body position and movement. The eustachian tube can provide some fluid drainage and also can open to balance air pressure inside and outside the ear.

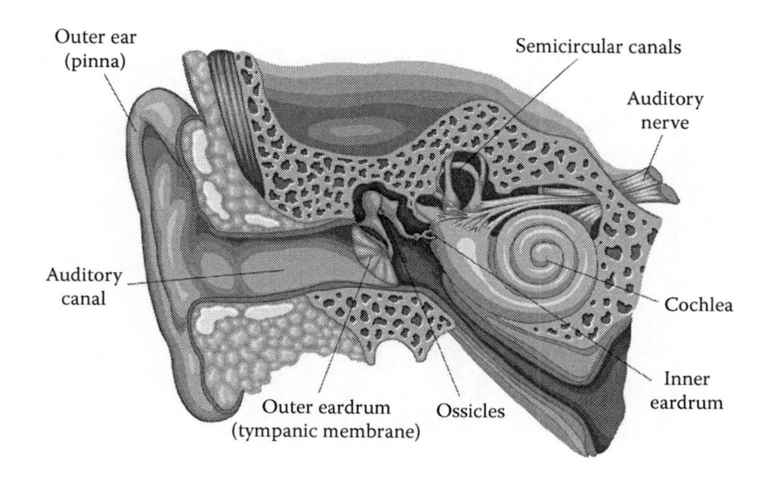

FIGURE 146
Ear anatomy.

Staffing

Physician, possibly assistant.

Equipment and Supplies

Myringotomy tubes, scalpel.

Preparation

Any fluid that leaks from the ear spontaneously is collected and analyzed in the lab. Blood tests are done to determine the cause of inflammation.

Procedure

A small incision is made in the outer eardrum, and an ear tube is inserted. The tubes are of various sizes and shapes to suit different situations. They are held in place by friction.

Expected Outcome and Follow-Up

Relief of pressure within the ear, drainage of fluid, and clearing of underlying infections. The tubes may come out on their own after a few days, or the physician may remove them when they have served their purpose. Antibiotics may be used to eliminate or prevent infections. The ear must be kept dry until healing is complete.

Complications

Further infection, permanent damage to the tympanic membrane.

Pacemaker Application

Alternate names—Pacemaker implantation.

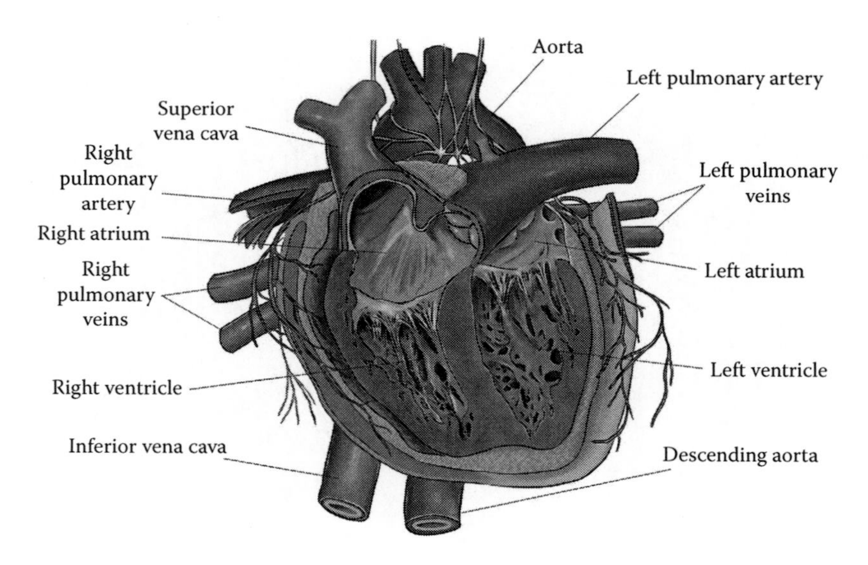

FIGURE 147
Heart and associated vessels.

Purpose

To install a permanent cardiac pacemaker that will regulate heart rate.

Indications

Irregular heart rhythms (arrhythmias) that may cause significant health problems if untreated, and which have not responded to other therapies. The specific cause of the arrhythmia is determined by a combination of special ECG studies and diagnostic imaging.

Anatomy

The heart is a muscular organ consisting of left and right atria, left and right ventricles, four valves, and associated blood vessels (Figure 147). These blood vessels carry the blood pumped by the heart out to various body parts and also carry blood to the heart itself. The atria are thinner walled than the ventricles. Nerves and other pathways in the heart carry electrical signals to control pumping contractions.

Pathology

Heart pumping effectiveness can be compromised by various problems within the heart. The muscles of the atria and especially the ventricles can be damaged by disease or lack of blood supply (cardiac infarction). The cardiac valves can be damaged by disease or age, thus allowing backflow of blood. Finally, the conductive pathways that help produce and coordinate contractions may be damaged (heart block), so that contractions are either weaker or less coordinated, or both. With some arrhythmias, the signals are intermittently blocked so that there are periods of normal heartbeats and periods of arrhythmia.

Physiology

The heart, being mostly muscle that is active all the time, requires an excellent blood supply to provide oxygen and nutrients and to remove wastes.

Signals from a center in the heart produce electrical signals that are carried throughout the heart to produce contractions. Various components provide delays to the signal so that contractions are coordinated and allow effective pumping.

Staffing

Normal surgical; the surgeon is a cardiologist.

Equipment and Supplies

Standard surgical, pacemaker. Pacemakers are battery-powered devices that can generate signals to help regulate contractions in either the atria, the ventricles, or both. They contain sensors and analysis circuitry that can decide when a pacing pulse is necessary and may be able to determine if a change in rate is required, for example during exercise. Diagnostic imaging equipment to guide electrode placement.

Preparation

Thorough testing is done to determine the exact cause of arrhythmia. This gives the cardiologist information so that the type of pacemaker and sites of electrode placement can be defined. A sedative is administered.

Procedure

Pacemaker implantation is usually done under local anesthesia. An incision is made in the skin of the chest, usually in the area just above the pectoral muscles. A suitable vein (usually the subclavian) in the area is accessed, and the electrode cable is inserted into the vein. Guided by fluoroscopy the cable is moved through the vein to the vena cava, through the right atrium and tricuspid valve, and finally to the point where the electrodes are to be implanted (Figure 148).

A special mechanism releases the electrode tips, and they are embedded in the heart muscle. The pacemaker is placed in a pocket of tissue and activated. Proper function is tested, and if all is well, the incision is closed.

Expected Outcome and Follow-Up

Regulation of cardiac rhythms and the resumption of normal activities. The pacemaker is programmable, and programming can be done with the use of special magnets acting through the skin. The patient is monitored to ensure the pacemaker is functioning correctly. Normal activities are resumed slowly. Since the pacemaker will set off metal detectors in security checks, a medic-alert bracelet confirming the presence of the pacemaker is provided. The bracelet also serves to notify medical staff about the pacemaker. Some electrical devices can affect pacemaker operation. Usually there are prominent signs in the vicinity of such equipment, but care must be taken to avoid them.

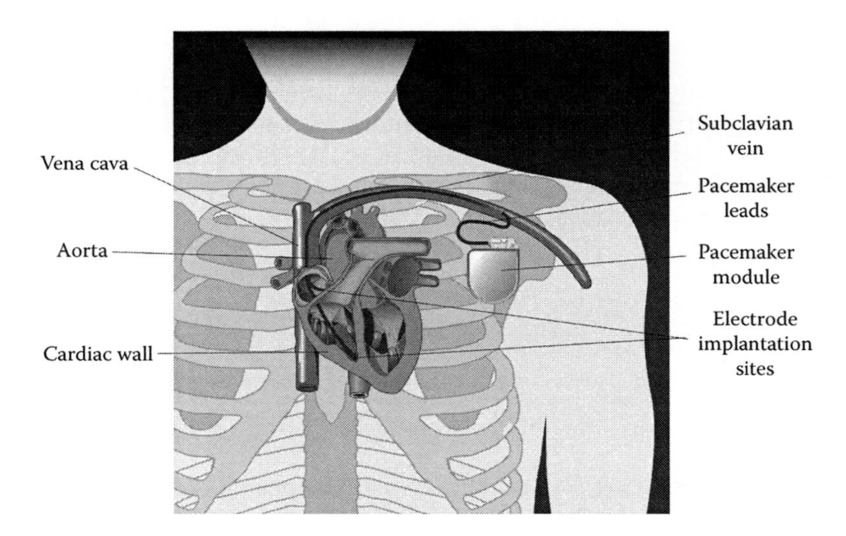

FIGURE 148
Pacemaker and electrode placement.

Complications

Collapsed lung, generator or lead failure.

Patent Ductus Arteriosus Repair

Alternate names—n/a.

Purpose

To repair the congenital defect of patent ductus arteriosus (PDA).

Indications

The presence of patent ductus arteriosus as shown by symptoms and diagnostic imaging.

Anatomy

The heart is a muscular organ consisting of left and right atria, left and right ventricles, four valves, and associated blood vessels (Figure 149). These blood vessels carry the blood pumped by the heart out to various body parts and also carry blood to the heart itself. The atria are thinner walled than the ventricles. Nerves and other pathways in the heart carry electrical signals to control pumping contractions.

Pathology

During gestation, a vessel called the ductus arteriosus connects the aorta and the pulmonary artery of the fetus. This is because the fetal lungs are not functional, and full blood flow through them would be counterproductive. Following birth, the ductus arteriosus normally closes off and eventually disappears. In some cases, however, the ductus

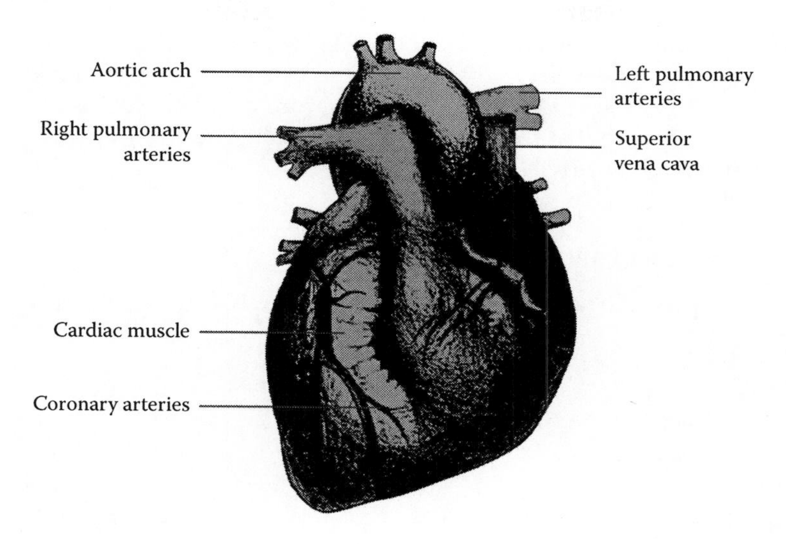

FIGURE 149
Heart anatomy.

arteriosus remains open, reducing the effectiveness of heart pumping and causing the heart to beat more powerfully in order to try to compensate. Adverse pressure in the pulmonary arteries results as well. All of this can lead to an enlarged heart and fluid buildup in the lungs. Severe cases can cause damage to other organs.

Physiology

The heart, being mostly muscle that is active all the time, requires an excellent blood supply to provide oxygen and nutrients and to remove wastes.

Signals from a center in the heart produce electrical signals that are carried throughout the heart to produce contractions. Various components provide delays to the signal so that contractions are coordinated and allow effective pumping.

Staffing

Normal surgical team; the surgeon is a pediatric cardiologist.

Equipment and Supplies

Standard surgical.

Preparation

Standard surgical.

Procedure

PDA repair is done under general anesthesia. It is usually performed when the infant is about 6 months to 2 years old, depending on the severity and the presence of other factors.

A small incision is made in the chest wall between two ribs, and the opening is retracted. This provides access to the heart and allows the structures to be visualized. The surgeon reaches in to the site of the PDA and ties sutures around it. The ductus may then be severed, or it may be left with just the sutures closing it off. The chest opening is allowed to close, and the incision is sutured.

Expected Outcome and Follow-Up

Relief of PDA symptoms. The patient is monitored in an ICU for a few days.

Complications

n/a

Peritoneal Dialysis

Alternate names—PD, ambulatory PD, intermittent PD, cyclic PD.

Purpose

To remove toxins from the blood when disease or damage prevents the kidneys from doing so. Ideally the process allows the kidneys time to resume normal function, but this occurs only rarely. Dialysis may allow the patient to survive until a kidney transplant can be performed.

Indications

Failure of kidney (renal) function as indicated by urine and blood tests.

Anatomy

The kidneys are a pair of organs located near the back wall of the abdominal cavity. They are well supplied with blood. The interior of the kidney is an open space called the pelvis, which collects urine and passes it to the ureters. The ureters connect to the bladder, which collects urine and holds it until it can be released via urination. Sphincter muscles are at the exit of the bladder, and the urethra leads from the bladder to the outside (Figure 150).

Pathology

Various disease processes or physical trauma can compromise renal function to the point where fluids and toxins build up in the body to intolerable levels.

Physiology

Kidneys take blood in from the body and filter out water and toxins, mainly urea. They also regulate electrolyte levels in the blood and produce hormones that are involved in blood pressure regulation.

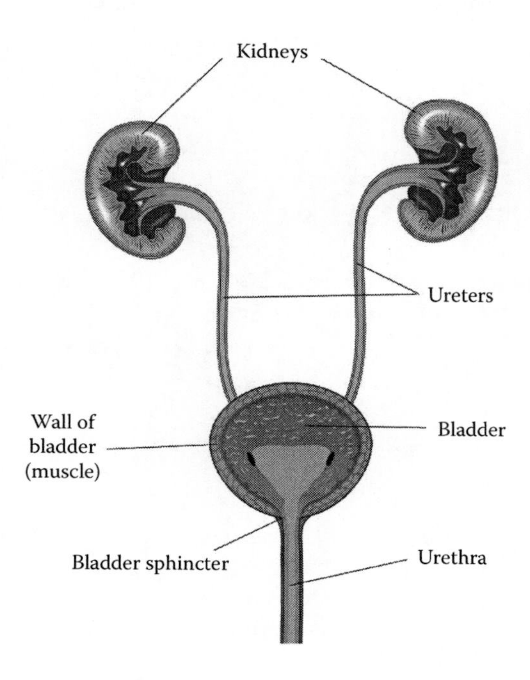

FIGURE 150
Urinary system.

The abdominal cavity has a large membranous surface area that is well vascularized. This means that fluids in the cavity can exchange chemicals with the blood, depending on relative concentrations. Substances that are more concentrated in the blood than in the fluid will move into the fluid, and vice versa.

Staffing

Nurse specialist, dialysis technologist. Peritoneal dialysis may also be performed by the patient or a family member.

Equipment and Supplies

PD machine (Figure 151) and dialysate, or just dialysate, depending on the type of process.

Preparation

PD is an ongoing process, and the patient is essentially always prepared. Initially a catheter is installed in the abdomen to allow filling and draining of the abdominal cavity with dialysate. Fluid balance is critical in dialysis, so the patient is weighed before and after each session to ensure balance.

Procedure

Three basic types of PD are used.

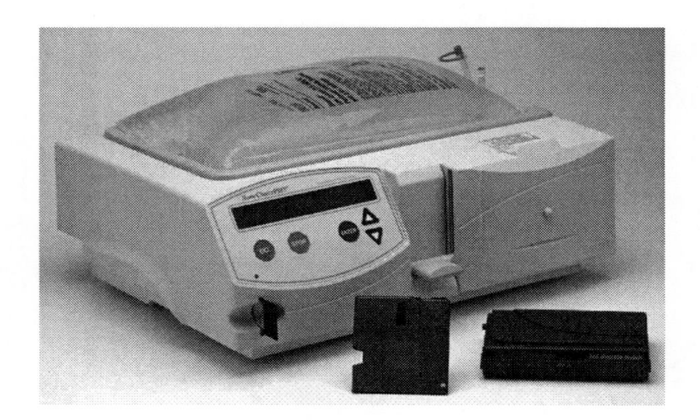

FIGURE 151
Peritoneal dialysis controller.

CAPD, or continuous ambulatory PD, involves the infusion of a specific volume of dialysate solution (usually by the patient or family member) into the abdomen via a permanent catheter. The dialysate is left in the abdomen for a few hours while going about normal activities, after which the fluid is drained out and replaced by fresh fluid.

With CCPD, continuous cyclic PD, a machine infuses dialysate at bedtime. The machine cycles fluid in and out of the abdomen continuously throughout the night. In the morning the fluid is drained, the machine is disconnected and the abdomen left free of fluid during the day. The process is repeated every night.

Intermittent PD (IPD) is similar to CCPD, but is usually performed in the hospital. A machine cycles dialysate through the abdomen repeatedly. It is done for 12 to 24 hours several times a week.

Expected Outcome and Follow-Up

Removal of toxins from the blood. The patient is monitored with regular physical exams and blood and urine lab tests to ensure the effectiveness of dialysis. Care must be taken to avoid infections at the catheter site. Blood pressure must be monitored carefully, because the kidneys no longer provide a pressure regulating function.

Complications

n/a

Prostate Surgery

Alternate names—A variety of terms are used, depending on the specific procedure used, including: transurethral prostate resection/resection of the prostate (TUPR/TURP); transurethral needle ablation (TUNA); transurethral microwave thermotherapy (TUMT); and water-induced thermotherapy (WIT).

Purpose

To remove excess and/or cancerous prostate tissue.

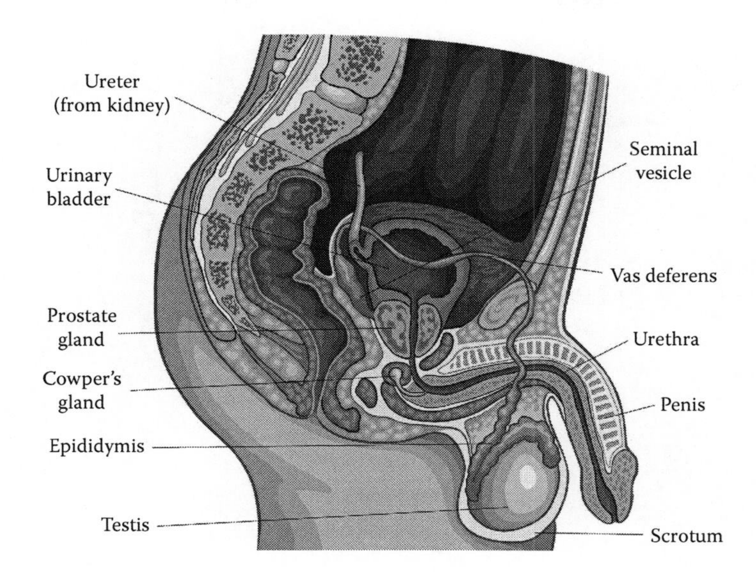

FIGURE 152
Male reproductive system including prostate gland.

Indications

Enlarged or cancerous prostate as evidenced by physical symptoms, blood tests, and physical examination, and confirmed by diagnostic imaging, cystoscopy, or prostate biopsy.

Anatomy

The prostate is a gland that surrounds the urethra just below the bladder in males (Figure 152).

Pathology

Normally the prostate is small, but disease processes can cause it to become enlarged, which can cause problems with urination and sexual function. Prostate cancer is potentially fatal and must be detected and treated early for the best chance of survival. It is the most common cancer in males (Figure 153).

Physiology

The prostate contributes components of seminal fluid.

Staffing

Normal surgical team; the surgeon is usually a urologist.

Equipment and Supplies

Standard surgical, special urological equipment.

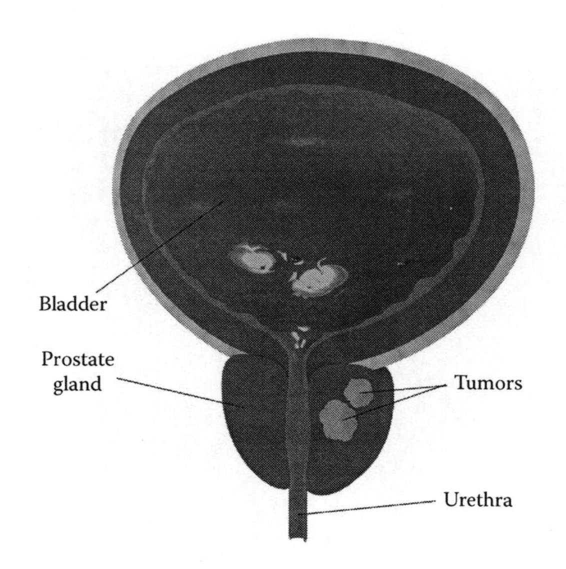

FIGURE 153
Prostate tumors.

Preparation

Standard surgical.

Procedure

A number of techniques may be used, under either local or general anesthesia.

In the past, open abdominal surgery was used to access and excise the prostate tissue.

A less invasive method was developed called transurethral prostate resection (TURP). In this procedure, a tube device is inserted into the urethral opening and passed up the urethra to the prostate. Cutting tools are used to excise as much of the excess or diseased tissue as possible. This tissue is flushed and aspirated from the site, and samples retained for lab analysis. A urinary drainage catheter is installed.

Other methods may be used to excise prostate tissue, with a similar access method to TUPR. Lasers, heated needles, heated water, and microwave probes all can destroy or burn away the tissue.

Expected Outcome and Follow-Up

Relief of symptoms. If cancer is confirmed by lab examination of excised tissue, chemotherapy and/or radiation therapy may be used.

Complications

Nerve damage, impotence, reduced control of urination.

Pulse Oximetry

Alternate names—Blood oxygen monitoring, pulseox.

Purpose

To measure the relative saturation of oxygen in the blood on a spot or continuous noninvasive basis.

Indications

Vital sign measurement.

Anatomy

Blood flows throughout the body (Figure 154). The digits and the earlobe are usually well vascularized with capillary beds and are relatively thin.

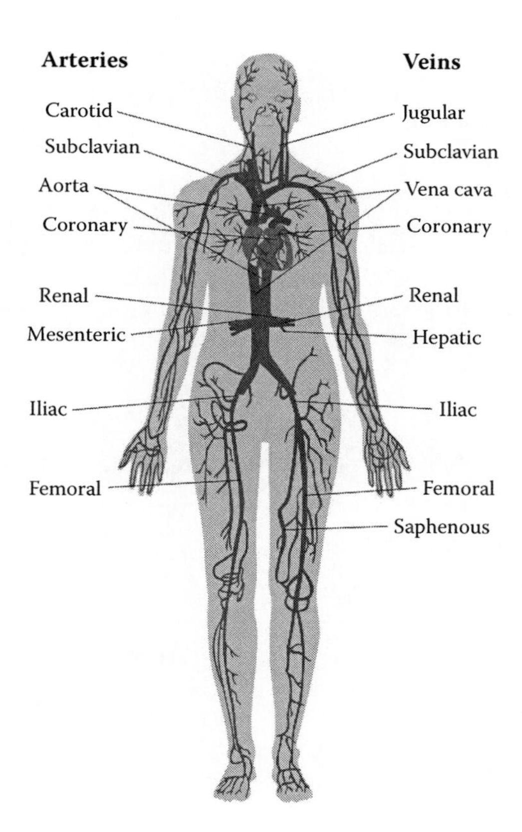

FIGURE 154
Circulatory system.

Pathology

Many disease factors can reduce the oxygen content of the blood, including congestive heart failure, emphysema, blood disorders, and lung cancer.

Physiology

Red blood cells carry oxygen using hemoglobin molecules, which also carry carbon dioxide for disposal. Depending on how much oxygen the blood is carrying, its color is different. These color differences mean that light of certain frequencies will be absorbed to a greater or lesser degree when passed through blood or tissue containing blood. By measuring how much light is absorbed, a value for oxygen saturation can be calculated. This value is given as a percentage of the maximum possible saturation. As a side product, pulse rate can be determined because blood flows intermittently due to the pulsation of the system. Tissues can be well vascularized, meaning there are many blood vessels in the area. If lots of blood is flowing through the area, it is well perfused.

Staffing

Nurse.

Equipment and Supplies

Pulse oximeter probe and monitor. The monitor may be stand-alone, or it may be integrated into a multifunction monitor.

Preparation

A well-vascularized, well-perfused location is selected and the probe applied. A limb on which NIBP measurements are to be performed should be avoided, because blood flow will be interrupted during those measurements. Access of ambient light to the probe should be restricted because it can interfere with measurements.

Procedure

The probe is connected to the monitor, and measurements are begun (Figure 155).

Expected Outcome and Follow-Up

Measurement of blood oxygen saturation.

Complications

Some patients may be sensitive to the materials used in the probe.

Radiation Therapy

Alternate names—Radiotherapy, radiation oncology.

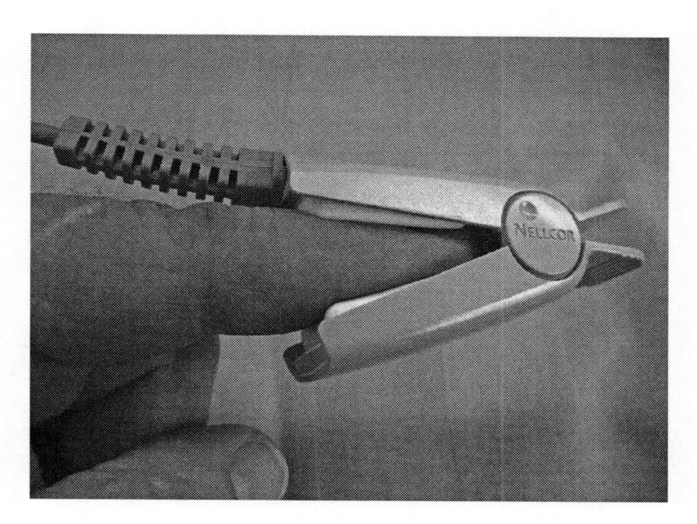

FIGURE 155
Pulse oximeter sensor on finger.

Purpose

To destroy cancerous cells in the body in order to prevent recurrence of cancer or reduce the symptoms of cancer that cannot be cured. Radiation therapy may also be used to shrink tumors in order to make their surgical removal or treatment by chemotherapy easier or more effective.

Indications

Cancer.

Anatomy

Cancer can occur in almost any part of the body.

Pathology

Cancer is the uncontrolled growth of cells in certain tissues in the body. These cells may separate from the initial site and travel to other parts of the body where they can start to grow a new tumor, a process called metastasization. The tumor growth may damage vital organs, or it may simply draw more and more of the body's resources, leaving less and less for normal functions. In either case, death may result. Cancer may be the result of spontaneous changes in the genetic structure of cells, or these changes may be initiated by radiation or chemicals. Cancers are different depending on the type of tissue in which they originate.

Physiology

n/a

Staffing

Radiation therapy technologist, oncologist.

Equipment and Supplies

Radiation therapy machine.

Preparation

The location of the target tumor is precisely determined by diagnostic imaging, or visualization if it is close to the surface.

Procedure

Radiation affects rapidly growing cancer cells more than slower-growing normal ones. This means that it will selectively kill cancer cells while causing minimal damage to healthy tissue.

Radiation may be given over the whole body to help reduce the chance of cancer spreading to tissues other than where it originated, but usually the radiation is applied selectively. Lead shields may be placed over the body so that radiation is only delivered to a small area, and the radiation may be aimed from two different directions so it is concentrated in the tumor area.

Radiation may also be delivered to tumors by using radioisotopes that are concentrated in the specific tissue of the cancerous organ. Once there, the material delivers low doses of radiation directly into the tumor.

Expected Outcome and Follow-Up

Reduction or elimination of cancer cells in the body. Patients are monitored to try to determine how effective the treatments have been in reducing the number of cancer cells. This measure is used to help decide on the frequency and level of subsequent treatments.

Complications

Radiation burns, edema, infertility, hair loss, fatigue, secondary malignancies induced by the radiation.

Septoplasty

Alternate names—Nose job, nose surgery, deviated septum surgery.

Purpose

To correct a deviated nasal septum or other septal defects.

Indications

Deviated septum as determined by visualization and possibly diagnostic imaging.

Anatomy

There are two nostril openings, and the passageways join further back in the nasal cavity. A wall of tissue called the nasal septum separates the passageways and also forms part of

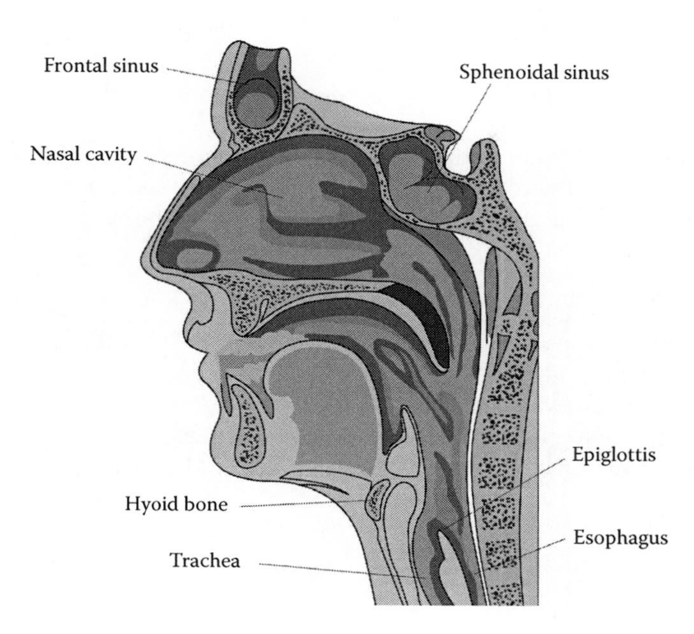

FIGURE 156
Vertical section of mouth and nose.

the structure of the nose. Nasal mucosa on the septal surfaces humidify inspired air and trap foreign particles in the air (Figure 156).

Pathology

The septum can become displaced by disease or injury. This narrows the nasal passage on one side, sometimes causing it to become inflamed. This inflammation can extend into the other nasal passage, making breathing through the nose very difficult. Septal defects may also cause or be related to excessive nosebleeds, chronic sinusitis, and sleep apnea. Tumors or polyps may develop on the septa.

Physiology

The nose allows air to pass in and out of the body during breathing.

Staffing

Normal surgical team; the surgeon is a specialist in nose or facial surgery.

Equipment and Supplies

Standard surgical.

Preparation

Standard surgical.

Procedure

Depending on the extent of the procedure, septoplasty may be performed under local or general anesthesia.

For a basic septal deviation correction, incisions are made along the septum, and some of the septal cartilage is removed while adjusting the position of the septum to a more symmetrical arrangement. The incision is closed and packing placed in the nostrils to control bleeding.

Expected Outcome and Follow-Up

Relief of symptoms. The packing is removed after a day. The patient is instructed not to blow his nose for several days despite congestion, because this can initiate nosebleeds and damage the surgical site. Mucosal crusts will be passed from the nose for a few weeks. Saline irrigation can help clear the crusts and other material and help reduce infections. Ice packs applied to the area can help reduce swelling and ease pain.

Complications

n/a

Shunt for Hydrocephalus

Alternate names—n/a.

Purpose

To relieve the pressure of cerebrospinal fluid (CSF) inside the brain in patients with hydrocephalus.

Indications

Excessive cerebrospinal fluid pressure along with hydrocephalus. Hydrocephalus is diagnosed via diagnostic imaging, especially CT or MRI scans.

Anatomy

The brain is the center of the nervous system, providing cognition, sensation, motor control, and reflex activity. Cerebrospinal fluid in and around the brain serves to cushion the brain from shocks and also helps maintain brain structure. The brain has some spaces, called ventricles, in its interior that do not contain brain tissue but are filled with CSF (Figure 157).

Pathology

Hydrocephalus is a condition in which there is an excess of CSF inside the skull. This causes the ventricles to expand, and if skull bone formation is not complete, the skull size may be increased abnormally. Hydrocephalus can be caused by brain tumors, infection in the brain, bleeding into the CSF, or congenital defects in the brain. Hydrocephalus symptoms may include headaches, balance disturbances, irritability, dementia, and reduced consciousness.

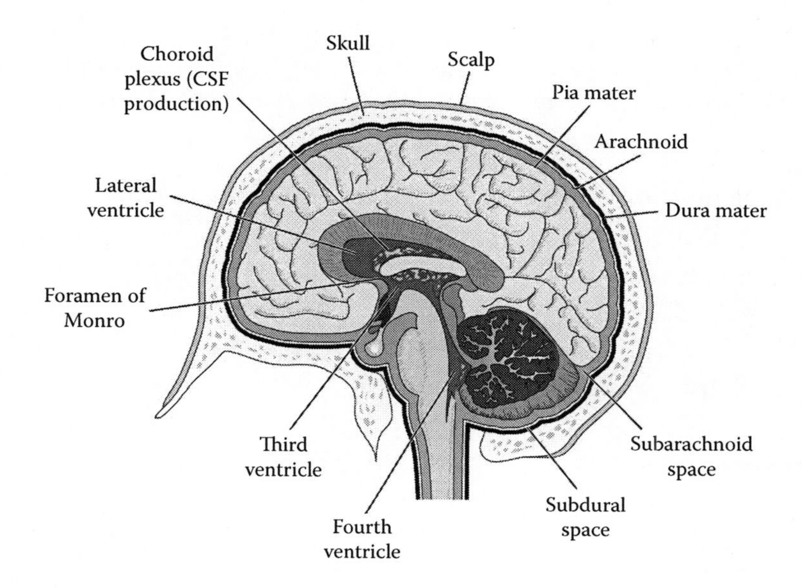

FIGURE 157
Brain anatomy showing fluid-filled ventricles.

Physiology

The brain functions as an electrochemical processor, handling input information from nerve receptors and providing output to muscle and other body systems control. CSF also helps provide immunological protection to the brain.

Staffing

Normal surgical team; the surgeon is a neurosurgeon.

Equipment and Supplies

Standard surgical, shunt, bone drill or saw. The shunt consists of a catheter with a built-in, pressure-adjustable, one-way valve to prevent reverse flow.

Preparation

Standard surgical.

Procedure

Incisions are made at the top and back of the skull and skin flaps moved aside. Another is made just below the tip of the sternum. A catheter is passed from the back incision under the scalp until it emerges from the top incision. The other end of the catheter is then manipulated under the scalp, the skin of the neck and chest and out the sternal incision. A special tool is used under diagnostic imaging guidance to insert the head end of the catheter into the cerebral ventricles. The abdominal end is inserted into the abdomen, where it vents into the abdominal cavity. Alternatively the end of the catheter can be inserted into the thorax and attached to the aorta. When the system function is confirmed, all incisions are closed. The pressure setting of some valves may be adjusted externally using a magnetic device.

Expected Outcome and Follow-Up

Reduction and control of CSF pressure, reduction in size of the ventricles. Shunted CSF is resorbed by the body. The function of the shunt is checked regularly.

Complications

Stroke, shunt failure.

Splenectomy

Alternate names—n/a.

Purpose

To remove a damaged or diseased spleen.

Indications

Cancer of the spleen, various diseases that result in blood cell damage when blood passes through the spleen, enlarged spleen, infected spleen.

Anatomy

The spleen is a small organ that is situated near the stomach on the right rear side of the abdomen (Figure 158). It is well vascularized.

Pathology

Cancer can form in spleen tissue. Liver cirrhosis, certain cancers, and other diseases can cause the spleen to become enlarged (splenomegaly). The enlarged spleen then may break down red blood cells more rapidly than normal leading to anemia. An enlarged spleen can also cause abdominal pain.

Physiology

The spleen stores red blood cells and also removes them from circulation when they are damaged. It helps regulate blood flow to the liver and is part of the immune system.

Staffing

Normal surgical team.

Equipment and Supplies

Standard surgical.

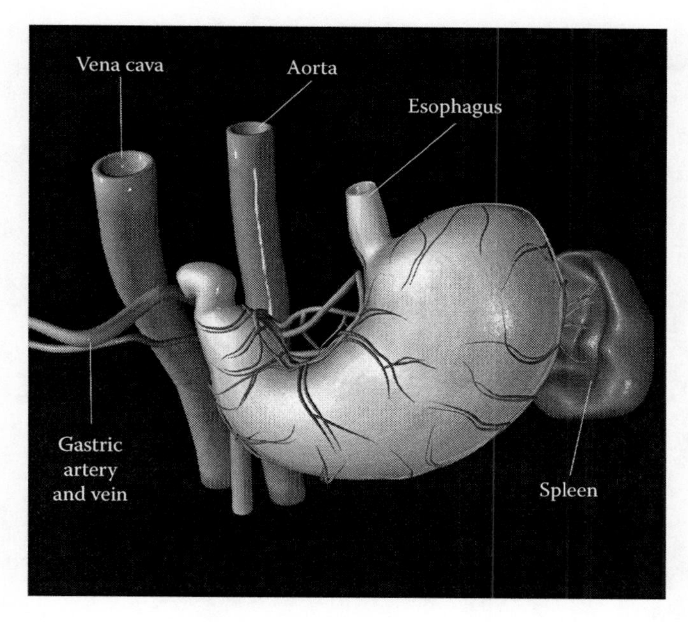

FIGURE 158
The spleen in relation to the stomach and blood vessels.

Preparation

Standard surgical.

Procedure

An abdominal incision is made to provide access to the spleen. The artery supplying the spleen is tied off, and some of the blood in the spleen is allowed to drain away through the vein. The blood vessels are then cut, and the ligaments supporting the spleen are removed. The spleen can then be removed and the incision closed.

In some situations, only part of the spleen may be removed to reduce its size but leave some tissue for normal spleen function.

Splenectomy can also be performed laparoscopically.

Expected Outcome and Follow-Up

Relief from symptoms. Some immune function will be reduced, making infection more likely.

Complications

Overwhelming infection, pancreatic inflammation, collapsed lungs.

Stress Test

Alternate names—Exercise test, treadmill test, cardiac stress test.

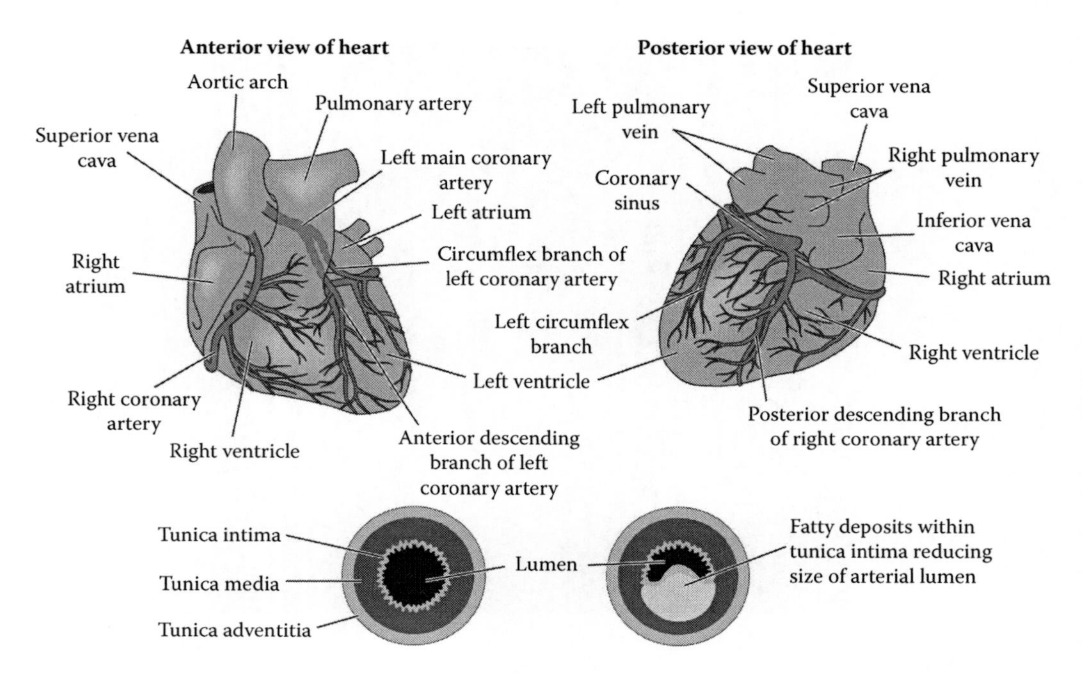

FIGURE 159
Heart anatomy and arteriosclerosis.

Purpose

To obtain an ECG recording during controlled exercise.

Indications

Known or suspected cardiac problems.

Anatomy

The heart is a muscular organ consisting of left and right atria, left and right ventricles, four valves, and associated blood vessels (Figure 159). These blood vessels carry the blood pumped by the heart out to various body parts and also carry blood to the heart itself. The atria are thinner walled than the ventricles. Nerves and other pathways in the heart carry electrical signals to control pumping contractions.

Pathology

Heart pumping effectiveness can be compromised by various problems within the heart. The muscles of the atria and especially the ventricles can be damaged by disease or lack of blood supply (cardiac infarction). The cardiac valves can be damaged by disease or age, thus allowing backflow of blood. Finally, the conductive pathways that help produce and coordinate contractions may be damaged, so that contractions are either weaker or less coordinated, or both. Exercise places an increased load on the heart and may exacerbate certain symptoms of heart disease.

Physiology

The heart, being mostly muscle that is active all the time, requires an excellent blood supply to provide oxygen and nutrients and to remove wastes.

Signals from a center in the heart produce electrical signals that are carried throughout the heart to produce contractions. Various components provide delays to the signal so that contractions are coordinated and allow effective pumping.

Staffing

Cardiology technologist to instruct the patient, set up the system, apply electrodes, initiate the test, and obtain recordings; a cardiologist to interpret the results and sometimes supervise the test.

Equipment and Supplies

Exercise machine, usually a treadmill, a system to control treadmill speed and inclination, and a cardiac monitor and recorder (Figure 160). The controller and monitor/recorder are usually integrated into one unit. A defibrillator should be readily accessible in case a heart attack is induced by the exercise.

Preparation

Electrode sites are cleaned and electrodes applied. The cable connecting the electrodes to the monitor is supported so that there is no strain on the electrodes during the procedure. The monitor is check to make sure an adequate ECG signal is present. The specific exercise

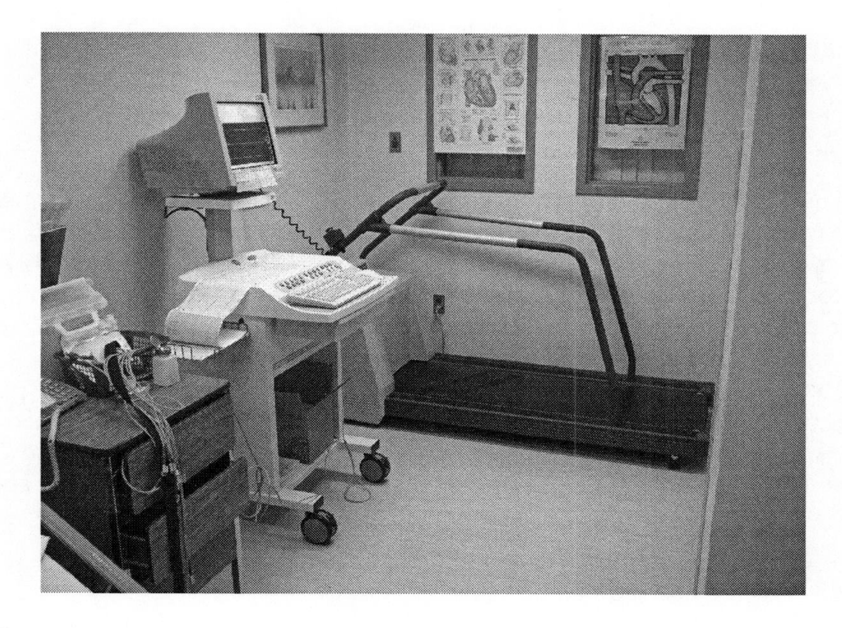

FIGURE 160
Cardiac stress test system with analyzer, including recorder and display screen on cart; treadmill in background.

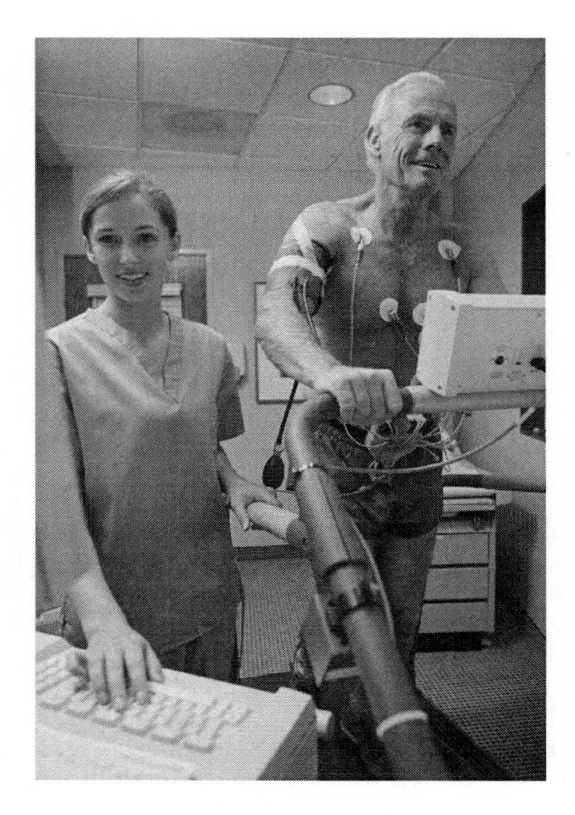

FIGURE 161
Patient undergoing stress testing.

protocol to be used is determined by the cardiologist according to the patient's condition and the results of previous tests.

Procedure

The patient stands on the treadmill, and the procedure is initiated (Figure 161). Exercise protocols usually include a warm-up session, a series of varying intensity levels, and a cool-down session. The protocol may include steadily increasing intensities until the patient feels discomfort, after which the cool-down session is begun. A number of standard protocols are preprogrammed, and custom protocols can also be developed. A panic button is easily accessible to the patient, either on the handle of the treadmill or in a handheld trigger. The operator can access the buttons as well; they will be observing the patient very closely during the procedure and will stop it if there are signs of distress. When activated, the exercise session is immediately halted (though not so suddenly as to cause a fall).

Upon completion of the session, ECG recordings can be generated for all or specific portions of the test. Some systems provide suggested diagnoses.

Expected Outcome and Follow-Up

A useful recording of ECG activity is obtained. Depending on the results, a course of treatment for cardiac disease may be initiated.

Complications

Heart attack. Some patients may be sensitive to the adhesive used in the electrodes, and skin irritation may occur.

Thyroidectomy

Alternate names—n/a.

Purpose

To remove the diseased thyroid gland from the patient.

Indications

Tumors of the thyroid; other processes that cause excess production of thyroid hormones and cannot be resolved with medication (Figure 162).

Anatomy

The thyroid gland is a butterfly-shaped organ that lies on either side of the trachea just below the larynx. Parathyroid glands are partially embedded in the surface of the thyroid glands.

Pathology

Thyroid tumors and other conditions can result in an enlarged thyroid and excessive production of thyroid hormones. An excessively enlarged thyroid is called a goiter. High production of hormones results in a higher than normal metabolic rate, which in turn can lead to excessive weight loss, irritability, excessive perspiration, nervousness, and muscle weakness.

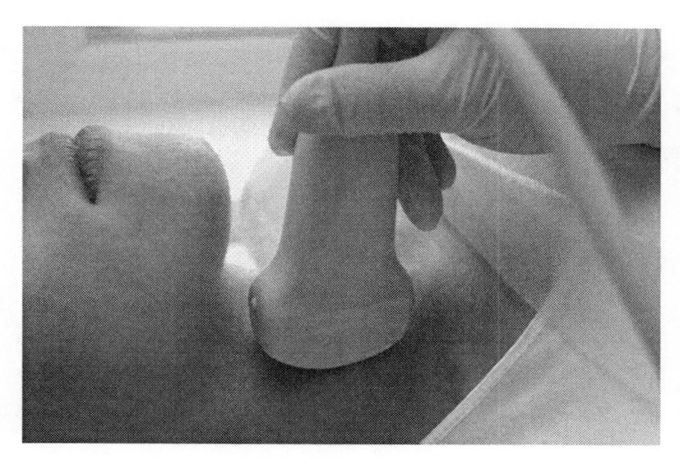

FIGURE 162
Patient having thyroid ultrasound scan.

Physiology

Thyroid hormones stimulate protein synthesis in the body and also increase oxygen consumption by tissues.

Staffing

Normal surgical team.

Equipment and Supplies

Standard surgical.

Preparation

Standard surgical. General anesthesia is used.

Procedure

An incision is made just above the clavicles, exposing the thyroid. The part of the thyroid to be removed is carefully dissected out, being careful not to damage the associate nerves or the parathyroid glands. The blood vessels supplying the excised tissue are sealed off as the process continues. Tissue samples are retained for lab analysis. When all of the targeted tissue has been removed, the incision is closed.

Expected Outcome and Follow-Up

Alleviation of symptoms. In the case of cancer, removal of all cancerous tissue. Sutures are removed after a week. If much or all of the thyroid was removed, hormone supplementation may be required.

Complications

Nerve damage, hypoparathyroidism if the parathyroids were damaged or removed, hypothyroidism if most or all of the thyroid was removed.

Tonsillectomy

Alternate names—n/a.

Purpose

To remove enlarged, inflamed tonsils or ones involved in recurrent infections.

Indications

Enlarged, inflamed tonsils that do not respond to medical treatment.

Anatomy

The tonsils are masses of lymphatic-like tissue located on either side of the back of the throat.

Pathology

Tonsils can become inflamed and enlarged due to infection. Recurrent infections can cause them to become permanently enlarged. The enlarged tissues can cause severe throat pain, difficulty swallowing, sleep apnea, speech impairments, and fever.

Physiology

The tonsils apparently are involved in the immune response, particularly in situations relating to the upper respiratory tract.

Staffing

Normal surgical team.

Equipment and Supplies

Standard surgical, tonsil removal tool.

Preparation

Standard surgical.

Procedure

General anesthesia is administered. The tongue is depressed to improve access to the tonsils, and the tissue is removed with a scalpel, an electrosurgery machine, or a combination of forceps and a wire snare. Bleeding is controlled with electrocautery and/or suturing.

Expected Outcome and Follow-Up

Alleviation of symptoms, prevention of recurrent infections. The patient is kept on their side to prevent aspiration of any bleeding that might occur in the back of the mouth. Antibiotics are given. The throat is usually very sore following surgery, and only liquids and later soft foods can be tolerated.

Complications

Voice changes.

Total Parenteral Nutrition

Alternate names—IV feeding, TPN.

Purpose

To provide hydration and sustenance to patients who are unable take food into their digestive system, or to give parts of the digestive system a rest to allow them to heal.

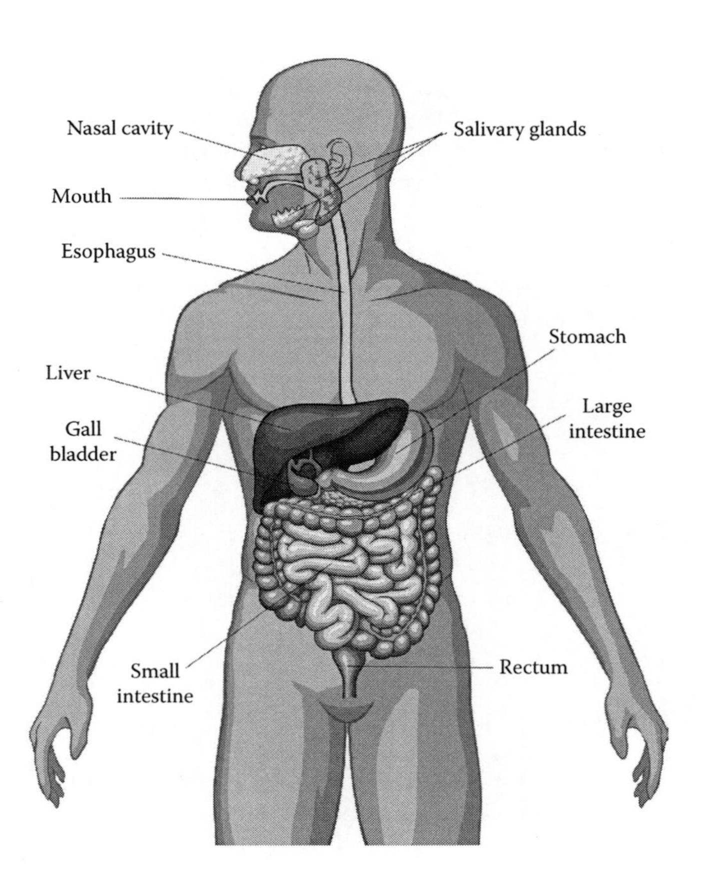

FIGURE 163
Digestive system.

Indications

Any of a number of conditions that preclude use of the digestive system, either temporarily or permanently, including severe malnutrition, certain stages of Crohn's disease or of ulcerative colitis, congenital abnormalities, or trauma.

Anatomy

Food enters the digestive system (Figure 163) via the mouth and esophagus and through the esophageal sphincter, a ring of muscle that helps prevent food from being pushed back up the esophagus, into the stomach. Another sphincter, the pyloric, keeps food in the stomach until it is ready to be passed into the small intestine. The stomach walls are muscular to help move food around inside and promote digestion, but there is little absorption of nutrients in the stomach. From the stomach, food passes into the small intestine, which consists of three main sections: the duodenum, the jejunum (the longest section), and the ileum. The colon, or large intestine, is shaped like an inverted U. It connects the lower end of the small intestine, or ileum, and the rectum. It has a relatively large diameter, and its walls are composed of smooth muscle and a lining with mucous glands and structures for absorbing liquid from the digestive tract. The rectum is the final portion of the digestive tract. It is a pouch-like organ that is closed at the distal end by the anus, a sphincter muscle.

Pathology

Severe malnutrition can impair the function of the digestive system so that it is unable to receive or process food.

Crohn's disease and ulcerative colitis are inflammatory diseases of the colon. When the inflammation is severe, the colon may not be able to accept any material from the upper parts of the digestive system.

Congenital abnormalities or trauma may make it physically impossible for food to enter or be used by the digestive system

Physiology

Glands in the lining of the stomach produce digestive enzymes and hydrochloric acid to start digestion of food. They also produce mucous material to prevent the enzymes and acid from digesting the stomach itself. Further digestive enzymes are added in the small intestine and digestion continues. Nutrients from digested food are absorbed into the bloodstream from the small intestine before it moved into the colon. The main functions of the colon are reabsorption of water and some electrolytes from the material passed into the colon from the small intestine and storage of waste products until they can be eliminated. The rectum receives fecal material from the colon and holds it until it can be evacuated. The anal sphincter keeps the rectum closed until evacuation.

Staffing

Physician, nurse, dietician.

Equipment and Supplies

TPN solution, IV set, IV pump.

Preparation

The nutritional needs of the patient are evaluated by physical examination and the results of blood and urine tests. The physical condition of the digestive system is ascertained using diagnostic imaging, unless the cause of the problems is already known.

Procedure

A central venous line is inserted, and a special TPN solution is administered; flow and volume are controlled with an IV pump. The makeup of the solution is modified to suit the specific needs of the patient. Regular blood and urine tests monitor the effectiveness of the therapy and may indicate the need for modifications to the solutions used.

Expected Outcome and Follow-Up

Adequate nutrition for health. The damaged or diseased portion of the digestive system may be able to heal or be repaired. The catheter insertion site must be kept clean and checked for signs of infection.

Complications

Glucose level fluctuations, electrolyte imbalances, bone demineralization, lipid metabolism irregularities, gallstones.

Tracheostomy

Alternate names—n/a.

Purpose

To provide an opening for air to reach the lungs when the upper airway is damaged, obstructed, or diseased, or when the patient will be unable to breathe on his own for an extended time.

Indications

Damage to or disease or obstruction of the upper respiratory system (mouth and nose, trachea, and larynx) that prevents the passage of air; quadriplegia or other conditions that prevent the patient from breathing on his own.

Anatomy

The trachea carries air from the mouth and nose to the lungs (Figure 164). The larynx is a cartilaginous structure surrounding the trachea. Air is moved into the lungs by the action of the diaphragm muscle in the base of the thoracic cavity and by the contraction of the intercostal muscles of the ribcage. The trachea is made up of rings of cartilage that allow movement but maintain an open passageway.

Pathology

Physical trauma or burns can cause significant damage to the upper respiratory system. Diseases including cancer can block or damage the system, and surgery for these diseases can result in the removal of critical components, preventing airflow. An object might be swallowed and lodge in the throat intractably, preventing breathing.

Physiology

Air is required for life.

Staffing

Normal surgical, though in emergency situation possibly a single individual.

Equipment and Supplies

Standard surgical; tracheostomy (trach) tube.

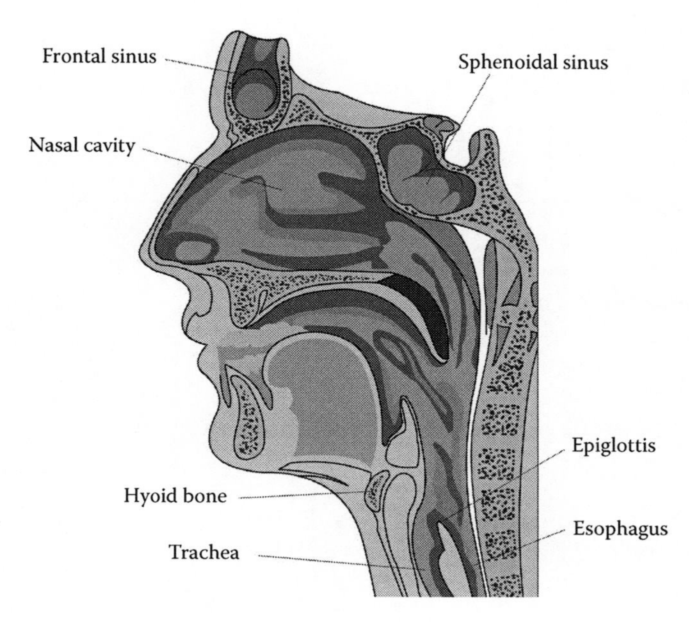

FIGURE 164
Mouth and nose anatomy.

Preparation

Standard surgical, except almost none for an emergency procedure.

Procedure

For emergency tracheotomies, the space just below the larynx is located, and an incision is made through the skin and into the trachea. A tube is inserted to keep the airway open; this may be any accessible, clean tube if necessary, though a proper tracheostomy tube is much preferred.

For scheduled tracheotomies, general anesthesia is used. An incision is made in the skin over the trachea, exposing the neck muscles and thyroid gland. The muscles are moved aside and the thyroid is cut down the middle to clear the trachea. The cartilage rings of the trachea are identified and an incision is made between adjacent rings. A tracheostomy tube is inserted in the opening. If the patient cannot breathe on his own, a ventilator is connected to the tube and artificial ventilation commences. Sutures are used to close part of the incision and hold the tube in position.

Expected Outcome and Follow-Up

The ability of air to reach the lungs. Since the upper respiratory tract normally provides humidification to inspired air, this function must be replaced by a humidifier. The tube is suctioned at intervals to keep it open. If the larynx is intact, the patient may be able to speak by blocking off the end of the trach tube.

If the tracheostomy is temporary, then when the condition necessitating it is cleared, the tube is removed, and the openings in the trachea and the skin are sutured closed.

Complications

Collateral damage to adjacent structures, collapsed lung, scarring of the trachea.

Tubal Ligation

Alternate names—Tube tying, tubal sterilization.

Purpose

To stop ova from reaching the uterus, thus preventing pregnancy.

Indications

The desire of the patient to avoid future pregnancies.

Anatomy

Two ovaries are located on either side of the uterus and are loosely connected to the uterus via the fallopian tubes. The uterus is a pear-shaped smooth-muscle organ in which a fetus develops in pregnancy. The uterus opens into the vagina through the cervix (Figure 165).

Pathology

n/a

Physiology

During the menstrual cycle, the ovaries produce an ovum, which travels down the fallopian tubes to the uterus. The lining of the uterus thickens and increases in vascularization

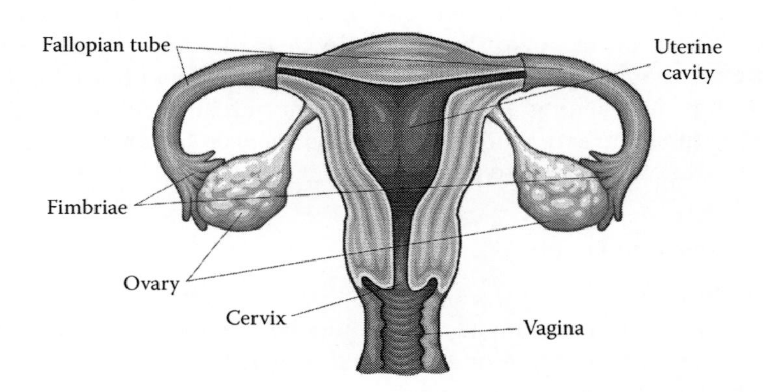

FIGURE 165
Female reproductive system.

in preparation for implantation of a fertilized ovum and thus a possible pregnancy. If no pregnancy occurs, the lining is shed as menstrual bleeding, and the cycle is repeated.

Staffing

Basic surgical team, or normal surgical team if done concurrently with another surgery.

Equipment and Supplies

Standard surgical, laparoscopic equipment.

Preparation

Standard surgical. If done on its own, tubal ligation is almost always done laparoscopically.

Procedure

If the procedure is done concurrently with another surgery, such as a caesarean section, the fallopian tubes are simply located and tied off or cauterized, or both. The other surgery continues as normal.

For laparoscopic tubal ligation, several small incisions are made in the abdomen to allow access for the laparoscopic instruments. The abdomen is insufflated and the fallopian tubes located. They are then tied off or cauterized, or both, via the lap tubes. The insufflating gas is released, the instruments are removed and the incisions closed.

Expected Outcome and Follow-Up

Prevention of pregnancy. Since there is some small chance that a fertile ovum is in the system below the point of tubal blockage, other forms of birth control are used until after the next menstrual cycle.

Complications

Very rarely the ends of the tube may reconnect spontaneously, allowing pregnancy to occur.

Uterine Ablation

Alternate names—Balloon ablation, endometrial ablation.

Purpose

To help reduce excessive menstrual bleeding.

Indications

Excessive menstrual bleeding.

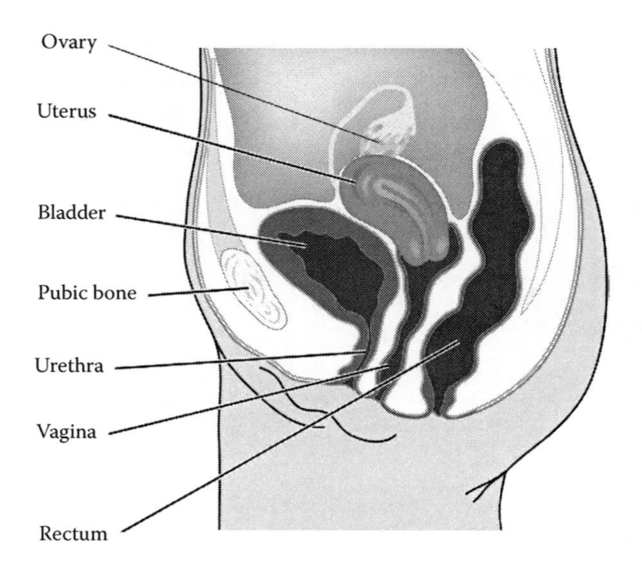

FIGURE 166
Uterus and adjacent organs.

Anatomy

The uterus is a pear-shaped smooth-muscle organ in which a fetus develops in pregnancy. The uterus opens into the vagina through the cervix (Figure 166).

Pathology

Excessive buildup of the uterine lining during the menstrual cycle can result in excessive bleeding and cramping.

Physiology

During the menstrual cycle, the lining of the uterus thickens and increases in vascularization in preparation for a possible pregnancy. If no pregnancy occurs, the lining is shed as menstrual bleeding, and the cycle is repeated.

Staffing

Basic surgical team; the physician is likely a gynecologist.

Equipment and Supplies

Balloon ablation equipment.

Preparation

The patient may be sedated.

Procedure

A tube is inserted into the uterus, and a balloon at its tip is inflated to fill the uterine interior. Hot water is circulated in the balloon for a prescribed period. The water is drained and the balloon withdrawn.

Expected Outcome and Follow-Up

Alleviation of symptoms. Cramping is likely to occur following the procedure, but should abate within a day or two. Some bleeding may persist for a week or two. Sexual intercourse and tampon use should be avoided for 2 weeks, but other activities can be resumed almost immediately.

Complications

Collateral damage to adjacent organs, uterine scarring.

Vasectomy

Alternate names—Male sterilization.

Purpose

To prevent sperm from being able to exit in ejaculation, preventing pregnancy in the man's partner.

Indications

The desire of the patient not to initiate pregnancy in the future.

Anatomy

The male reproductive system consists of the testicles, the vas deferens, the prostate gland, the penis, and the urethra. The testicles and vas deferens are contained within the scrotum, or scrotal sac (Figure 167).

Pathology

n/a

Physiology

Sperm are produced in the testicles and stored in the vas deferens. On ejaculation, the tract contracts rhythmically to expel the sperm and fluids added by the prostate through the urethra/penis and out.

Staffing

Physician.

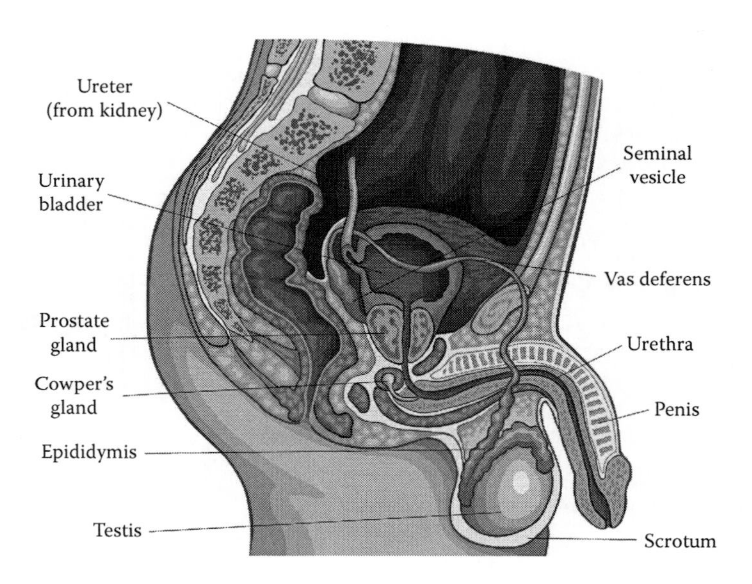

FIGURE 167
Male reproductive system.

Equipment and Supplies

Scalpel, sutures, cautery unit.

Preparation

A local anesthetic is administered.

Procedure

A small incision is made in the scrotal sac. The vas deferens are located and either ligated or cauterized, or both. The incision is sutured.

Expected Outcome and Follow-Up

Prevention of pregnancy of the partner. Swelling may be alleviated with ice packs. Activities are restricted for a day or two. Since a few sperm may be present in the portion of the tract distal to the closure, unprotected sexual intercourse should be avoided for a few days.

Complications

Inflammation in the area.

Appendix A: High-Technology Equipment

Physiological Monitors

Patients in serious condition must have their vital signs monitored continuously. Physiological monitors bring a number of different critical measurements together, so that a comprehensive picture of the patient's condition can be obtained. The number of parameters measured is different for various situations, and monitors are designed to meet these different needs.

There are two general formats of physiological monitors: configured, in which the unit is built with various measurement capabilities and cannot be changed; or modular, in which various modules that measure different parameters can be added or removed as needed (Figure 168). Configured monitors are typically more compact and less expensive; modular monitors are more flexible in terms of the types of measurements they can perform and have the advantage that, if a particular module fails (or a newer model becomes available), it can simply be unplugged and exchanged with another in minutes.

Monitors can measure electrocardiogram (ECG), respiration, temperature, blood oxygen saturation (SpO2), blood pressure (either noninvasively or invasively), other pressures (such as intracranial or spinal), temperature, cardiac output, exhaled carbon dioxide levels (capnography), blood carbon dioxide saturation, and blood chemistry (POC blood analysis). In addition, monitors may be able to interface with other devices such as ventilators and nurse call systems. They may have a recorder module included, or they may connect to a central monitoring system and/or patient charting system.

The display portion of a monitor shows the results of the various measurements, either as a graph or as a numerical value, or both. Monitors can usually display a number of different parameters simultaneously and can store values even when they are not displayed.

Most parameters monitored can be set up so that an alarm sounds if the values go beyond certain predetermined levels; these alarm levels are set to generic values when the monitor is turned on, but can be changed depending on individual circumstances. Alarms may be audible or visual or both, and they may be set to initiate a recording, signal a central monitoring station, or trigger a nurse-call system.

Monitors may record various measurements, which can later be called up as tables or graphs so that changes can be tracked more easily; some systems allow marking of specific events, such as physical activity or medication administration, so that these can be correlated to measurements. Tables and graphs may be available for printout on a built-in recorder or on a larger-format printer at a central station.

Some monitors have a built-in battery so they can continue to function if there is a power failure or during patient transport.

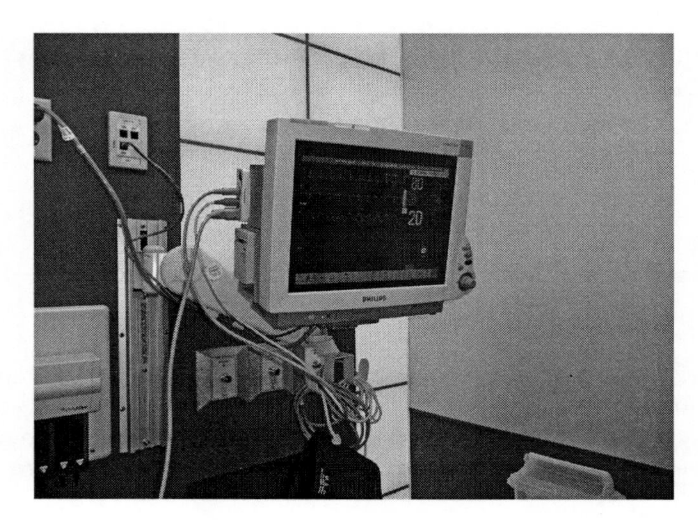

FIGURE 168
Physiological monitor with parameter modules.

Telemetry Systems

Cardiac patients are often able to get out of bed and walk around; this may be part of their recovery process, or it may simply be a convenience. In either case, it is important for medical staff to be able to continue monitoring the patient's ECG signals. Telemetry systems allow patients to be mobile without unwieldy and possibly dangerous long cables.

ECG signals are picked up from the patient's skin by electrodes and wires, just as with regular ECG monitoring. The signals are then amplified and processed, again by circuitry similar to that in a bedside monitor. Within the small module carried by the patient, however, is a radio transmitter that broadcasts signals carrying the ECG information. These signals are picked up by an antenna system and processed to extract the original ECG waveform, which can then be displayed on a central monitor. This monitor can usually display signals from several patients simultaneously, and may be stand-alone or part of central monitoring system that handles information from bedside monitors as well. The central station handles all recording, trending, patient admission information, and alarms.

Older telemetries broadcast the ECG information as an analog signal, much like an AM or FM radio station. Newer systems turn the analog signal into a stream of digital information before broadcasting it, which gives better resistance to interference and allows either lower transmitter power (thus prolonging battery life) or else greater range.

Most telemetry transmitters have a nurse-call button that the patient can use to signal for assistance or to mark any unusual feelings or symptoms he might have associated with his condition.

Some systems also have the capability of obtaining and transmitting blood pressure and blood oxygen saturation (SpO2) information as well as ECG, which can help give a more complete picture of a patient's condition (Figure 169).

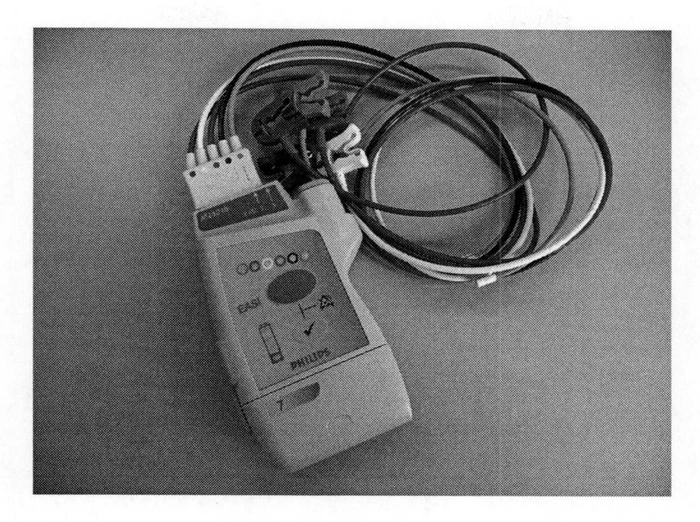

FIGURE 169
Telemetry transmitter packs, with ECG leads and SpO2 sensor.

Central Station Monitoring System

In any critical care area, medical staff needs to be able to monitor each patient, but not necessarily from the bedside at all times.

Physiological monitors and telemetry units are associated with each patient and send data to a central location where important signals can be displayed. Most central stations display detailed information about a particular patient as necessary, while normally displaying a single waveform and numeric values (usually ECG) for all patients simultaneously.

Patient information can be entered (usually with a standard computer-type keyboard) via the central station at admission and removed on discharge. Staff may be able to adjust alarm levels from the central station, but usually cannot cancel alarms without going to the bedside.

Recordings of information from the various bedsides can usually be made at the central station, either on a strip-chart recorder or on a larger-format recorder or printer (usually used for tables and graphs or more comprehensive reports). Central monitors can be integrated into or connected to patient data recording and analysis systems (Figure 170).

Central monitoring also can give the capability of displaying information from one beside monitor while working at another.

FIGURE 170
Dual-display central station monitor with keyboard and strip chart recorder.

Ambulatory ECG Recorders

It can be important to track the ECG signals of patients for an extended period, as they go about their daily activities. Ambulatory ECG recorders perform mobile ECG measurement, by putting the input and processing circuitry for ECG signals into a small, battery-powered package, along with a means of recording the signal (Figure 171). In the past, this

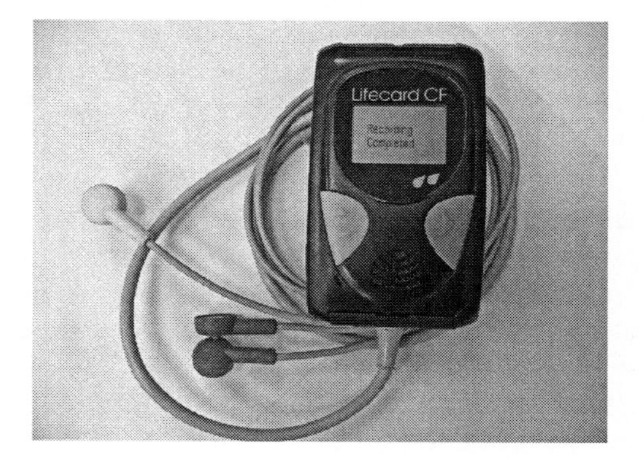

FIGURE 171
Digital ambulatory ECG recorder with leads and LCD waveform display screen.

was done with magnetic tape (often a standard cassette, geared down to run for 24 hours), but newer units use electronic memory modules which have the advantage of needing no moving parts or mechanical alignment, are thus lighter, more durable, and require less maintenance. Picking up the ECG signals from the patient involves the same type of skin electrode and lead wires as other ECG monitors and recorders, but because there is no need for display or printout of the signal in the device itself, the large and high-power-consumption monitor/recorder can be eliminated. Also, because the signal is not analyzed on the machine and there are no alarms or other parameters to be considered, the electronic circuitry can be relatively simple.

These units have a clock that records a coordinating signal along with the ECG, and there is normally a button that the patient can use to mark times of exertion, chest pains, meal times, or other circumstances as requested by his physician.

ECG Recorder/Machine

An ECG machine makes precise measurements of the heart's electrical activity from several different perspectives, by using an array of electrodes (usually either five or twelve) placed in carefully selected positions on the patient's chest. The circuitry of the machine then translates these signals into a graph and prints them out on chart paper, with one trace for each of several combinations of electrodes (or "leads"). Because the signals are very small (measured in thousandths of a volt), compared to much larger electrical interference signals, the circuitry must be very sophisticated to provide useful results.

Most newer ECG machines provide annotations on the chart, giving lead labels, time and date, patient information, etc. Some provide precise values for various measurements of the ECG signal, and some also give a possible interpretation of any abnormalities present in the signals. The more complex machines usually have some kind of display screen and a keyboard for entering information (Figure 172). Some have data ports for connection to computer systems and/or to allow sending the results through a modem to a remote location.

ECG electrodes are usually either an adhesive patch with a conductive area in the center, or a silver cup with a rubber suction bulb that literally sucks the cup onto the patient's skin. The suction types are used with a conductive cream and are quicker and easier to apply and remove; adhesive electrodes are less obtrusive and stay in place better for longer-term applications.

Most ECG machines have batteries that allow them to be used when access to a power outlet is difficult; many also have data storage so that records can be printed out at a later time, because they are often used in the midst of an emergency situation and need to be removed from the scene as quickly as possible.

Stress Test Systems

The ECG of a patient is usually measured with the patient in bed. In diagnosing many cardiac problems, however, the physician needs to see how the heart reacts to exercise.

Stress test systems allow the opportunity of testing ECG during exercise. They consist of a mechanism for exercise in a fixed location, such as a treadmill, and a component

FIGURE 172
12-lead ECG machine with integrated data entry keyboard and LCD display.

that combines control circuitry for the exercise device with physiological monitoring. The monitor looks at the patient's ECG waveforms, usually on multiple leads, and displays the signals on a screen and on a chart recorder. Some systems may also monitor other parameters, such as oxygen saturation (SpO2, pulse oximetry) or respiration.

The system controls the level of exercise according to a preprogrammed pattern. For example, by varying speed and/or slope of the belt of a treadmill, exercise levels can be changed. The exercise levels can also be controlled directly by the operator; however, the preset programs are usually used. These are developed by cardiologists who have determined the optimum exercise patterns for various patients and conditions, in order to maximize the chance of obtaining clinically useful results while minimizing risk to the patient. The operator can stop the exercise program at any time if it appears that the patient is becoming over-stressed, and systems have safety mechanisms to stop if the patient moves off the equipment during exercise (Figure 173).

ECG signals are picked up from the patient's skin just as with normal ECG monitors, but with certain modifications (Figure 174). Because the electrodes will be applied for a short time, but must be able to withstand (sometimes vigorous) patient movement and possible perspiration, they are designed somewhat differently than those used for longer-term monitoring. Suction cup electrodes may be used, with rubber bulbs to create suction and silver cups to make contact with the patient's skin. A conductive cream is usually used with this type of electrode. A harness of some sort may be used to keep the electrode lead wires and cables in place, and, of course, the cable must be quite long to allow adequate patient movement.

Heart–Lung Machines

During some cardiac surgery, the patient's heart must be stopped so that surgeons can perform their work. Blood flow and oxygenation must be continued for this time, which may be several hours, and heart–lung machines perform this function.

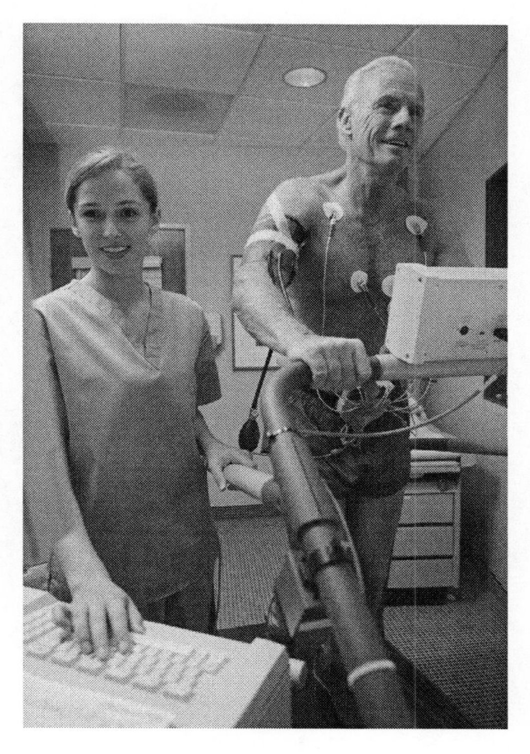

FIGURE 173
Patient undergoing cardiac stress test.

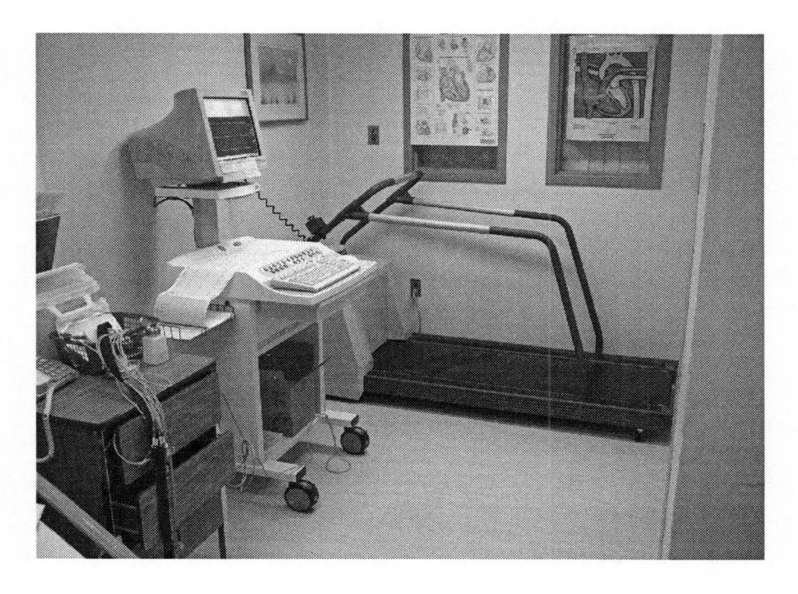

FIGURE 174
Cardiac stress test system with analyzer, data entry keyboard, chart recorder, and display on cart, and interconnected treadmill in background.

FIGURE 175
Heart–lung machine being operated by a perfusionist during open heart surgery.

The majority of total blood flow is drawn from the patient's venous system, usually the superior vena cava or the right atrium. It is then pumped into a membrane system that allows the escape of carbon dioxide from the blood and the intake of oxygen to the blood. The blood is then pumped into the aortic arch where it continues to circulate through the patient's body. The heart and lungs are bypassed, and the heart can be made to stop to allow surgical procedures to take place.

An anticlotting agent, usually heparin, must be added to the blood to prevent clots forming in the system and passing to the patient. This agent must be neutralized immediately following surgery so that normal clotting can take place.

Temperature of the blood must be controllable in order to maintain, raise, or lower patient body temperature as required by specific circumstances. It may be advantageous to induce hypothermia in order to minimize tissue damage during certain procedures.

Heart–lung machines are operated by perfusionists, specially trained technologists who monitor and control system functions working in coordination with anesthetists, surgeons, and the rest of the surgical team (Figure 175).

Electroencephalographs (EEG)

Electrical activity in the brain can give important information about its function and possible disease conditions. By taking signals from a large number of electrodes on specific points on the patient's scalp, amplifying and processing these signals, and then comparing and combining the various signals and analyzing the results, the electroencephalograph can present a recording of the signals that originate from various points within

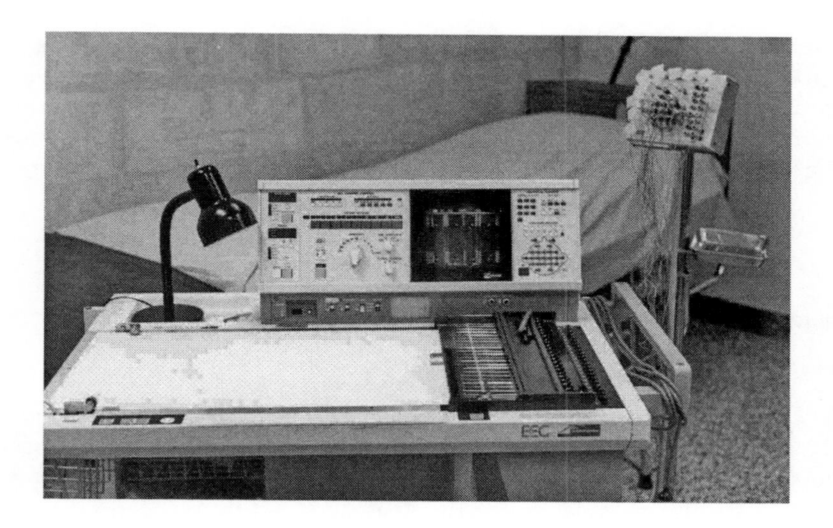

FIGURE 176
Electroencephalograph machine with multichannel recorder and control panel in foreground, and patient electrode interface box in background.

the patient's brain. The recording can be printed out on a chart, or recorded electronically for later examination (Figure 176). Most EEG machines incorporate a means of providing various stimuli to the patient, such as light flashes or sounds, which can then be related to changes in the electrical signals from the brain.

Invasive Pressure Monitors

It is often important to know pressure values within the patient's body. This is most commonly the pressure in limb arteries, but can also include blood pressures in the venous system, at various points of the arterial system close to the heart, or cerebrospinal fluid pressures in the skull or spinal column. Blood pressures are pulsatile, and the maximum (systolic), minimum (diastolic), and mean pressure values are also important.

Blood pressure can be measured at the limbs noninvasively (by noninvasive blood pressure monitors, NIBPs), but this method is susceptible to errors, it is not continuous, and it cannot give values for anything but limb arterial pressure.

By introducing a catheter with an electronic pressure transducer at its tip, exact measurements of pressure can be made continuously, at whatever point the catheter is positioned. Transducers are small and can be guided from a convenient insertion point to various locations within the circulatory system, either by simply watching the pressure values and knowing the typical measurements from various locations, or by viewing the catheter and vessels with X-rays.

Because of the continuous nature of these pressure measurements, a waveform of the values can be displayed on an associated monitor; the shape of the wave signal can be significant as well as the various pressure values. Systolic, diastolic, and mean values can be displayed; many monitors also record measurements over time so they can be displayed in tabular or graphic form. High and low alarm levels can be set, sometimes for systolic, diastolic, and mean values independently.

Pressure transducers are sensitive and must be calibrated prior to each use; they must also be "zeroed" regularly by exposing them to atmospheric pressure in order to compensate for slight variations

Noninvasive Blood Pressure Monitors

A patient's blood pressure is a vital measurement for diagnosis. When absolute accuracy is needed, and/or monitoring will be over an extended period, and/or when continuous, instantaneous measurements are required, a catheter is inserted into the patient's blood vessels, and pressure is measured directly by an attached transducer. This invasive technique is often not desirable, however, and is generally impractical to take blood pressure readings manually at frequent intervals over an extended period of time.

Noninvasive blood pressure monitors (NIBPs) allow blood pressure measurements to be made automatically.

Most NIBP systems use a technique similar to that of manual blood pressure measurement. An inflatable cuff is placed over the patient's limb (usually the arm but sometimes the thigh). The cuff is inflated until arterial blood flow is occluded, and then pressure is slowly released while monitoring the return of partial and then complete blood flow. The cuff pressure at which partial flow commences is equivalent to the systolic blood pressure, while the point at which complete flow is restored is equivalent to the diastolic pressure.

Two basic types of NIBPs are in common use. In one, a microphone placed near the patient's artery, usually under the inflatable cuff, determines the presence or absence of blood flow. Associated electronic circuitry processes the sound signals and analyzes the resulting waveforms to determine the systolic and diastolic points.

The second type of NIBP measures small variations in air pressure within the cuff caused by blood pulsing within the patient's arm. The pattern of these variations is used to determine blood pressure measurements.

In both systems, readings may be initiated by the operator, or may occur at regular intervals, the timing of which is determined by the operator. There is some kind of display for results; systolic and diastolic values may be displayed alternately in a single window, or they may each have separate displays. Mean pressure may be calculated and displayed as well.

Some systems have a "stat" button, which causes the machine to take several measurements in quick succession. Many also have programmable alarm limits to notify staff when pressures are too high or too low. Another common feature is a recorder module, which allows printouts of results, along with the time they were taken, and sometimes a graph of pressure trends over time.

A new technology uses motion of the blood vessel wall to determine pressures, and will allow continuous, instantaneous blood pressure measurements, without requiring the insertion of a catheter.

NIBPs may be stand-alone; they may be built into devices that also measure ECG and/or SpO2, or they may be modules in a general physiological monitoring system (Figure 177).

FIGURE 177
Vital signs monitor with temperature, SpO2, and NIBP capability. A strip chart recorder is on the side of the unit.

Electronic Probe Thermometers

Probe thermometers use a tiny electronic device to measure the temperature of the body part to which they are applied. The temperature is then displayed on a digital readout, which may be on a small, handheld box or on the screen of a larger, multifunction monitor. Handheld units usually have the sensor on the end of a thin rod, which can be placed under the tongue, under the arm, or rectally. To ensure patient safety, there are separate, color-coded probes for oral and rectal use, and the probes are covered with a smooth, tough disposable cover, which is changed for each patient (Figure 178).

Often, there is a timer in the handheld part that beeps when the temperature reading has had time to stabilize, or the unit can be set to continuously measure the patient's temperature. For longer-term monitoring, sensors on a small, flexible wire are available; these can be placed on the patient and not interfere too much with movement.

Units usually have a switch to select between Celsius (centigrade) and Fahrenheit degrees.

Glucometers

Devices for measuring blood glucose are small, reliable, accurate, and simple enough to be used by patients themselves, though of course medical caregivers use them as well. The devices rely on a chemical reaction in which glucose in the blood causes a color change on a small paper or plastic strip. A small drop of blood is applied to the strip, and, after

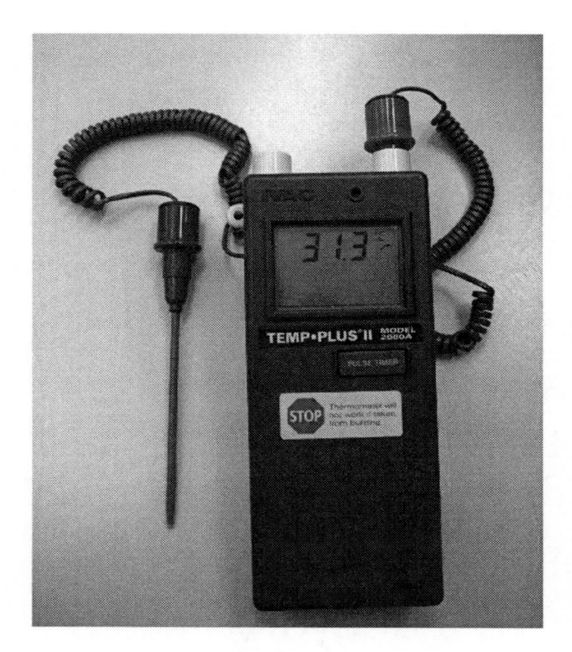

FIGURE 178
Electronic probe thermometer with oral (outside unit) and rectal (in holder) probes.

a preset time (usually signaled by the testing device), the strip is inserted into a port in the unit. Electronic circuitry measures the color change of the strip and from these data, determines the blood glucose level and displays it on a screen.

Because of the critical nature of this test, glucometers must be very accurate and stable. They have means by which the user compensates the unit to each batch of test strips, to account for slight variations from batch to batch. They also have calibration strips, which have precisely controlled colors, to check for proper operation.

Blood glucose levels can change quickly following meals or exercise, and measurements must be made with this in mind. Factors such as timing and type of food or drink taken and patient activity level must be noted, along with the reported blood glucose values.

Fetal Monitors

The time of labor and delivery is critical for the baby, and it can be very useful to be able to monitor its condition, particularly its heart rate. This can be done on a short-term basis with a stethoscope or Doppler unit, but for longer-term monitoring, a fetal monitor is preferred. Fetal monitors can also measure and record other parameters (Figure 179).

Fetal heart rate is picked up in one of two ways. Most commonly, a special transducer is placed on the mother's abdomen. This transducer has several crystals, which produce a beam of ultrasound. The beam is focused at approximately the depth of the fetal heart. When it hits the beating heart, some of the signal is bounced back to the transducer, which picks up this reflection. The frequency of the reflected signal varies as the fetal heart moves

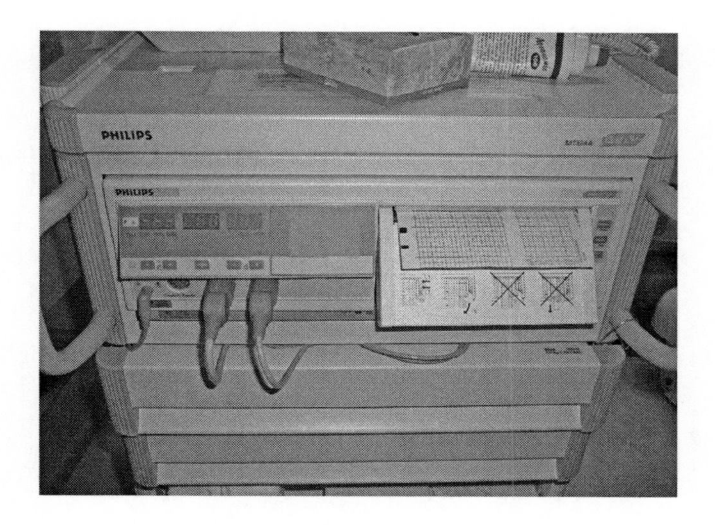

FIGURE 179
Fetal monitor with numeric displays for fetal heart rate (dual), relative strength of contraction (toco), cables for ultrasonic fetal heart sensors, tocograph sensor, and patient-controlled marker button, plus chart recorder for the various parameters.

in its beating, and this change in frequency can be processed and analyzed to produce a signal corresponding to the fetal heart beat. The rate is displayed as a numerical value, and the sound of the beat is also available at a speaker so that the caregiver can hear if the rate is changing, or if the beam is no longer focused on the baby's heart. The signal is also recorded on a graph chart, which allows caregivers to see trends in the rate over time. In order to allow maximum transmission of the ultrasonic signal, a gel is placed between the transducer and the mother's skin. The transducer is usually held in place by an adjustable elastic belt (Figure 180).

The ultrasound transducer is quick and easy to apply and can give very valuable information. However, it can lose the signal through movements of either the fetus or the mother.

The second method of picking up the fetal heart rate is through an electrode. A curved section of fine silver wire is inserted through the vagina and pierces the surface of the fetal scalp. A second plate electrode on the mother's skin provides a reference, and electronic circuitry can pick up the electrical signals of the baby's heart. This ECG signal is somewhat more stable that that from an ultrasonic transducer, but it requires that the fetus be in the head-down position and that dilation is sufficient to allow placement. The electrical ECG signal can also provide certain information not available with the ultrasound signal. The ECG signal is used to display a numerical value and tone, as well as record on a chart, just as with the ultrasound signal.

Another function of the fetal monitor is to measure the relative strength of uterine contractions. This is important since the fetus is most likely to be distressed during contractions. The uterine contractions are measured by a transducer disk (called a TOCO transducer) that presses a central tab against the mother's abdomen. During a contraction, the muscles in the uterus become much harder, and the disk is pushed back into the transducer. These variations are measured and processed to produce a graph of relative contraction strength, which is displayed as a numerical value on the monitor, and also graphed on the same chart as the fetal heart rate. This allows a good correlation to be made between contractions and any possible fetal distress.

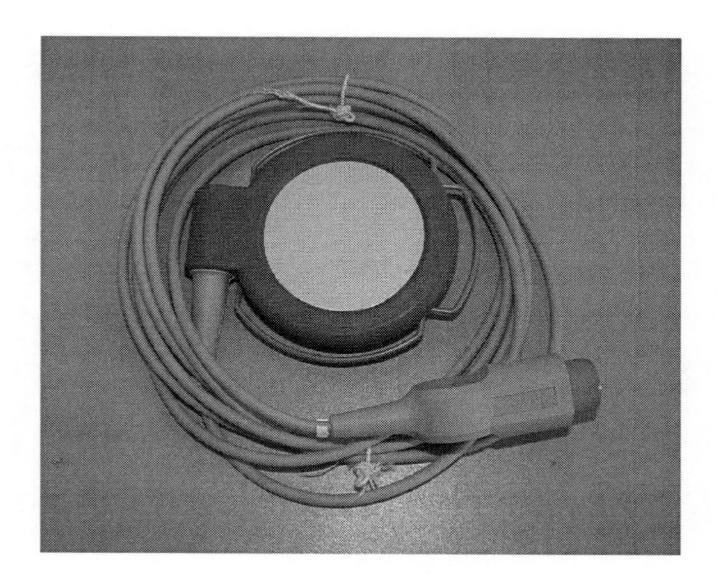

FIGURE 180
Fetal monitor ultrasound probe. Piezoelectric crystals are embedded in the face of the sensor. They produce an ultrasonic signal that is focused on the fetal heart, and also receive reflections of these signals from the heart in order for the monitor to analyze them and calculate the fetal heart rate.

Some fetal monitors also measure maternal blood oxygen saturation (SpO2), because low oxygen levels can be very harmful to the fetus.

Another option for a monitor is the ability to handle twins; this simply requires a second, independent ultrasound portion, one for each fetus. Graphs are both on the same chart.

Some fetal monitoring systems have telemetry, with which the transducers are plugged into a small, battery-powered transmitter. The transmitter sends the signals back to the monitor; this has the benefit of allowing the mother to walk around.

Infant Incubators

Neonates, especially if they are premature, need an environment in which factors such as temperature and humidity are controlled, and in which they are easily observable. Caregivers must be able to access the infant to change diapers, administer medication, feed, and simply provide physical contact, preferably without disturbing the controlled environment.

Infant incubators are made up of a clear plastic chamber, electronic and mechanical systems to monitor and control the environment, and a stand to bring the chamber up to a comfortable working height. The stand is usually on lockable wheels and has storage compartments for diapers and other supplies.

The plastic chamber is designed to give maximum visibility from various angles while providing reasonable insulation from exterior conditions and access to the infant. There is a larger hatch for moving the infant in and out of the incubator when necessary, as well as several "portholes" with flexible sleeves to allow caregivers to reach in to the baby, with minimal air leakage (Figure 181). The portholes are usually covered by a clear door when not in use (Figure 182).

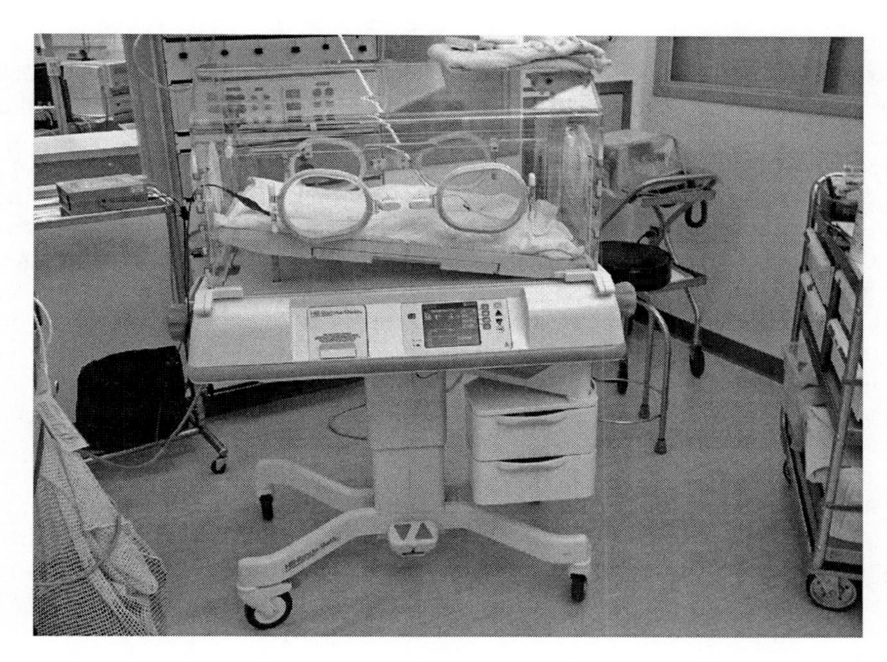

FIGURE 181
Infant incubator with Plexiglas cover, tilted infant bed, access ports, display and control panel, and height-adjustable stand.

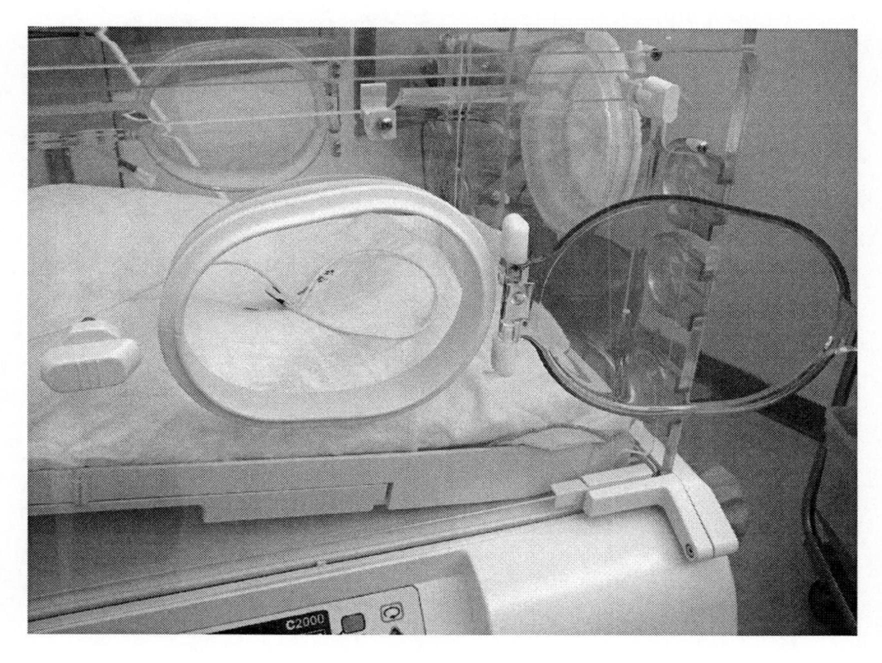

FIGURE 182
Infant incubator access door.

The outer wall of the chamber may be double layered to provide better insulation and also to control air flow; warm air may be directed between the outer layers and distributed into the interior at several points for more even heating. The chamber material must be strong enough for safety, but it must also be very clear for visibility, and allow penetration of ultraviolet rays for bilirubin therapy, if that is needed.

Temperature is the most critical environmental parameter to be controlled in an incubator. Neonates are often unstable in their internal temperature control; being small, and often having little body fat, their temperature can change quickly. It is important, though, for caregivers to be able to see as much of the infant's skin as possible, to watch for changes in color and texture that may indicate problems. Skin surface exposure is also important for bilirubin therapy.

Temperature may be controlled in one of two ways. Most commonly, a sensor measures air temperature; then, the output of the internal heater is adjusted after comparing the measured temperature to the desired temperature set by the caregiver. This sensor may be built into the unit, at some point in the airflow, or it may be a plug-in type with the actual temperature pickup suspended in the air above the infant.

The second method of temperature control involves placing a temperature sensor directly on the baby's skin. Again, this measurement is compared to the desired temperature set by the caregiver, and heat output is adjusted accordingly. To be effective, this method requires that the sensor be placed carefully; limb temperatures may vary considerably from torso temperature; the sensor must not be covered, and it must not interfere with treatments. Additionally, the adhesive used may irritate the baby's skin, and the sensor and wire may hinder removal of the baby for feeding, bathing, cuddling, or treatment outside the incubator. Given these considerations, incubators may not have skin temperature monitoring/control.

Temperature measurements are displayed on a front panel, which also shows the settings for desired temperature (Figure 183). There are alarms for over and under temperature, and usually a redundant high temperature alarm set somewhat above the primary high temperature alarm. Alarms are usually visual and audible.

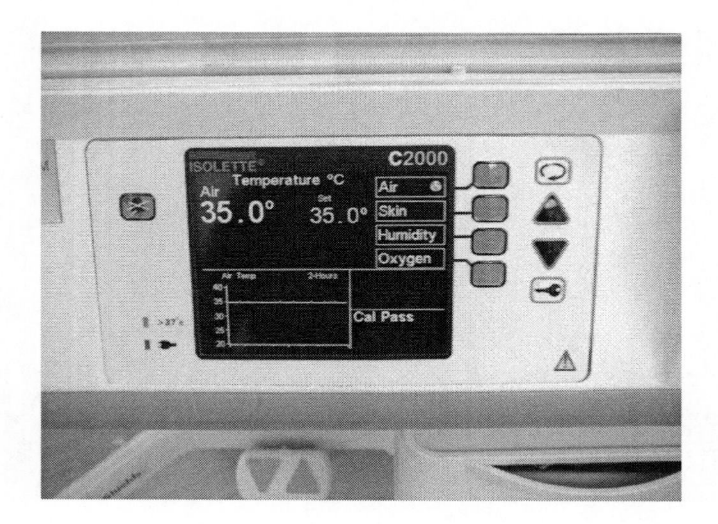

FIGURE 183
Display and control panel for infant incubator. Parameters including air and skin temperature, humidity, and oxygen concentration can be displayed in numeric or trend graph formats.

Humidity control is usually simple, with a supply of sterile distilled water being placed in the airflow so that some of it can evaporate. Care must be taken to keep the water reservoir clean, because bacteria or fungi may grow in it, with potentially harmful effects.

Air must be filtered to remove as much dust as possible, and these filters must be changed regularly.

Incubators normally have a port by which oxygen can be added to the interior air. Because oxygen levels must be high enough to be effective but not so high as to cause problems with infant retinal development, the oxygen concentration in the incubator should be monitored and alarmed.

It is important that air be well circulated within the incubator to maintain even and controlled heat. A fan, usually located near the heating element, accomplishes this. Because of the many hard surfaces within the incubator chamber, and because infants have sensitive hearing, fan noise must be designed to be as low as possible. In addition, because proper airflow is critical, most incubators have some kind of detector that will initiate and alarm if airflow is lower than normal.

Finally, because incubator function is critical, most have a power failure alarm, so that staff can take appropriate measures if power is interrupted.

Bilirubin Therapy Systems

Neonates sometimes have an excess of bilirubin in their system, which causes a jaundiced appearance. It is desirable to reduce these levels, and the simplest method of doing so is to expose the skin to ultraviolet (UV) light, which causes the bilirubin to break down.

Bilirubin therapy systems are simply light sources that produce ultraviolet light of the optimum wavelength and intensity. Too much exposure can cause burns, while too little is ineffective. Some systems use overhead fluorescent lamps that are designed to emit the correct type and amount of UV light. Several factors affect the amount of exposure the baby receives, such as the skin surface available, material such as the Plexiglas of an incubator between the light source and the infant, the distance between the light and the baby, and the age and condition of the bulbs. Because of these variables, it is important that the UV levels at the baby's skin be measured at the start of treatment, and at intervals if the treatment time is prolonged. In order to prevent excessive exposure, it is also desirable that a timer mechanism be used to either cut off the UV light after a predetermined time or to remind staff to turn the light off.

Overhead systems often have a tape measure built in to aid in placing the light at the correct distance from the baby (Figure 184). The lights produce some heat, and because much of the baby's skin is exposed to facilitate treatment, it is important to monitor the baby's body temperature during treatment.

Another type of bilirubin therapy system avoids some of the problems associated with overhead lamp types. This method uses a "blanket" of material in which fiber optic strands are embedded. These strands carry UV light from a central source and distribute it evenly, so that the baby's skin is exposed to a consistent illumination wherever it is covered by the blanket. The factors of distance and of intervening materials are eliminated, and the light sources used tend to be more stable than fluorescent bulbs. The blanket system is also smaller and less cumbersome than an overheard lamp system. Of course, output levels must be checked regularly and exposure times limited.

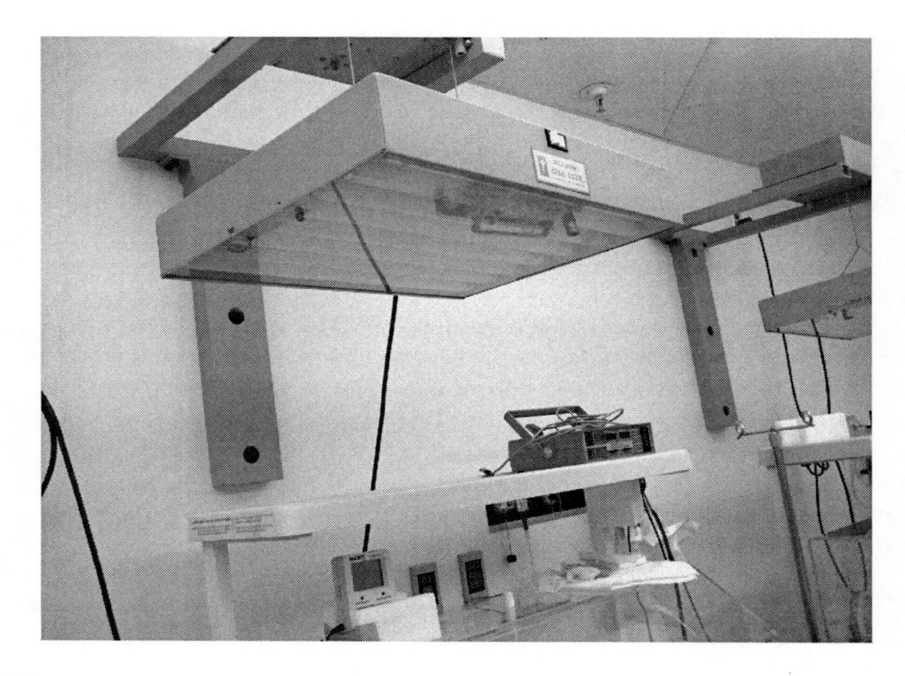

FIGURE 184
Overhead ultraviolet lamp for treatment of bilirubinemia. Height is adjustable to allow for different exposure levels.

Spirometers

Measuring lung function is critical in determining the state of health of many patients, especially those with respiratory disorders. Various parameters can be measured, including those of normal breathing and the extremes of inspiration and expiration.

Both the volume and the rate of airflow during breathing can be measured by a spirometer; each can be graphed against the other to produce a roughly circular chart, beginning and ending at the same point, or volume can be graphed against time to give a linear chart. Measurements are usually repeated a few times for each level of breathing (normal and maximal effort). The shape of the charts, including minimums and maximums, as well as any irregularities, can tell the caregiver a great deal about the health of the patient's lungs. In addition, tests performed both before and after medications can give a good indication as to the effectiveness of the medication. Long-term tracking of test results provides a means of following disease and/or recovery progress.

The simplest spirometers are strictly mechanical, with the patient breathing into a bellows that has a chart pen attached, which in turn marks rotating or moving graph paper, producing a linear chart of volume vs. time. More advanced versions use some kind of electronic sensor to measure flow; once the flow rate and cross-sectional area of the breathing tube are known, volumes can be calculated. These data can then be used to produce the circular graphs mentioned previously, either on graph paper or on video monitor. Electronic circuitry can perform all relevant measurements and give a printout of the results, as well as allowing pre- and postmedication tests to be compared directly, on a graph and/or as numerical values.

Spirometers may also form part of a system that can monitor such parameters as expired CO_2 levels, body temperature, ECG, blood oxygen saturation (SpO2), and others.

The prime considerations of spirometer design are that they be accurate and repeatable, and that they provide an absolute minimum of obstruction to the patient's breathing.

Pulse Oximeters

Many disease conditions can reduce the amount of oxygen present in a patient's blood. Physical effects of low oxygen levels may not be apparent until after the optimal time for intervention, therefore a means of measuring oxygen content (or oxygen saturation) of the blood is important.

A pulse oximeter measures the amount of oxygen in the patient's blood by sending both red and infrared light beams through tissue, and measuring how much of each is transmitted. Blood in the tissue transmits light differently depending on its oxygen saturation levels; by comparing the values for infrared and red light transmission, the unit can calculate oxygen percentage saturation.

Most oximeters also give a readout of pulse rate, in beats per minute. Some have a tone that sounds with each pulse beat, which may change in pitch depending on the oxygen concentration. Some may also have alarms for high and low oxygen levels and pulse rates; the alarm levels may be fixed or adjustable, though adjustable alarms are usually only used in more critical settings (Figure 185).

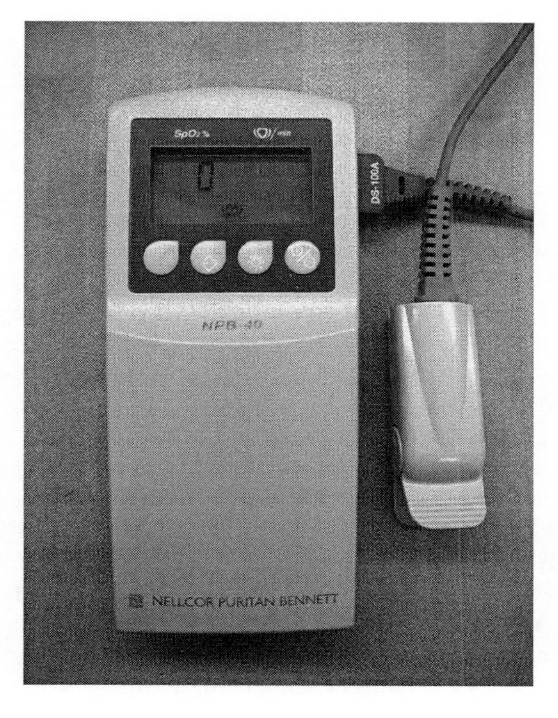

FIGURE 185
Pulse oximeter with finger probe.

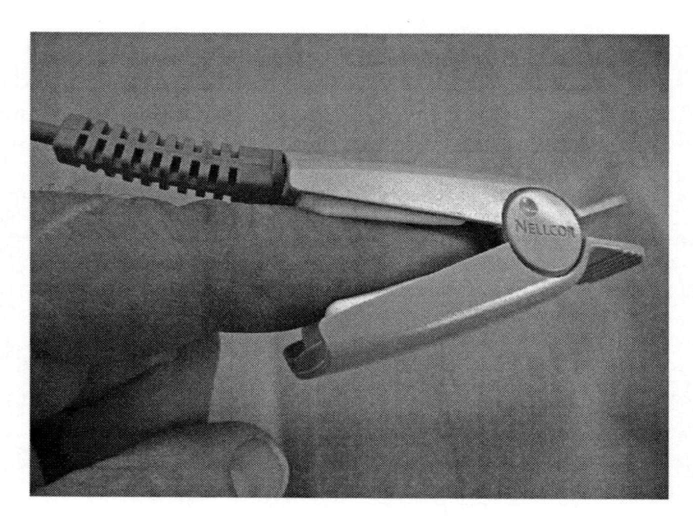

FIGURE 186
Pulse oximeter probe on patient's finger.

A sensor, either wrap-around or clip-on, is placed on a small part of the body that has good blood flow, such as a finger, toe, or earlobe (Figure 186).

The sensor has an emitter that produces the red and infrared signals, and a receiver that picks up what is left of each after they pass through the tissue. The unit then processes the measurements and gives a digital readout of the saturation value.

Laparoscopy Systems

Traditional abdominal surgery is traumatic for the patient. Large incisions are a direct shock to the system. They require a high degree of anesthetization, which is itself hard on body systems, and healing times and risks of infection are both significant. The large scars left are a cosmetic problem for some patients, as well.

Laparoscopic surgery reduces these negative aspects of surgery by reducing the incision size from 10 to 30 centimeters or more, to as little as one or two centimeters. This is made possible mainly by advances in fiber optic technology; optical cables provide a flexible pathway for illumination as well as for conveying images of the surgical target to the surgeon.

In order for the surgeon to be able to see to target, the abdomen must be filled with gas (insufflated). This requires a gas control system that regulates and measures flow rates, delivered volumes, and sometimes inflation times. To provide a constant degree of inflation, the gas supply must be able to sense and control the intra-abdominal gas pressure (Figure 187).

Some simple procedures are done with a simple lens and eyepiece, which allows the surgeon to directly observe the internal structures and instruments (Figure 188).

Most procedures, however, are visualized by an electronic camera pickup attached to the fiber optic cables. The camera feeds an electronics module that conditions the signal and sends it to one or more high-resolution color video monitors, which the surgeon observes. This allows other personnel to see the progress of the operation for teaching purposes and to allow assistance that is more effective. Images are often recorded for future examination.

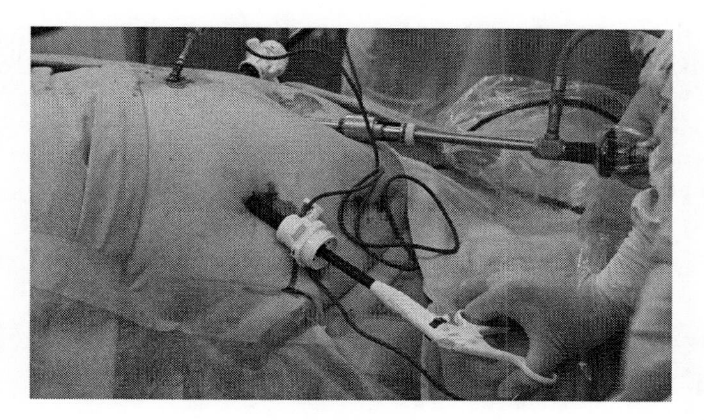

FIGURE 187
Laparoscopic procedure—note how the abdomen is distended by the insufflation process.

Because color is sometimes critical in identifying structures and diseased tissues, both the illumination and display components of the system must have high color accuracy. Light sources must provide high-intensity illumination, which can be either manually or automatically controlled; light levels may interact with the video system, as well, to provide optimum values.

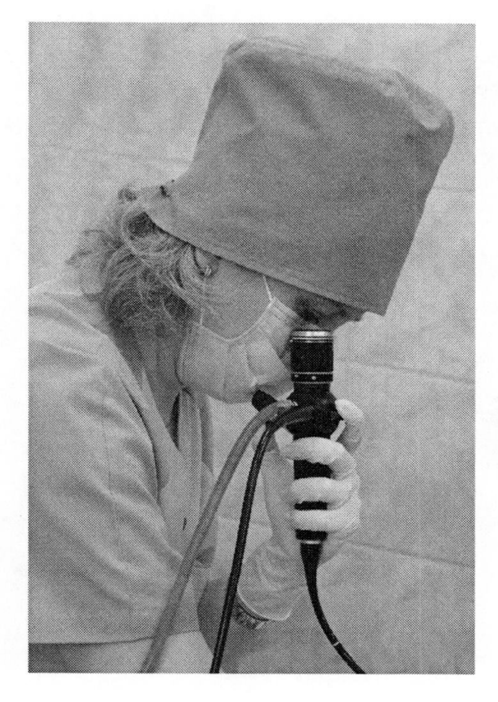

FIGURE 188
Direct observation through lenses and fiber optic channel.

A channel in the access tubing allows insertion of the actual surgical instruments; these may be mechanical cutting or dissection devices, specially designed electrosurgery probes, or attachments to break up kidney or gallstones.

Sometimes one tube is used for observing the site while another provides access for surgical instruments.

Suction must also be available, for removing excess fluids from the site and for aspirating excised tissues.

Endoscopy Systems

It is often necessary to examine the inner structures and surfaces of a patient's digestive or respiratory systems. Any means of performing minor surgical procedures (such as excisions of growths or cauterization of wounds) requires the ability to visualize the tissues in question.

Endoscopy systems consist of several components.

- The central component is a tube, called an endoscope or scope (Figure 189). The endoscope has various parts, including:
  - Fiber optic light guides for illumination and observation.
  - Channels for supplying air to inflate (insufflate) the body cavities being examined and to remove such air after the procedure.
  - Another channel for irrigation and aspiration of the area.
  - If surgical procedures are to be performed, a channel for devices used for these procedures.
  - In some cases, a set of cables and bands that allow the operator to turn the tip in various directions.

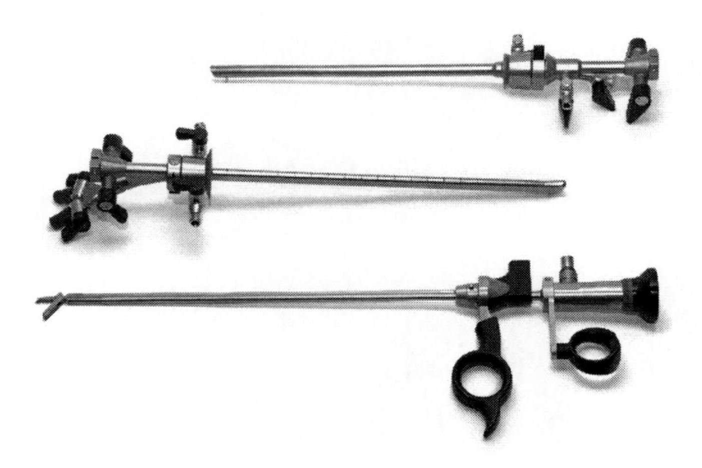

FIGURE 189
Rigid endoscopes and tools.

FIGURE 190
Endoscopy fiber optic light source. A slide control adjusts brightness. The fiber optic cable connection is on the right.

- The scope must be very smooth, both to prevent damage to tissues during insertion and removal, and to reduce the opportunity for infectious organisms to remain on the surface of the tubing. Scopes are generally marked with clearly visible rings that indicate the depth of insertion.

- Another component of the system is the light source, which provides high-intensity, adjustable illumination for procedures (Figure 190). The light source may incorporate other components, such as an insufflator and irrigation aspiration pumps or simple electrosurgery unit circuitry, or these components may be separate.
- Insufflators supply and control the gas that is used to inflate (insufflate) the surgical site, with adjustable gas flow rates and pressures (Figure 191). Because the insufflating gases are absorbed by abdominal tissues, a constant flow of gas is required to maintain insufflation levels.
- Finally, the system requires a means of displaying images. This may consist of a simple lens for viewing by a single operator, or, more commonly, a video camera, which produces video images for viewing on one or more video monitors. These images can also be recorded on videotape or other media (Figure 192, Figure 193, Figure 194).

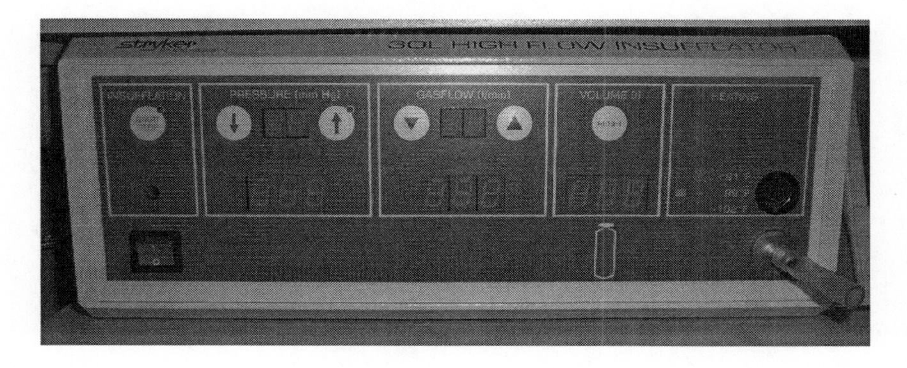

FIGURE 191
Endoscopy insufflator. Controls allow adjustment of pressure and volume as well as gas flow rate.

FIGURE 192
Endoscopy camera head with mechanism for attachment to endoscope. The camera head contains lenses, one or more photosensors, and initial processing electronics.

Oto/Laryngo/Ophthalmoscopes (OLOs)

The ear (oto), throat (laryngo), and eye (ophthalmo) are body parts that often require close examination. To do so effectively, equipment must provide bright, adjustable lighting that can illuminate the subject while allowing a clear view.

- Otoscopes typically have a tapered tube that can be inserted into the ear canal, with a magnifying lens (sometimes interchangeable for various magnifications) over the large end (Figure 195). Light is supplied by a small bulb in a handle that is either battery or AC powered. The light in most otoscopes is channeled from the bulb into the ear by a set of optical fibers embedded within the viewing tube, so

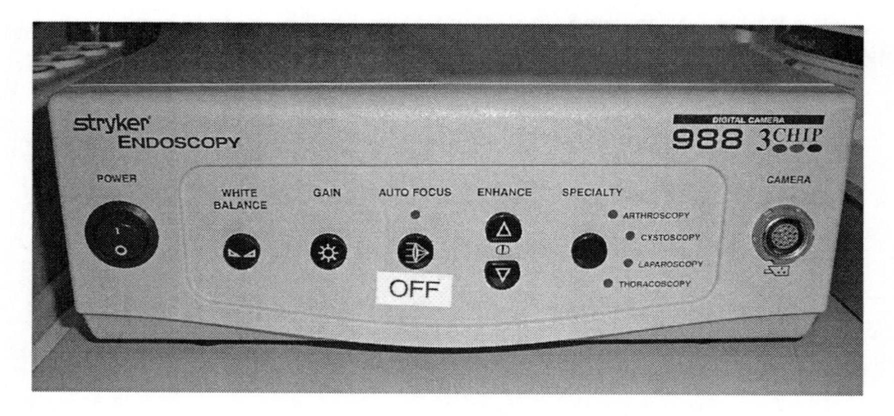

FIGURE 193
Endoscopy camera console. This component receives video signals from the camera head and provides further image processing, as well as controls for the camera head. A variety of video outputs are available, and the unit can interface with the light source to allow for automatic light intensity adjustments.

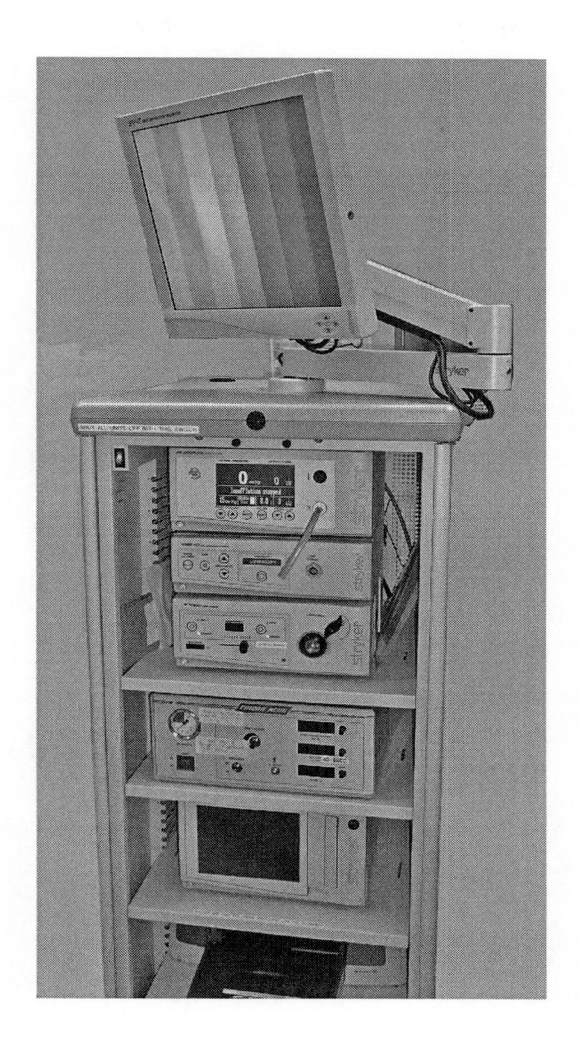

FIGURE 194
Endoscopy system rack with (from top) video display, insufflator, camera console, light source, specialized uterine inflation unit, and video recorder.

that the light emerges in a ring around the end of the tube, thus providing light without interfering with viewing. Some otoscopes use a simpler arrangement, with a bulb shining directly down the tube and blocked from view by a shield. This shield, of course, obstructs the field of view somewhat. Disposable covers are usually used to help prevent the spread of infection from one patient to another.

- Laryngoscopes need a means of holding the soft tissue in the patient's mouth (mainly the tongue) out of the way for viewing (Figure 196). They are usually equipped with a curved metal blade for this purpose. The blade has rounded edges and is made of metal so that it will not break if the patient inadvertently bites down. A bright light (either a bulb or the end of an illuminated fiber optic channel) is placed at the base of the blade to illuminate the area being viewed. Laryngoscope blades must be sterilized between patients.

FIGURE 195
Medical staff using an otoscope to examine a patient's ear.

- Ophthalmoscopes have a bright light source as well, but because structures of concern in the eye are generally smaller than those in the ear or throat, ophthalmoscopes have lenses of various powers to aid in observations (Figure 197). Some eye conditions are visualized better with colored light, so filters are often available for the observation light. In addition, some situations require a narrow line of light to illuminate the eye, so mechanisms are provided to make such lines, usually of various widths.

OLO scopes can be powered with rechargeable or disposable batteries, or they may be connected to a power supply fed from a wall outlet. Because they are required for many emergency cases, the units in the ER are often wall mounted and line powered, so that dead batteries are not a problem.

Diagnostic Ultrasound Units

Certain tissues cannot be visualized well with X-rays, and some situations (such as pregnancy) require that X-rays be avoided if at all possible, but visualization of internal structures is still needed.

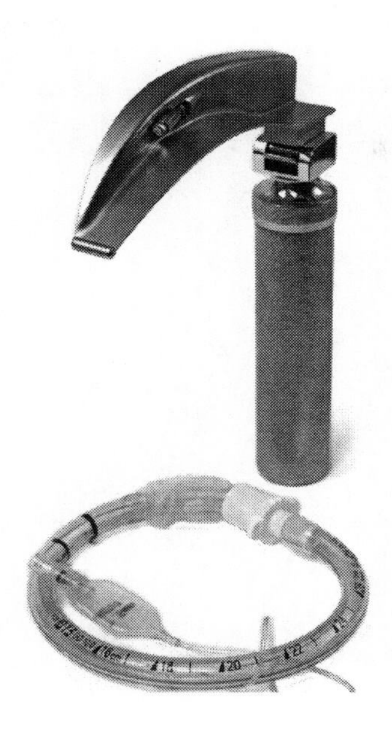

FIGURE 196
Laryngoscope. The handle contains a battery and head for connecting different blades, while the blade includes a high-intensity light bulb.

Diagnostic ultrasound machines produce beams of ultrasonic sound waves, which can be directed into the patient's body (Figure 198). The sound waves are reflected by tissues of different densities, and by the boundary layers between different tissues. By picking up and processing the reflected signals, an image of the internal structures can be obtained. Various frequencies of ultrasound are used for different circumstances (such

FIGURE 197
Ophthalmoscope being used in eye examination.

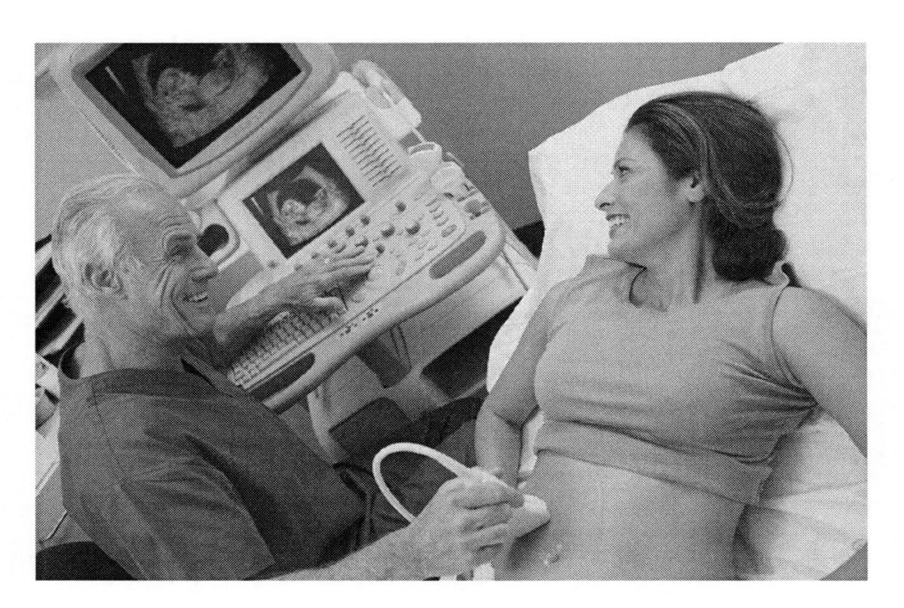

FIGURE 198
Ultrasound unit in use with a pregnant patient.

as the depth of the target organs), and probes are available that give wider or narrower beams. Visualization of a heart valve, for example, requires a narrower beam than one used for visualizing most or all of a developing fetus. The beam is scanned back and forth electronically to produce a full image.

Because of the nature of the wave generation and pickup, raw ultrasound images represent a two-dimensional or "slice" view of the tissue in question. While this may be sufficient for many applications, more sophisticated units can take the results of many such slices and combine them to give a three-dimensional view.

Devices may be general purpose, or they may be specifically designed for particular applications, such as cardiology or maternity.

The patient is positioned as required, and the ultrasound probe is applied to the skin. Because sound waves are partially blocked when they encounter a boundary such as between the probe surface and the skin, especially if there are any air gaps, a special gel that helps couple the sound waves more efficiently is used. This also helps lubricate the skin to make positioning the probe easier. Particularly for fetal imaging, because images are taken from a variety of angles, large quantities of coupling gel are required. Operators usually keep the gel in a heated chamber so that it is warm when applied.

Room X-Ray Units

X-rays of the appropriate power will penetrate tissues. If an appropriate detector is placed on the opposite side of the body from the source of the X-rays, a "shadow" image is obtained which can give vital information about various structures within the body, because different body parts and tissues block the X-rays to a greater or lesser degree.

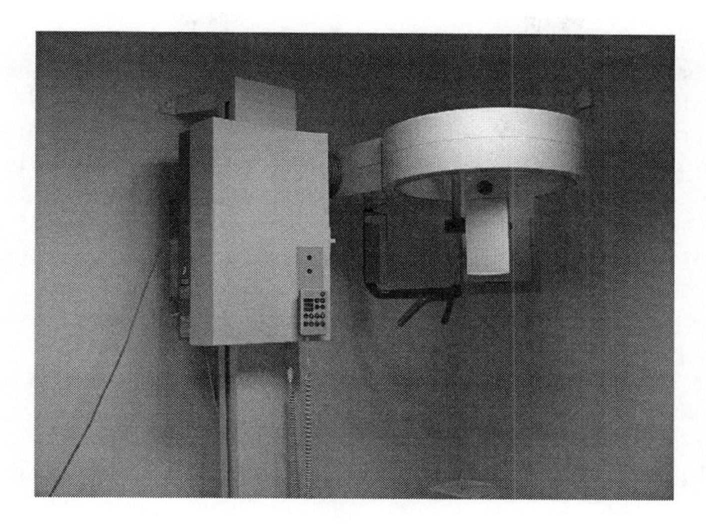

FIGURE 199
X-ray unit. X-rays are generated in the head on right, while the column allows for vertical adjustment.

By focusing a high-energy beam of electrons onto a spinning metal disc, the atoms of the metal can be excited to a point where they give off X-rays, an extremely high-frequency form of electromagnetic radiation (Figure 199). Other examples of electromagnetic radiation are light rays, infrared or heat radiation, and radio signals.

Traditionally, the most common means of detecting and recording the images produced by an X-ray machine is silver halide–based photographic film. This film is similar to everyday film used in cameras, but is generally much larger. It is optimized for X-ray exposures rather than light exposures, and it is very fine grained so that maximum detail can be seen in images.

Newer systems employ an electronic detector system that uses the same general technology as the detectors in digital cameras, modified to receive X-ray images.

Electronic components act to focus the X-ray beam at the plane of the film or detector, and thus care must be taken to place the receiver and the X-ray source at the correct distance from each other.

Because the X-rays themselves cannot be seen, a visual system of aiming is used, with a light shining from the same point as the X-rays and making a shadow of cross-hairs. This part of the unit also has adjustable edges so that exposure can be limited to the specific target area.

Adjusting the power and time of the X-ray emission controls exposures. Power is measured in units called kilovolt-amps, or kVA. Different tissue densities and thickness, and varying requirements for detail and speed of exposure, determine these parameter settings.

Certain structures can be better seen on X-rays if a material that blocks the rays is placed into spaces within the body. Examples of this include a liquid form of barium that the patient may drink to help define parts of the upper digestive tract, or which may be given by enema when the lower parts of the digestive system are being investigated. Radiopaque dyes may be administered intravenously to show either blood vessels or the urinary system (after the dye is removed from the bloodstream by the kidneys).

Some X-ray units have mechanisms to automatically and rapidly change several film cassettes, in order to follow the progress of barium or dye that has been introduced into the patient's body as it moves through. Others have a roll of film, as in a movie camera, to record a greater number of frames (though in a smaller format). Digital systems simply load images from the detectors into a computer system so that successive images can be taken easily and rapidly.

Excessive exposure to X-rays can be harmful, especially to reproductive organs, and so patients are normally draped with flexible lead aprons and/or sheets to limit exposure to the target area. Staff members use this equipment over long periods of time and so must take precautions to avoid overexposure. Lead aprons are worn if they must be in the vicinity of the patient during exposures, otherwise they move behind a leaded shield for the time of exposure. Film badges are worn which can help evaluate the total dosage received over a period of time, and if levels reach certain threshold levels, staff may be reassigned for a while.

Patients are asked to remove any clothing from target areas of the body, and to replace it with a thin gown in order to minimize blocking of the X-rays. They are then placed in a position (either on a table or standing) that will allow the best view of the target, and distances and areas to be exposed are set. Because exposure times are often measured in relatively large fractions of a second, or sometimes even multiple seconds, the patients must remain still during exposures. If the target area is in or near the chest, the patients are asked to hold their breath for the exposure.

Barium is usually administered before the X-ray procedure starts, though it may be given while making exposures if, for example, swallowing actions are to be examined. Radiopaque dyes are usually injected immediately before X-rays are taken; preinjection images may be taken for comparison purposes.

CT Scanners

Conventional X-rays produce a "shadow" image of the tissues through which the X-rays pass. Many disease conditions or abnormalities can be visualized better when a "slice" view is available, so that structures are not obscured by tissues above or below, as may be the case with regular X-ray images.

CT (computerized tomography) scanners pass a series of narrow-beam X-rays through the target area of the body in a circular pattern, and then combine the information from each exposure with computers to develop a "slice" image. With enough individual exposures and accurate computer analysis, very detailed images can be produced. Because the system uses narrow beams and a combination of many subimages, each total image exposure can be relatively low, and the total radiation dose is acceptable. Newer devices use more sensitive X-ray receptors and faster computers to reduce the total radiation exposure and both the time required to acquire an image and the time to process the image and have it available for viewing. Images may be presented as an X-ray-like transparency, a printed picture, or on a video display; they can be stored as computer files for later retrieval and analysis. By taking a series of CT images, each adjacent to the next, a three-dimensional representation of structures can be developed.

Patients are moved by a powered table surface into the scanning chamber, where exposures are made (Figure 200). Because exposure times can be somewhat prolonged, the patient must remain as still as possible; newer systems have reduced exposure times, but this is still a factor. The range of "slices" that can be obtained is limited by the size of the

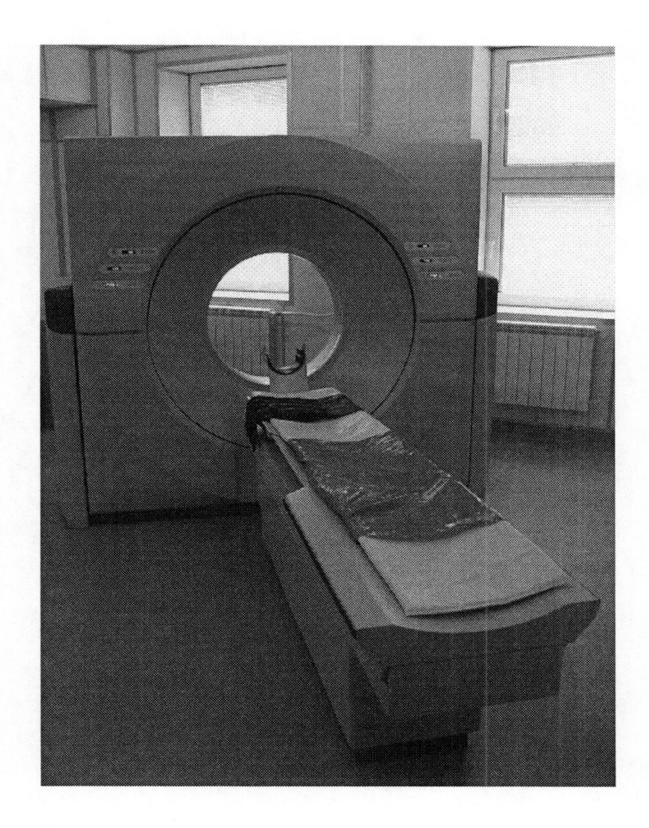

FIGURE 200
CT scanner.

scanning ring and the shape of the human body. Most systems have a ring not much larger than a human torso.

Magnetic Resonance Imaging (MRI) Scanners

Because the human body is composed largely of water, a means of imaging the distribution of water in the body provides a picture of various structures giving a different kind of information than that provided by X-rays or other scanning technologies.

Water molecules consist of two atoms of hydrogen and one of oxygen. When they are within a powerful magnetic field, radio waves of a precise frequency can interact with the hydrogen atoms and cause them to resonate and emit their own radio signals. These signals can then be detected and analyzed, and an image of hydrogen (and thus water) distribution within the body can be obtained. With sufficiently advanced detection and analysis, very detailed images of body tissues can be obtained. These images can then be displayed on a video monitor and stored and/or printed out for later examination.

Because extremely powerful magnetic fields are used in MRI procedures, all magnetically active metallic objects must be removed from the patient. This means that patients with stainless steel implants or pacemakers may be excluded from such studies. The operation

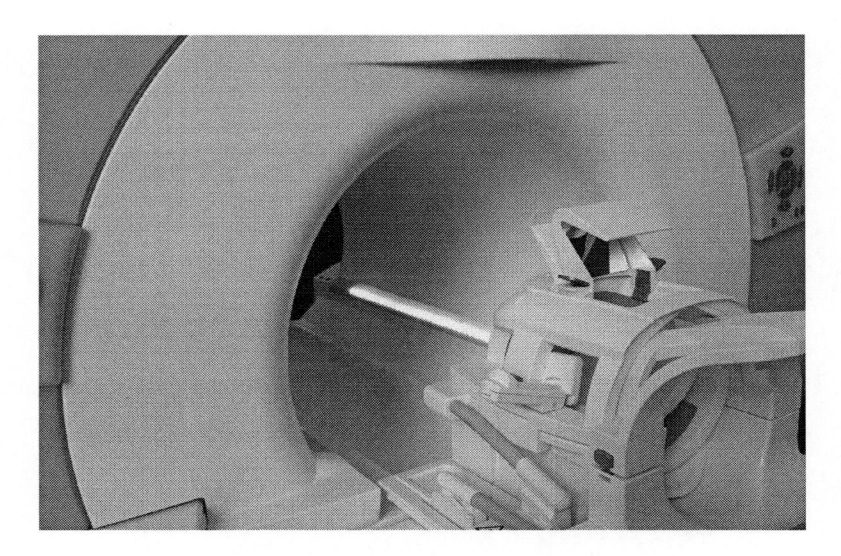

FIGURE 201
MRI scanner.

of pacemakers is also affected directly by magnetic fields, so that further restricts MRI use with these patients. Dental fillings and crowns are not usually affected.

The patient is positioned on an examination table, which can be moved around to facilitate imaging of different areas (Figure 201). The operator moves to a shielded location, and the magnetic field is initiated. To obtain the images desired, patient position is adjusted by the table, under both operator and program control. Typically, two to six images are taken during a procedure, and because each image takes a few minutes for positioning and exposure, the whole procedure may take 15 to 45 minutes. This time may be longer for especially detailed or complex studies. Patients must remain still during the actual imaging, but can shift to a limited degree between exposures.

A contrast medium may be injected into the patient for some studies, in order to improve image quality.

Proton Emission Tomography (PET) Scanners

It can be important to know which areas of the body are active metabolically in determining disease or injury conditions in a patient. Because specific compounds are involved in certain metabolic functions, a means of identifying the location and rate of uptake of these compounds will give information about metabolic activities in the tissue or organ in question.

Various molecules utilized in metabolic activity within the body can be "labeled" with radioactive elements that either replace the normal atoms of that same element within the molecule or attach themselves to the molecule in such a way that chemical activity is effectively unchanged. The resultant molecules are produced in a huge, high-tech machine called a cyclotron, and are referred to as "radiopharmaceuticals." These radioactive atoms are then carried through the body along with the whole molecules and are concentrated within organs or tissues that utilize these molecules and are currently active.

One example of such a molecule is fluorodeoxyglucose, which acts like normal glucose and is absorbed by any tissue in the body that is active metabolically. This is particularly useful in the brain, where tumors, injury sites, or areas involved in mental processes typically are more active than other nearby areas.

A second radiopharmaceutical is radioactive iodine, which is concentrated in the thyroid gland, especially in tumors or abnormally active regions. Conversely, iodine is less concentrated in abnormally inactive areas.

There are many other such radiopharmaceuticals, each of which is specific to certain types of metabolic activity in particular tissues or organs. They are used to help diagnose disease conditions in the target areas, or to perform research studies involving those areas.

Once the radiopharmaceutical has been concentrated in the target tissue, the actual PET scan can be performed. Radioactive atoms give off protons in the form of gamma rays, which can be detected by a device called a gamma camera. By arranging as many as 180 of these gamma ray detectors in a circular pattern around the target area, gamma ray activity can be measured. Combining signals from the various detectors and processing with a computer system gives "slice" images, which are colored artificially to show different degrees of metabolic activity in the image area. Each color assigned to the image by the computer corresponds to a particular level of activity; this makes it easy to differentiate between active and inactive regions.

Some PET scan systems can take a series of adjacent slice images and combine them to give a three-dimensional view.

Images are viewed on a video display and can also be stored or printed for later examination.

The radiopharmaceutical chosen for the particular study is administered to the patient intravenously. After a predetermined time, which varies with each type of examination, the patient lies on a moveable table, which slides them into the detector ring. Movements of the table are controlled by the operator and/or the computer system to position the patient in order to obtain the desired images. Images may be obtained in a series of slices to produce a three-D picture, and several exposures may be taken over time to observe changes in metabolism.

Radiation produced by the chemicals involved is relatively low level, and the molecules either decay to harmless levels in a short time, or they are excreted by the body, or both. Because operators are potentially exposed to more radiation than individual patients, they must take adequate precautions to avoid overexposure.

Mammography Units

Early detection of breast cancer is critical in reducing harm from the disease, including mortality. X-ray imaging is one means of aiding such detection. Because cancerous tissue is relatively close in X-ray density to normal breast tissue (as compared to bone, for example), and because tissue nodes that may be clinically significant are often small, means that can increase the effectiveness of imaging must be adopted. A reduction in the thickness of over- and underlying tissue surrounding possible tumors is an important factor.

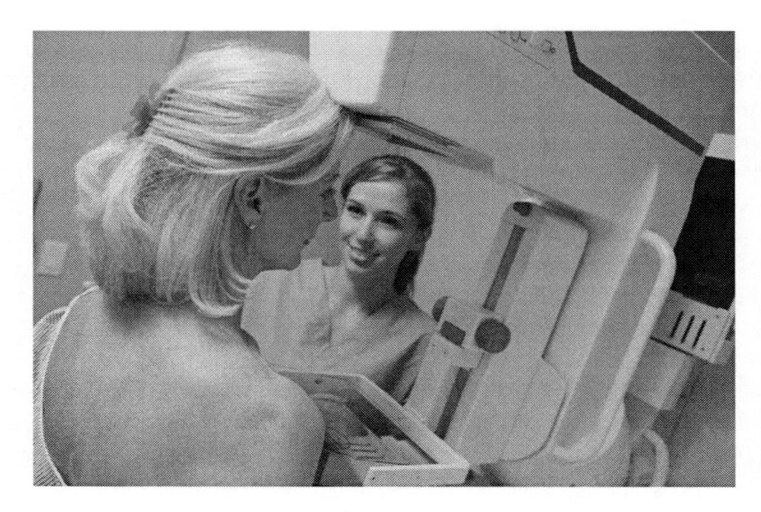

FIGURE 202
Patient undergoing mammography examination.

Mammography units are similar to normal X-ray units, with X-ray characteristics optimized for breast tissue. They have attachments that compress the breast, horizontally and vertically, in turn. With the breast compressed, thus having a much thinner profile, X-ray exposures are taken. If lumps have been detected by other means, these areas may be examined more closely.

The operator sets power and exposure times for the procedure, and then adjusts the X-ray head to the patient. The patient, with the assistance of the operator, places one breast on a clear plastic plate. A motorized mechanism then lowers a second plastic plate onto the opposite surface of the breast until the tissue is adequately compressed. The operator then moves to the control area (shielded by lead glass and walls) and makes the exposure. The breast is then compressed along the other axis and exposures taken again (Figure 202).

Patients are shielded appropriately to reduce X-ray exposure to nontarget areas of the body.

As the compression required for good images is significant, this procedure can be quite uncomfortable for the patient.

Radiation Therapy Units

Some types of cancer cells can be killed by exposure to radiation; they tend to be more sensitive to radiation than normal cells because their rapid reproduction rate is disrupted by the radiation.

By focusing the radiation from one or more sources on the area in which cancer cells are likely to be present, a radiation therapy unit allows the selective removal of cancer cells, while leaving most of the surrounding tissues relatively undamaged. A block of radioactive material (produced in high-technology nuclear facilities) is enclosed in a lead container, which has a window that can be opened to allow radiation to escape in a narrow beam.

The patient is positioned so that the radiation beams can reach the target area with minimal penetration of healthy tissue. Because treatment times are quite long, the patient must also be made as comfortable as possible while remaining motionless. When the patient has been positioned and nontarget areas shielded, an operator remotely opens a window in the container holding the radioactive source, allowing the target area to be exposed to radiation. During the course of a treatment session, the angle of the radiation may be changed so that healthy tissue above and below the tumor site is irradiated less than the tumor.

Some tissues will selectively absorb particular radioactive materials, in which case (if there is a tumor in that particular tissue) relatively low concentrations of the material can be injected intravenously; the patient's body then concentrates the substance in the target tissue where it can deliver effective doses of radiation until it dissipates. An example of this is the thyroid gland's ability to concentrate iodine, including its radioactive isotopes. Obviously, this method does not require a radiation therapy machine, but many of the same precautions in handling the radioactive material are required

IV Pumps

Medications, blood, or fluid often must be delivered into a patient's bloodstream at an exact rate for a relatively long time. In the old days, caregivers had to measure how fast the liquid was dripping and calculate the rate, adjusting the flow often. The intravenous pump makes this much easier by pumping the fluid in a controlled and accurate manner (Figure 203).

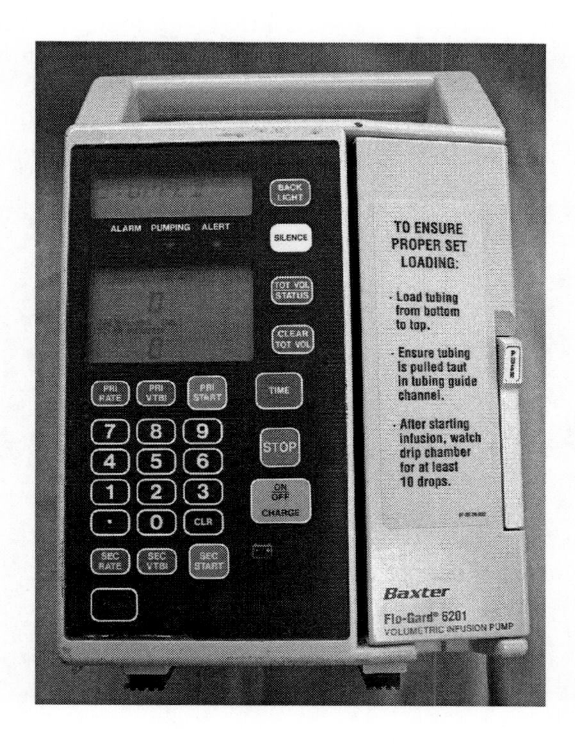

FIGURE 203
Infusion pump.

There are several different techniques of pumping. One of the most common is "peristaltic," in which either a set of "fingers" or some rollers on a drum squeeze the fluid tubing, one after another, pushing a wave of fluid through the tube. Because the tubing is precisely manufactured, the speed at which the pump operates determines the flow rate of the fluid. The other common pumping method uses a cassette that fits into the pump and is compressed by a piston. The size of the pumping chamber in the cassette is very precise, and by varying the frequency of compression, the flow rate can be controlled. This method is generally more accurate than the peristaltic method. With either method, an internal computer calculates and displays the flow rate, and also the total volume to be infused, how much has been infused, and how much is left.

Pumps have very accurate sensors which can detect air bubbles in the line and stop the pump, sounding an alarm before the bubbles can get past the pump and into the blood stream. They also can detect blockages, either before or after the pump.

Enteral Feeding Pumps

For a variety of reasons, a patient may not be able to swallow food adequately. In such cases, food must be administered through a tube inserted into the stomach. Obviously, the food has to be liquid, and since such liquids are generally quite thick, they may not flow easily enough to pass through the tube by gravity alone.

Feeding pumps provide pressure to make the liquid food flow through the feeding tube and into the patient's stomach (Figure 204). The most common mechanism for these pumps is a wheel, around part of which an elastic section of tubing is stretched. Cylindrical roller bearings at intervals around the wheel compress the tubing as the wheel rotates, squeezing the fluid in the tube in a peristaltic wave. This pushes the liquid food through that section of tubing and toward the patient, while at the same time drawing more food from a

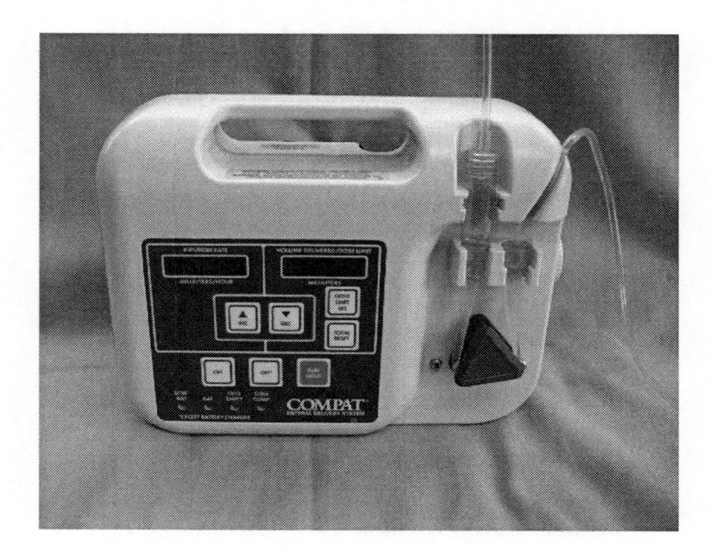

FIGURE 204
Feeding pump.

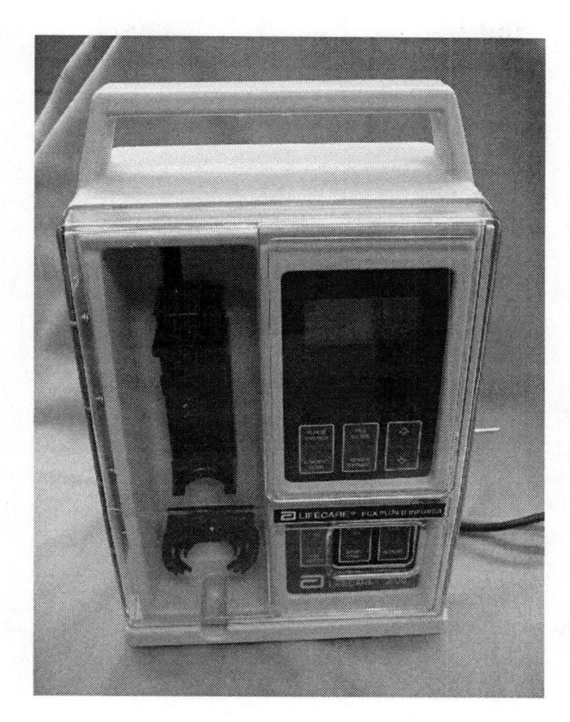

FIGURE 205
Patient-controlled analgesia (PCA) pump.

reservoir. Pumps generally have controls that allow flow rate and desired volume to be set, and have some method of signaling when the set volume is reached. They often have a means of detecting blockages, and alarms for when this occurs.

Patient-Controlled Analgesia Pumps

When patients are experiencing ongoing pain, either from surgery, injury, or disease, intravenous analgesia requirements are variable and subjective. Giving the patient some control over when dosages are administered both allows them better control over their pain and allows nursing staff to attend to other, more critical duties.

A patient-controlled analgesia (PCA) pump allows a patient to control his own pain-killer administration (Figure 205). To help prevent abuse of analgesic drugs, the solutions are usually packaged by the hospital pharmacy in sealed glass cartridges. These cartridges contain a piston and delivery port that is compressed by the pump mechanism to deliver the solution. This mechanism is precisely controlled to ensure accurate dose delivery.

Pumps can be programmed for continuous delivery rates, maximum total dosages, maximum patient-controlled dosage, and maximum patient-controlled dose frequency. Most units have preprogrammed values for specific drugs at specific concentrations, which can be selected directly, thus avoiding some potential calculation errors.

Most controls are located behind a lockable panel.

Blood Warmers

Blood must be kept refrigerated to prolong its storage life, but infusing it into a patient at this temperature can cause a serious drop in body temperature, especially if the patient is small and/or hypothermic, or if a large amount of blood is required in a short time.

A blood warmer is a device that allows blood to be heated to near body temperature before it is infused. This requires careful control, however, because overheating blood can damage it. Blood warming units use either a water bath or metal plate heaters to warm the blood. Because the temperature must be increased substantially in a short time, there must be a large surface area for sufficient heat exchange to take place. In metal plate heaters, the blood passes through a plastic pouch that has a long, back-and-forth passage through it, so the blood has a long distance in which to be warmed. Water bath units have the advantage of quicker heat exchange, often using a double-walled tube, with blood flowing toward the patient in the inside tube and warm water being pumped in the opposite direction in the outer jacket. The opposing flow means that the blood is close to the water temperature when it exits the double-walled section; the total length of tubing within the warmer is much less in this type, which means less wasted blood.

In both types of warmers, the exit of the warmer must be as close to the patient as possible so that it does not cool off too much before reaching the patient. With its somewhat moveable section of double-walled tube, the water jacket warmer allows for a shorter unheated section of tubing before it reaches the patient.

Both types also usually have a temperature display, as well as a double over-temperature cutout and alarm system, because overheating the blood can be harmful both to the blood and to the patient.

The section of tubing in the warmer is discarded after use.

These devices may be used in conjunction with an IV pump, which would be placed upstream of the warmer so that heat is not lost in the pump and associated tubing.

Hemodialysis Units

In cases where kidney function is extremely reduced or nonexistent, or where peritoneal dialysis is not appropriate, patients may have their blood purified by hemodialysis.

Blood is shunted from an artery in the patient's body and passed into a system that causes the blood to flow through a chamber where blood and a special dialysate solution are separated by a membrane film. The film allows some molecules to pass through but not others (and blocks blood cells), and the dialysate solution is formulated to encourage toxic substances and excess salts and water to move from the blood, through the membrane, and into the solution.

The system adjusts the temperature of the fluids involved so that the processed blood is close to body temperature. The cleansed blood is then passed back into a vein, usually adjacent to the artery from which it was removed.

Once the arterial and venous ports have been established, the hemodialysis unit (Figure 206) is filled with the various solutions required. Operation of the system is confirmed, and then blood flow is initiated. After the prescribed treatment time, the arterial

FIGURE 206
Hemodialysis machine.

flow is stopped and as much blood as possible is returned to the patient; then the venous line is closed.

In some patients, ports in the artery and vein can be semipermanent so that they can be used for repeated dialysis procedures.

Peritoneal Dialysis Units

When a patient is experiencing severely reduced kidney function, toxins and excess salts and water normally removed by the kidneys build up in the body and must be removed to prevent serious harm or death.

Peritoneal dialysis machines utilize two important principles. The first is that substances in solution tend to move from areas of higher concentration of that substance to areas of lower concentration; the second is that the inside of the abdomen (peritoneum) provides a large surface area that is well perfused. By filling part of the peritoneal space with saline solution, toxic materials in the blood will tend to move into the saline. The saline can then be drained, removing some of the toxic material as well.

A catheter is inserted into the patient's abdominal cavity; this may be semipermanent, and closed off when not in use. Saline solution is passed through the peritoneal dialysis

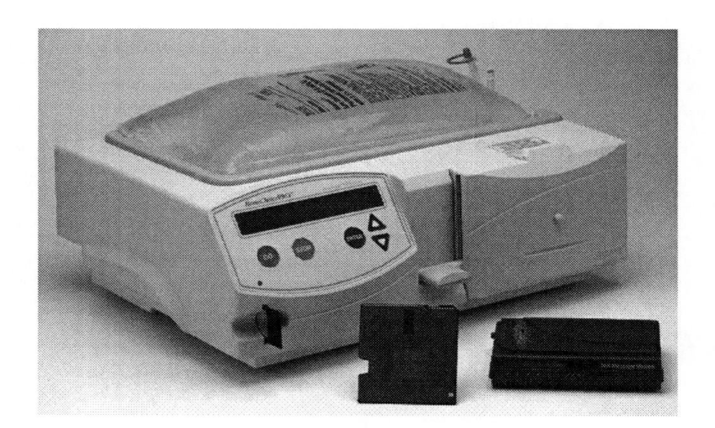

FIGURE 207
Peritoneal dialysis controller.

unit, which warms it to near body temperature and controls the flow rate (Figure 207). The abdomen is filled with a volume determined by body size. The unit also functions as a timer, to signal when the saline should be drained. The process is repeated several times in order to maximize toxin removal. Volumes are recorded to ensure that no significant amount of saline is left in the abdomen. This technique has a number of drawbacks, including the trauma of catheter insertion, the possibility of infection, and the fact that some toxins may not be removed effectively, while some beneficial substances may be inadvertently removed. Some systems use smaller volumes of fluid for a longer time, with fewer repetitions, while others use larger volumes for shorter times, repeated a number of times. The smaller-volume process can allow the patient to be at least partially mobile during treatment, and is thus referred to as ambulatory peritoneal dialysis. The large-volume technique requires that the patient be essentially immobile for the duration of the treatment, and is referred to as cycling peritoneal dialysis; it can be done while the patient is sleeping.

Lithotriptors

Kidney and gallstones (calculi) can be extremely painful, and often dangerous to the patient. Surgical removal is possible, but like any surgery, carries a degree of risk, as well as an extended recovery time. If these stones can be broken up into small-enough fragments, they can be passed with minimal discomfort, thus avoiding the drawbacks of surgery.

Lithotriptors (literally, stone breakers) use high-powered, high-frequency sound waves to shatter stones. Because the stones are much harder and more brittle than the surrounding tissues, the sound pulses affect them much more. When pulses are of sufficient magnitude and duration, they cause the stones to gradually break apart. To avoid soft tissue damage as much as possible, treatment times are generally extended. There are two general types of lithotriptors: direct contact and focused.

- Direct contact units consist of a special tip on a catheter. Circuitry passes a high-voltage pulse through the tip, causing a microburst of vaporized water. This burst in turn generates a sound shock wave, which begins to break the stone apart.
- Focused systems utilize a number of sources that produce narrow sound beams. These beams can be closely focused on the stone from outside the body, and the convergence of the multiple beams is sufficient to break up the stone(s). These systems are further divided into two types: immersion and coupled.
  - Immersion units have a bathtub in which the patient sits, and transducers in the water produce sound waves. The water transmits the sound waves well, and they are transferred into the patient's body, where they are focused on the stone.
  - Coupled units use a rubber bladder in which the sound waves are produced. The bladder is pressed against the patient, with a coupling gel between to aid in sound transmission.

Electrosurgery Units

One of the problems encountered in surgery is bleeding from the many blood vessels of various sizes that are severed while incisions are being made. This is more prevalent in some types of tissue, and results in not only an increased loss of blood for the patient, but also decreased visibility for the surgeon and support staff.

Electrosurgery units help solve the problem of excess bleeding by using high-intensity electrical signals to perform cutting procedures (Figure 208). A pencil-type probe with a blade on the end (which comes in a variety of shapes for different circumstances) applies the electricity to the cutting point, while a large conductive grounding pad placed on a fleshy part of the patient provides a return circuit. Power levels are adjusted for different

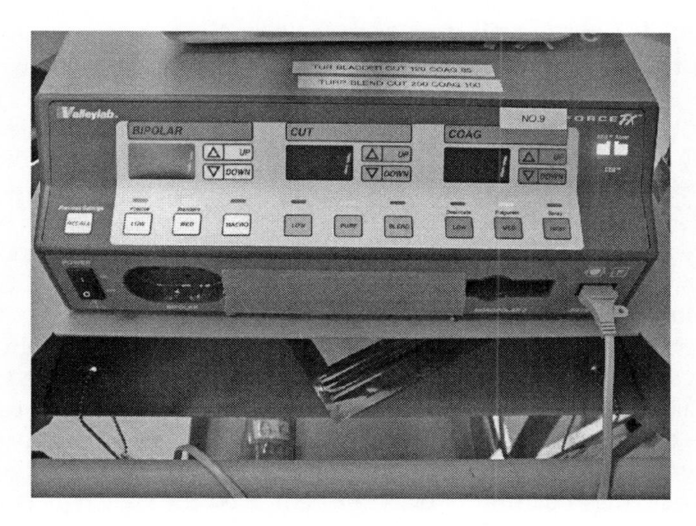

FIGURE 208
Electrosurgery unit, with controls for Cut, Coagulation, and Bipolar outputs.

procedures. The electrical energy vaporizes the tissue immediately around blade, and there is enough energy spreading into nearby tissues to heat them to a point where many of the smaller blood vessels are sealed off (cauterized). Larger vessels must still be tied off, but much of the excess bleeding experienced with scalpel incisions is avoided.

In some circumstances, a very localized effect is needed, such as when performing tubal sterilizations. In such instances, a special electrode is used which passes current directly from one side of the probe to the other, passing through the tissue to be cauterized. This function is called "bipolar" operation, as opposed to the more normal "monopolar" operation using a surgical pencil and separate ground pad.

Experimentation has shown that different shapes of electrical signals produce different effects: more concentrated cutting, more general cauterization, or a blending of both.

Some electrosurgery units utilize a highly charged beam of gas (usually argon) to perform the cutting and cauterization, instead of a metal blade.

In some circumstances, only cauterization is needed, and simpler devices which use the same principles as full-function electrosurgery units, but which only provide cauterization, are used. Simple cauterization generally requires less power and less precise control than cutting, so these units are usually smaller and less expensive.

Surgical Lasers

Certain surgical procedures require that tissues be vaporized in order to achieve the desired results, such as removal of skin abnormalities or cutting through tissue. Coagulation of tissues adjacent to the cutting site may also be desirable, or coagulation may be the goal.

Surgical lasers utilize very pure and controlled light beams to heat tissue. Depending on the width and power of the beam, this heating may be extreme, vaporizing the tissue, or less so, in which case cauterization is produced. The laser beam may be controlled so that the depth of its effect is very precise, allowing removal of skin blemishes or tattoos.

A power supply and optical system develop the laser beam. Its characteristics are determined in part by the material used to generate the laser light. Some such materials are CO_2, neodymium/yttrium/argon (Nd:YAG), and erbium YAG. The laser beam is directed by a series of mirrors to the target area.

A low-powered laser beam, utilizing the same pathway as the treatment beam, may be used to point the system at the correct location before applying power. Control circuitry allows for adjustment of beam power and duration of treatment.

In order to properly visualize the target area, and to observe the effects of the laser beam, a microscope system may be used.

Because the laser beams used in these systems are powerful, they can cause problems if they are directed to places other than the surgical site. They are capable of starting fires on flammable materials such as drapes or clothing, and they can cause skin burns or severe eye damage. Strict rules must be in place and enforced for areas where lasers are used, and signs and door interlocks must be used to prevent anyone from entering the area inadvertently.

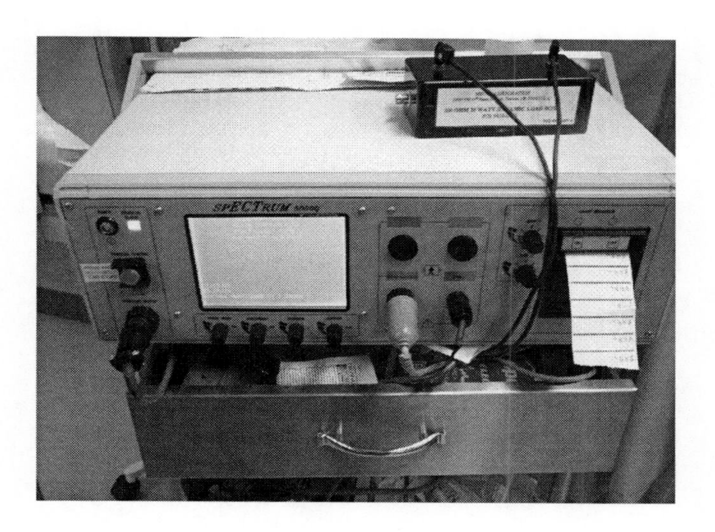

FIGURE 209
Electroconvulsive therapy unit, with controls for stimulus waveforms, display for ECG and EEG signals, connectors and cables for ECG, EEG, and stimulus, strip chart recorder, and stimulus delivery button.

Electroconvulsive Therapy Units

Certain psychiatric conditions seem to respond well to the administration of a controlled electrical signal to the brain.

An electroconvulsive therapy (ECT) machine provides precisely shaped and timed signals (both parameters under control of the operator) via special electrodes placed on the patient's scalp (Figure 209).

Phacoemulsifiers

Cataract surgery is performed often, and the techniques of removing the clouded natural lens from the eye and replacing it with an artificial one are finely tuned. One aspect of the procedure, that of removing the old lens, is aided by a specialized device called a phacoemulsifier (Figure 210).

A primary objective in surgery is to minimize trauma, and to this end, incisions are kept as small as possible. This is especially true for ophthalmic surgery, but the ideal incision size does not allow for removal of the lens in one piece during cataract surgery.

The phacoemulsifier aids in this step by breaking the lens material into very small pieces (emulsifying it) and extracting the resulting product. A fine tip delivers very-high-frequency vibrations that break up the solid lens material; a parallel duct then applies suction to the area, removing the lens particles, while a second duct supplies irrigation fluid as required. Some units utilize a specially tuned laser beam to break up the lens.

The unit requires controls for the various steps (these controls can usually be operated by a foot pedal assembly, leaving the surgeon's hands free), as well as displays showing suction and infusion levels.

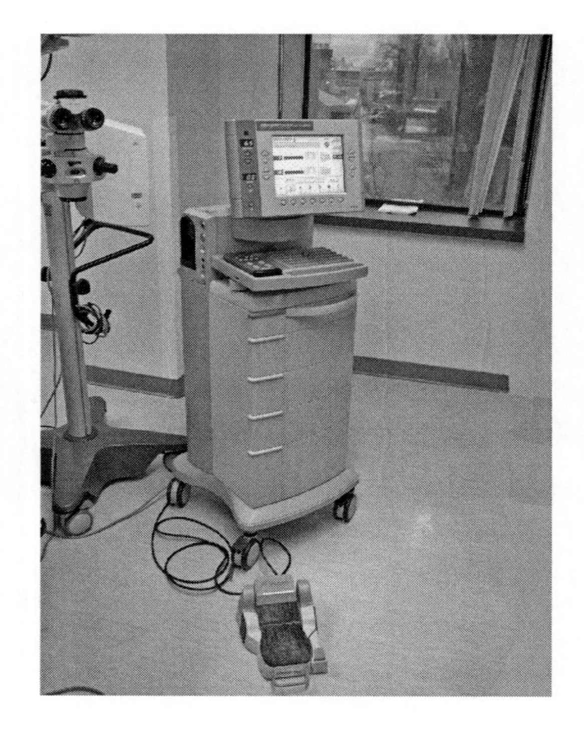

FIGURE 210
Phacoemulsifier machine with digital and video displays, connectors for tools, data entry keyboard, and foot control.

Defibrillators

The contraction of the heart is controlled by electrical signals, which must be of the correct size, timing, and distribution to cause coordinated and effective pumping. Many factors can disrupt these signals, such as blood electrolyte and/or gas levels, body temperature, various drugs, damage to the heart muscle or to the conductive pathways that distribute the signals, and physical or electrical shock. When the heart no longer beats effectively, blood flow is compromised, and serious harm or death can result. Note that the heart may continue to beat in some fashion, but not effectively (that is, it fibrillates or is in fibrillation), or the heart may cease to beat at all (asystole, "flat line"). Whatever the cause, it is essential that proper function be restored as soon as possible. This may be accomplished by medication, by physical intervention such as CPR or body temperature adjustment, or by applying an electrical signal to the heart that will restart or re-coordinate (defibrillate) its contractions. A defibrillator is a device that supplies such corrective signals (Figure 211).

Defibrillators may act on the heart directly, with electrodes applied to the cardiac tissue, or indirectly, with electrodes (called paddles) placed on the exterior of the patient's chest. The corrective signal, or shock, required is much, much greater when the electrodes are external, because a large portion of the signal will be blocked or redirected by the tissues

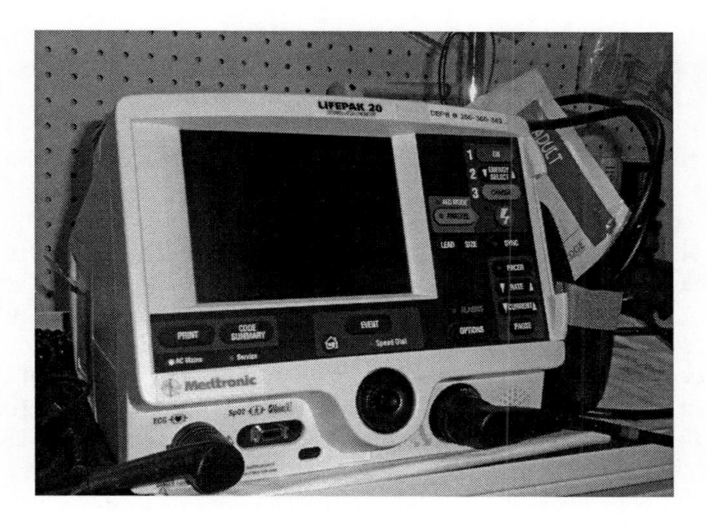

FIGURE 211
Defibrillator with controls for energy levels, EG functions, and discharge. A video screen provides functional information, as well as ECG waveforms.

between the electrodes and the heart itself. In most emergency situations, it is not possible to apply the electrodes directly to the heart, so most defibrillators are of the external type.

Internal defibrillators can be either implantable, in which case they must be capable of determining when shocks are necessary and administering them automatically, or manual, in which case the size and timing of the shock is determined by the operator. Most manual internal defibrillators are simply regular external defibrillators with specially designed spoon-like electrodes for direct application to the heart (for example, during open-heart surgery). These defibrillators must be capable of delivering the small-sized shocks required, and most have a circuit that detects when the internal paddles are being used to prevent the selection of higher shock levels.

Implantable defibrillators use wire electrodes that are embedded in the cardiac muscle at the appropriate location. They measure the ECG signal and analyze it to determine its characteristics. If these characteristics fall into a class that has been designated as abnormal but correctable, they will automatically administer an appropriately sized and timed shock to reestablish proper cardiac function. This is different from implantable pacemakers, which apply smaller signals in order to maintain proper rhythms. Generally, a pacemaker is used when the degree of damage to the heart that requires corrective action is less, whereas an implantable defibrillator is used when there is greater damage and more extensive intervention is required to restore proper function.

External defibrillators may also be automatic; this is usually a feature of a machine that can also be used in manual mode.

Manually controlled external defibrillators have some means of detecting and displaying and/or analyzing the ECG signals from the heart. This may be done directly through the paddles, or through separate ECG electrodes. Using separate electrodes generally produces more accurate ECG signals, but it takes time to apply them, and so the paddles themselves are used in situations that are more urgent. When separate electrodes are used, they must be located so as not to interfere with the optimum placement of the paddles. If the paddles are placed on top of the ECG electrodes, the shock can be reduced in effectiveness and/or burns can result to the patient's skin. Units often have a visual indicator of paddle contact quality.

Controls for the defibrillator include energy selection level, charge initiation, and discharge. The discharge buttons are usually a pair, one on each paddle handle, which must both be depressed to cause a discharge. Charge initiation and sometimes even energy selection controls may also be located on the paddles, as well as on the front panel.

These units have some kind of display of the ECG signal. This may be a simple visual or audible indicator of each beat, but usually it is a graphic display of the ECG waveform. Along with the graphic display, there may also be an audible beat indicator, whose volume can be adjusted or turned off. There is often a paper recorder that prints the signal and usually indicates the point at which the shock is administered and its size, as well as the date and time, which can be important for later medical analysis or legal considerations. The recorder often will have a memory function that allows it to print out several seconds of ECG signal from before the time the shock was applied, which gives a better picture of the event as a whole.

Defibrillation may be required to correct other abnormalities of the cardiac rhythm. The most common of these is when the atria of the heart are fibrillating but the ventricles are beating normally (atrial fibrillation or A Fib). In this case, the shock must be applied at a very precise time in relation to the ventricular beats. Because this would be almost impossible for a human operator to time correctly, defibrillators are often equipped with circuitry to detect the ventricular beats and apply the shock at the next correct time after the operator depresses the shock (or discharge) buttons. This procedure is called synchronized cardioversion. Because the proper timing requires an accurate ECG signal, the defibrillator will not enter this mode unless ECG electrodes are applied and used. The mode is normally called "sync" (synchronized), and there is usually a light flashing and/or tone sounding to indicate its selection. There is usually a "marker" signal placed on top of the normal contraction signals so that the operator can determine whether the timing is correct.

Because they are often used in emergency situations, defibrillators often have internal batteries that can provide power for all functions.

Defibrillators may have other functions built in, such as pulse oximetry (SpO2), in order to help caregivers better assess the condition of the patient.

Cardiac Pacemakers

Cardiac rhythm is normally controlled by a system within the heart, moderated by various parameters such as oxygen demand and levels of hormones such as adrenalin. The natural pacemaker system sends signals to the heart muscle in a pattern that produces coordinated contractions of the various parts of the heart, of a strength and rate appropriate to body state.

Various disease processes can disrupt this natural pacemaker to such a degree that cardiac contractions are no longer sufficient for the needs of the patient.

In these circumstances, an artificial pacemaker is used to provide proper pacing signals. Pacemakers may be temporary or permanent, and may be external or implanted. External pacemakers can be further divided into invasive and noninvasive types.

External pacemakers are typically used for short-term applications, either until the patient's natural pacemaker can resume normal function or until an implantable pacemaker can be installed, with noninvasive types being used for shorter times than invasive.

Noninvasive external pacemakers use electrodes placed in specific places on the patient's chest to pass electrical signals into the heart. These signals stimulate the heart to beat more effectively and are usually coordinated with whatever natural cardiac signals are present. They can be adjusted for rate and amplitude, and may either completely control cardiac contractions or act as a "booster," filling in for missing beats as required. The control signals have to be quite large for enough of the signal to reach the heart, and the signal passes through areas of the body where it is not needed. Long-term use of electrodes on the patient's skin can cause irritation or burns.

Invasive external pacemakers function in a similar way to noninvasive types, except their signals are carried to the heart by wires inserted into the patient's body and attached directly to the heart. They have the advantage of more precise control and require much less power to effect pacing compared to noninvasive types, but they take much longer to apply, as well as carrying the problems associated with any invasive procedure.

Implantable pacemakers (Figure 212) perform the same function as their external counterparts, but have many special restrictions. They are inside the patient's body and therefore relatively inaccessible. This means that they must have a long-lasting power source; even though the power requirements are small, they may be needed for years. Special battery technology has been developed, and nuclear power sources have been used. The units must have some means of being controlled without physical contact. In earlier devices, and many current ones, rate and signal strength could be adjusted using a powerful magnet placed near the implant site. Newer units may have mechanisms to measure body needs and adjust themselves automatically according to these needs. Implanting a foreign object within the body has its own set of considerations apart from the pacemaking functions, and these factors are an important part of the design criteria as well.

FIGURE 212
Implantable pacemaker module.

Ventilators

In response to disease conditions or trauma, the ability of a patient to breathe may be reduced or nonexistent. In these cases, ventilation must be provided artificially, either through mouth-to-mouth means, via a "breathing bag" which is pumped by hand, or by a mechanical ventilator. For situations where artificial ventilation is required for more than a few minutes, a mechanical ventilator is preferred.

Ventilators range from relatively basic devices that simply provide a properly timed boost in air pressure to assist patients in drawing air into their lungs, up to very complex units with a number of variable parameters, built-in compressors and oxygen blenders, monitoring and alarms systems, and battery backup power.

The goal of all systems is the same: to provide adequate ventilation to patients, while minimizing harm to their lungs and associated structures. Minimizing harm is especially important in situations where ventilation may be required for periods of days, months, or years.

Most full-featured ventilators can operate in a variety of modes, depending on the needs of each patient (Figure 213).

Breaths may be delivered by the ventilator at a preselected rate, when the patient goes too long without taking a breath unassisted, or whenever the patient makes an effort to draw a breath (within set limits).

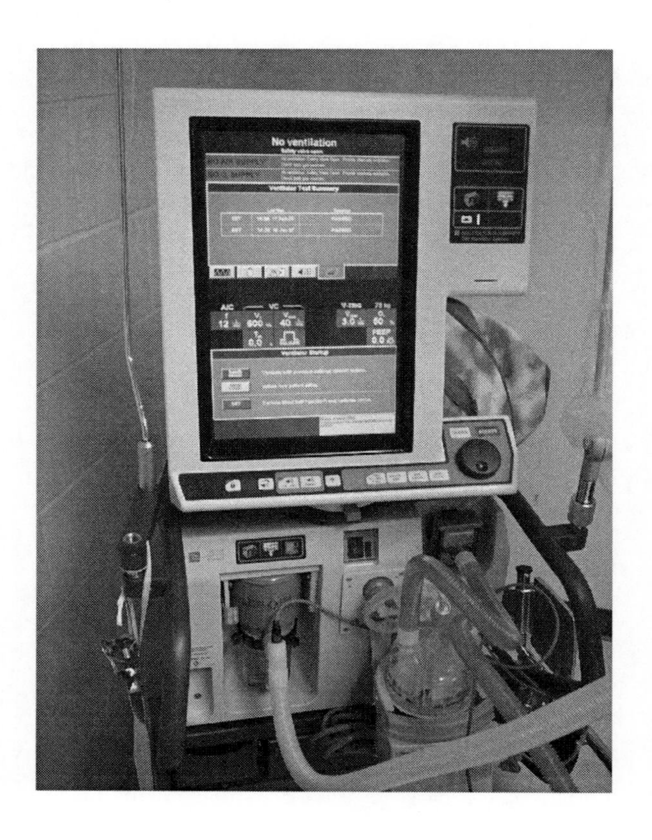

FIGURE 213
Full-function patient ventilator with video screen for patient information and settings display, parameter controls, filters and humidifier, and patient connection circuit.

Because ventilation is obviously a critical factor, ventilators must be designed to be highly reliable, in both normal and emergency situations. Ideally, they should be able to continue to function if line power or wall oxygen pressure should fail, and should have redundant critical components to minimize the risk of failures. Alarms must also be reliable, and must be designed so that they cannot be turned off.

When a patient is being ventilated, it may be important to monitor the oxygen content of delivered air, blood oxygen concentration, and expired CO_2 levels. These functions may be performed by separate devices or by ones integrated into the ventilator; CO_2 levels are usually measured by a separate device. All of these functions may be performed by a physiological monitor; some such monitors can interface with the ventilator, allowing for recording of ventilation parameters and integration of alarms into a central monitoring system.

Anesthetic Machines

During surgery, there are a number of important considerations for the well-being of the patient, many of which are addressed by the attending anesthesiologist using an anesthetic machine (Figure 214).

Primary among these considerations is the elimination of the sensation of pain, which is accomplished by the administration of one or more gases that render the patient unconscious (Figure 215).

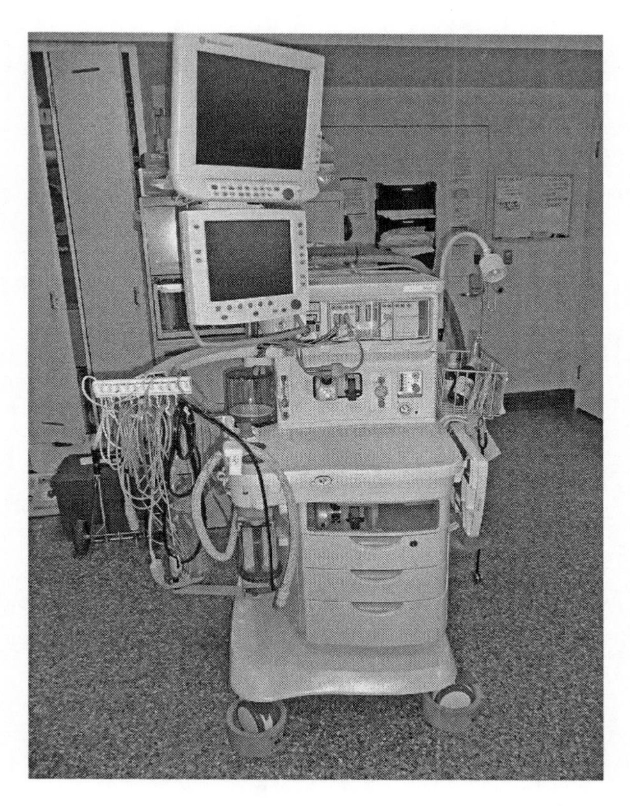

FIGURE 214
Anesthetic machine.

FIGURE 215
Sevoflurane vaporizer.

The anesthetic machine provides a source for these gases, regulates their pressure and flow, mixes them with oxygen, adds humidification if required, and delivers the final mixture to the patient. Because the levels of the various components are critical, machines often have systems to measure and display percentages and flow rates, with associated alarms for high or low levels. This function may also be performed by an auxiliary unit, either stand-alone or integrated into a multiparameter monitor (Figure 216).

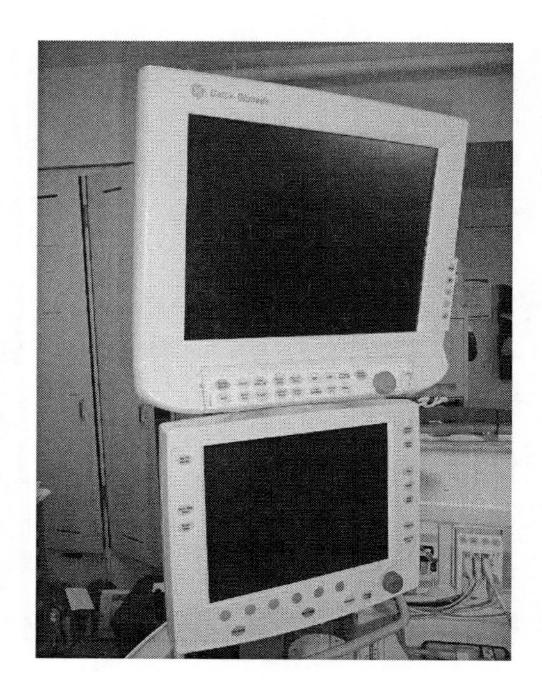

FIGURE 216
Anesthetic machine video displays. In this unit, the upper screen displays patient information including ECG and respiration numerics and waveforms, while the lower screen is concerned with system settings and performance.

Hypothermia is a common reaction during surgery, and so air provided to the patient may be heated; in this case, humidification is especially important to prevent drying of airway tissues.

Some surgical procedures require that the patient's muscles be immobilized to prevent unwanted contractions in response to either the physical trauma of surgery or to the electrical stimulation of electrosurgery units. Drugs to accomplish this are administered, but they paralyze breathing muscles as well, which means that the anesthetic machine must be able to provide artificial ventilation at a controlled rate and volume. This ventilation is delivered via an endotracheal tube, which in these situations also carries the anesthetic gases.

Gases exhaled by the patient consist of nitrogen, carbon dioxide, and unused anesthetic agent. In order to reduce the amount of agent required, the exhaled gases are recirculated, which means that the CO_2 must be removed and oxygen added. The CO_2 removal is accomplished by passing the gases through a canister containing soda lime granules, which absorb only the CO_2 (Figure 217). When the granules are saturated, they change color, at which time the canister can be reverse-flushed and made ready for additional use.

Anesthetic machines must also provide access to monitoring of vital signs such as ECG, blood pressure, temperature, blood oxygen saturation, and expired CO_2 levels. These functions may be built into the anaesthetic machine itself, or the machine may simply be a platform on which an independent physiological monitor is mounted.

FIGURE 217
Carbon dioxide absorber module.

Appendix B: Normal Values and ECG

Normal Values and ECG

This section is included to give perspective on some of the more common medical measurements. "Normal values" is the range of values found in healthy, normal individuals.

Hematology—Red Blood Cells (Erythrocytes)	
RBC (Male)	4.2–5.6 M/μL (million cells/μL)
RBC (Female)	3.8–5.1 M/μL
RBC (Child)	3.5–5.0 M/μL
Hematology—White Blood Cells (Leukocytes)	
WBC (Male)	3.8–11.0 K/mm^3 (thousand cells/mm^3)
WBC (Female)	3.8–11.0 K/mm^3
WBC (Child)	5.0–10.0 K/mm^3
Hemoglobin	
Hgb (Male)	14–18 g/dL
Hgb (Female)	11–16 g/dL
Hgb (Child)	10–14 g/dL
Hgb (Newborn)	15–25 g/dL
Hematocrit	
Hct (Male)	39–54%
Hct (Female)	34–47%
Hct (Child)	30–42%
General Chemistry	
Bilirubin, total	0.2–1.4 mg/dL
BUN	6–23 mg/dL
Calcium (total)	8–11 mg/dL
Carbon dioxide	21–34 mEq/L (milliequivalents/L)
Carbon monoxide	symptoms at ≥10% saturation
Chloride	96–112 mEq/L
Ethanol	0 mg%; coma at ≥400–500 mg%
Glucose	65–99 mg/dL
HDL (Male)	25–65 mg/dL
HDL (Female)	38–94 mg/dL
Potassium	3.5–5.5 mEq/L
Sodium	135–148 mEq/L
Urea nitrogen	8–25 mg/dL

Lipid Panel (Adult)	
Cholesterol (total)	<200 mg/dL desirable
Cholesterol (HDL)	30–75 mg/dL
Cholesterol (LDL)	<130 mg/dL desirable
Triglycerides (M)	>40–170 mg/dL
Triglycerides (F)	>35–135 mg/dL

Urine	
Specific gravity	1.003–1.040
pH	4.6–8.0
Na	10–40 mEq/L
K	<8 mEq/L
Cl	<8 mEq/L
Osmolality	80–1300 mOsm/L
Glucose	≥180 mg/dL)

Cerebrospinal Fluid	
Osmolality	290–298 mOsm/L
Pressure	70–180 mm H_2O

Hemodynamic Parameters	
Cardiac index	2.5–4.2 L/min/m^2
Cardiac output (Q)	4–8 L/min
Stroke volume	60–100 mL/beat
Systolic arterial pressure	90–140 mm Hg
Diastolic arterial pressure	60–90 mm Hg
Central venous pressure	2–6 mm Hg
Ejection fraction	60–75%
Left atrial pressure	4–12 mm Hg
Right atrial pressure	4–6 mm Hg
Pulmonary artery (PA) systolic pressure	15–30 mm Hg
PA diastolic pressure	5–15 mm Hg
PA mean pressure	10–20 mm Hg
PA wedge pressure	4–12 mm Hg
PA end diastolic pressure	8–10 mm Hg
Right ventricular end diastolic pressure	0–8 mm Hg

Neurological Values	
Intracranial pressure	5–15 mm Hg

Blood Gases—Arterial Values	
pH	7.35–7.45
$PaCO_2$	35–45 mm Hg
HCO_3	22–26 mEq/L
O_2 saturation	96–100%
PaO_2	85–100 mm Hg

Blood Gases—Venous Values

pH	7.31–7.41
$PaCO_2$	41–51 mm Hg
HCO_3	22–29 mEq/L
O_2 saturation	60–85%
PaO_2	30–40 mm Hg

Basic Vital Signs

- Cuff blood pressure—As measured by automatic or manual cuff-inflation methods at the arm. Normal for adults is considered to be 120 mm Hg systolic and 80 mm Hg diastolic, expressed as "120 over 80" or 120/80. Adult pressure is usually considered to be borderline high when pressure is above 140/90, and treatment is suggested for pressures over 160/100. Extreme hypertension can produce pressures as high as 230/135. Low-normal adult blood pressure is 100/65, though some athletes may have pressure in this range as a normal condition. Children tend to have pressures in this range as well. Hypotension is below about 90/60, and patients are likely to enter a coma state if pressures fall below about 50/30. Note that different methods of measuring cuff pressure, or even different people using the same method, can produce somewhat different values. Automatic noninvasive blood pressure machines usually give repeatable results, but these may not agree perfectly with results obtained manually. Values from both ways of measuring cuff pressure should be taken as approximate and used more to detect trends than to represent true, high-accuracy pressures.
- Pulse rate—Pulse rate varies over a large range depending on the individual, their state of health, and amount of exercise being performed. Normal resting rates for healthy adults can range from 60 to 100 beats per minute (bpm), with athletes generally having lower rates, in the 45 to 55 range. Children have higher resting rates than adults, and in utero, a fetus might have a normal pulse rate of around 200 bpm. Exercise increases pulse rates, as high as 150 or even 200 bpm, while rates often drop during sleep. Normal, nonathlete adult resting rates of below about 60 bpm is called bradycardia, while such rates above about 100 bpm is called tachycardia.
- Respiration rate—Like pulse rate, respiration rate can vary over a wide range, with exercise increasing rates and rates for children being generally higher than those for adults. Normal resting rates for adults are between 10 and 20 breaths per minute, while for infants rates may be between 20 and 40 breaths per minute. During exercise, rates may increase to about 45, with athletes reaching up to 70 breaths per minute.
- Temperature—Body temperature can be measured at a variety of points in the body, including the temporal area of the head, under the tongue, in the rectum or vagina, at the outer eardrum, or at internal organs during tests or surgery. Normal oral temperature is considered to be 36.8°C, plus or minus 0.7°, although some individuals may have healthy temperatures somewhat outside of this range. Core body temperature is usually about 1° higher than oral temperature. Body temperature follows a daily or diurnal rhythm, with lowest temperatures occurring during deep sleep. A temperature of over about 100°F is called hyperthermia

and can result in a severe medical crisis, while temperatures below about 34°C is called hypothermia and can again cause major problems. Exercise, emotional state, ambient temperature, various disease processes, and some drugs can all affect body temperature.

- Blood oxygen saturation—This parameter is measured with a device called a pulse oximeter, which passed different wavelengths of light through a well-perfused body part, such as a finger or earlobe. Analysis of the transmitted light provides a reasonably accurate measure of the percentage ratio of oxygen in the patient's blood, with 100% being the maximum possible.

ECG Measurements and Arrhythmias

ECG Signals

The electrical signals within the heart can give important information concerning the state of that vital organ. These signals can be detected on the skin of the patient's chest, where they can be recorded and measured via an electrocardiogram (ECG) machine or monitor. The waveform has a typical shape, and the parts of the waveform have been given labels: P, Q, R, S, and T (Figure 218). As muscles go through a cycle of contraction and relaxation, they develop an electrical signal that is either positive (polarization) or negative (depolarization).

The P wave is caused by the atria depolarizing. It is usually smooth and positive and has a short duration, less than 0.12 sec.

The time between the P wave and QRS portion is called the PR interval and is normally about 0.12 to 0.20 sec long.

The QRS complex is a result of depolarization of the ventricles. It normally is about 0.04 to 0.12 sec long.

The ST segment is the time during which the ventricles contract. The length and shape of this segment of the waveform in particular is often measured to give further information regarding cardiac health.

The QT interval begins at the onset of the QRS complex and extends to the end of the T wave. It indicates the time from which the ventricles depolarize until they start to repolarize.

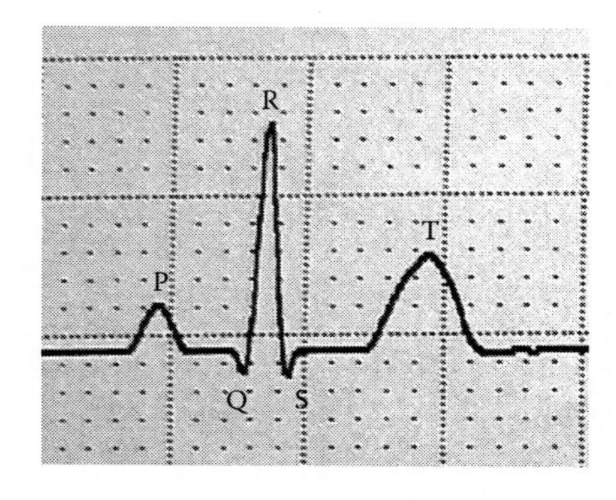

FIGURE 218
ECG QRS complex waveform.

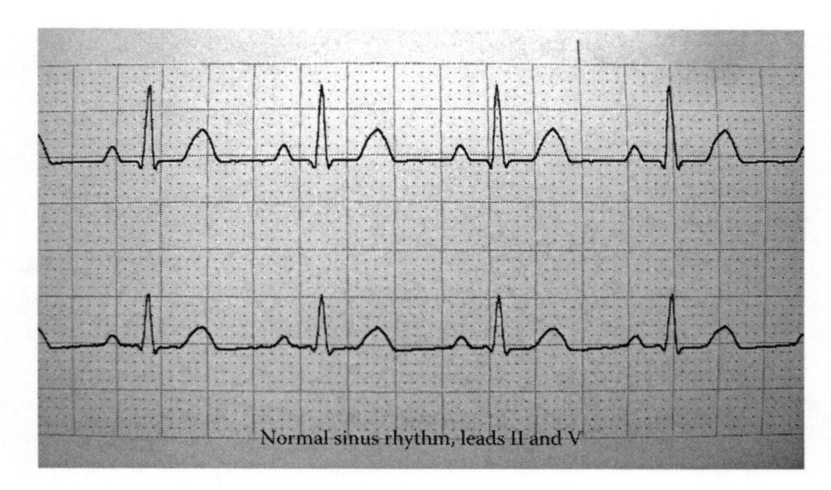

FIGURE 219
ECG waveform recording with traces from two different leads.

The T wave is a result of ventricular repolarization. It is normally rounded and positive (Figure 219).

Arrhythmias

Any variation from a normal ECG rhythm (or normal sinus rhythm) is called an arrhythmia. Arrhythmias range from occasional odd beats that do not affect the well-being of the patient and can be present for years, to critical patterns that, if not corrected immediately, will result in death.

Arrhythmias may also simply be a normally shaped waveform that, in a patient at rest, is too fast (tachycardia) or too slow (bradycardia).

Arrhythmias can arise from a variety of causes: an imbalance in the chemicals that are involved in cardiac function; damage or interruption to the signal pathways in the heart; faults in the systems that control heart activity; or damage to the myocardium, either through inadequate blood flow (ischemia) or complete loss of blood supply to a particular area of the heart, which results in muscle tissue death (necrosis), called a cardiac infarction.

Some of the more common arrhythmias include:

- Asystole—This is not really an "arrhythmia" because there is no rhythm at all. Asystole means the complete cessation of any contractions—the patient is "flat lined." There is no blood flow, so obviously it is a critical situation and must be treated immediately to prevent death, usually by using a defibrillator.
- Atrial Fibrillation (A Fib)—Fibrillation is an uncoordinated contraction of various groups of muscles in the heart, in this case those of the atria. A Fib is not immediately life threatening, since the ventricles continue to contract normally, pumping a significant amount of blood.

- Ventricular Fibrillation (V Fib)—Fibrillation of the muscles of the heart ventricles. This is an immediately life-threatening condition which must be treated within minutes, since there is almost no blood flow.
- Ventricular Tachycardia (VT or V Tach)—A rapid heart rate that is initiated within the ventricles, usually caused by serious heart disease. V Tach is potentially life threatening and must be treated promptly. Usually considered as five or more consecutive PVCs (see below).
- Premature Ventricular Contractions (PVCs)—These extra beats originate spontaneously within the ventricles and can give the sensation of the heart "skipping" a beat. PVCs are relatively common, especially in younger people, and often disappear with time.
- Bigeminy—This rhythm consists of alternating normal beats and PVCs.

Glossary for Normal Values

Bilirubin: A by-product of the breakdown of red blood cells, it is normally destroyed by the liver. High levels can indicate liver problems. High concentrations of bilirubin in the body can cause the skin to turn yellow (jaundice).

BUN: Blood Urea Nitrogen.

Cardiac index: The amount of blood pumped by the heart per unit of time divided by body surface area, usually expressed in $L/min/m^2$.

Cardiac output: A measurement of blood flow through the heart, expressed in L/min.

Central venous pressure: The blood pressure in the right atrium or veins near the heart, mainly the vena cava.

Cerebrospinal fluid: A clear, colorless fluid containing small amounts of glucose and protein. Cerebrospinal fluid fills the ventricles of the brain, spaces between the brain and the cerebral membranes, and the central canal of the spinal cord. It acts as a shock absorber as well as carries some nutrients and materials involved in immune response.

Cholesterol: A fatlike steroid alcohol found in animal fats and oils, in bile, blood, brain tissue, milk, yolk of egg, myelin sheaths of nerve fibres, the liver, kidneys, and adrenal glands. It is the main component of most gallstones and is involved in atherosclerosis (hardening of the arteries.)

Diastolic arterial pressure: The peak pressure reached in arteries close to the heart (such as the ascending and descending aorta), corresponding to the maximal contraction of the ventricles.

Ejection fraction: A measure of the ability of the ventricles to contract.

HCO_3: Bicarbonate. A chemical produced in red blood cells, liberated through exchange with chloride. The kidneys may then excrete it.

HDL: High-Density Lipoprotein ("good cholesterol").

Hemoglobin (Hgb): A molecule involved in carrying oxygen in the blood; an iron atom forms part of the molecule.

Hematocrit (Hct): The relative volume of blood occupied by red blood cells.

Intracranial pressure: The pressure exerted by cerebrospinal fluid within the skull.

Mole: A mole is the quantity of anything that has the same number of particles as are found in exactly 12 grams of carbon-12. That number of particles is referred to as Avogadro's number, which is 6.02×10^{23}.

O_2 saturation: A measure of the amount of oxygen carried by the blood compared to the maximum theoretically possible, expressed as a percentage.

Osmolality: The concentration of osmotically active particles in solution expressed in terms of osmoles of solute per kilogram of solvent.

Osmole: A unit indicating the number of moles of a chemical compound that contribute to a solution's osmotic pressure.

$PaCO_2$: Partial pressure of carbon dioxide in the blood.

PaO_2: Partial pressure of oxygen in the blood.

pH: A logarithmic measure of the activity of hydrogen ions in a solution, which also indicates relative acidity/alkalinity. pH 7 is neutral; pH 0 is extremely acidic; pH 14 is extremely alkaline.

Pulmonary artery wedge pressure: The pressure measured in the pulmonary artery when blood flow in the artery is occluded, usually by a small balloon on the tip of a catheter. This catheter also contains a pressure transducer.

Specific gravity: The density of a liquid compared to water, pure water having a specific gravity of 1.

Stroke volume: The amount of blood pumped out of one ventricle of the heart as the result of a single contraction.

Systolic arterial pressure: The low reached in the cycle of pressure in arteries close to the heart (such as the ascending and descending aorta), corresponding to the maximal relaxation of the ventricles.

Triglycerides: Glycerides in which the glycerol has three fatty acid molecules attached. They are the main constituents of vegetable oil and animal fats.

Appendix C: General Anatomy

Basic anatomical drawings of major body systems are provided for reference.

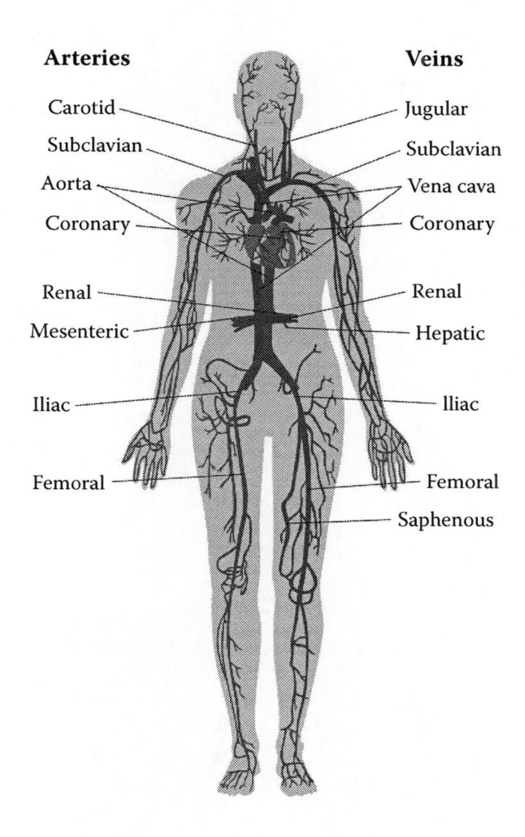

FIGURE 220
Circulatory system.

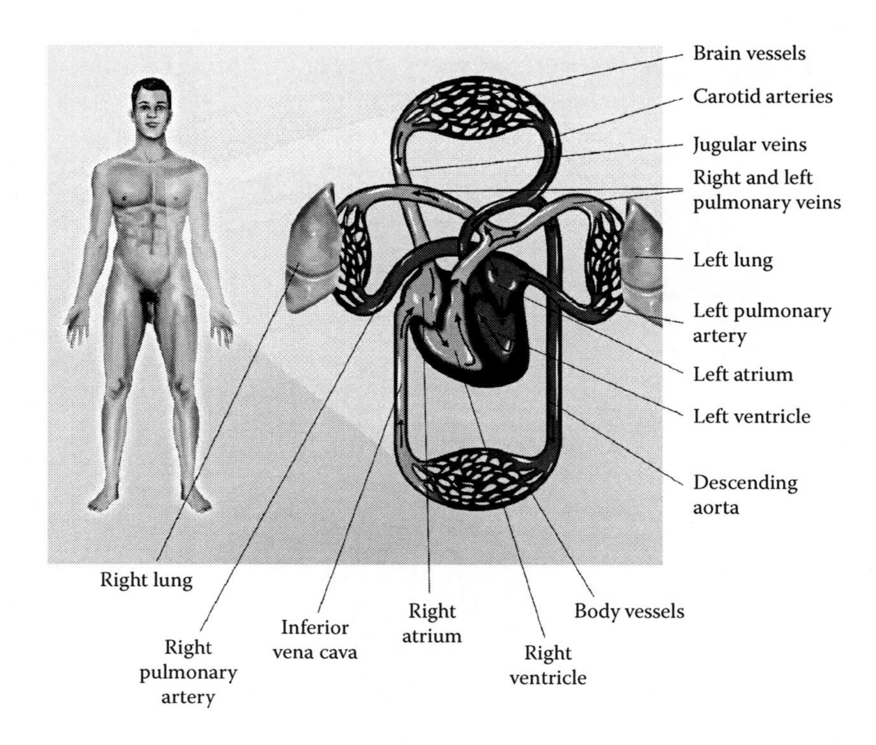

FIGURE 221
Circulatory system schematic.

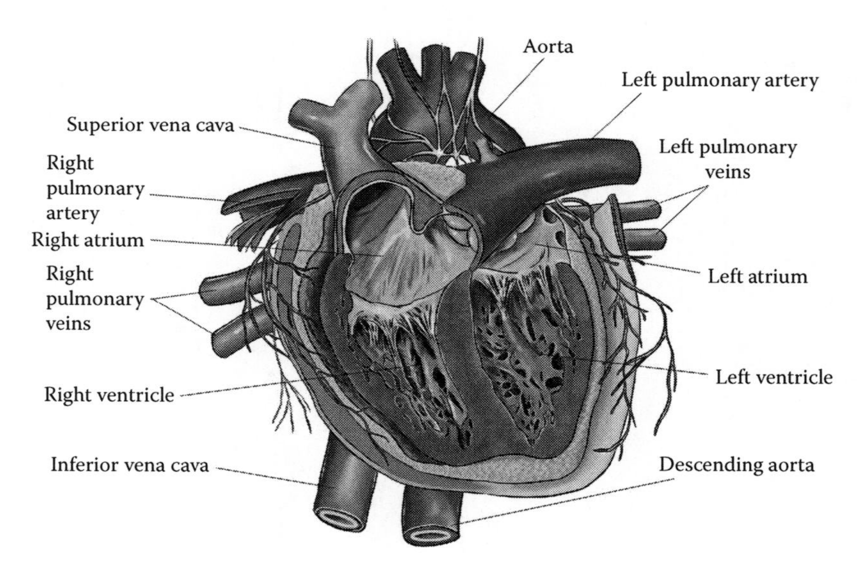

FIGURE 222
Heart cross section.

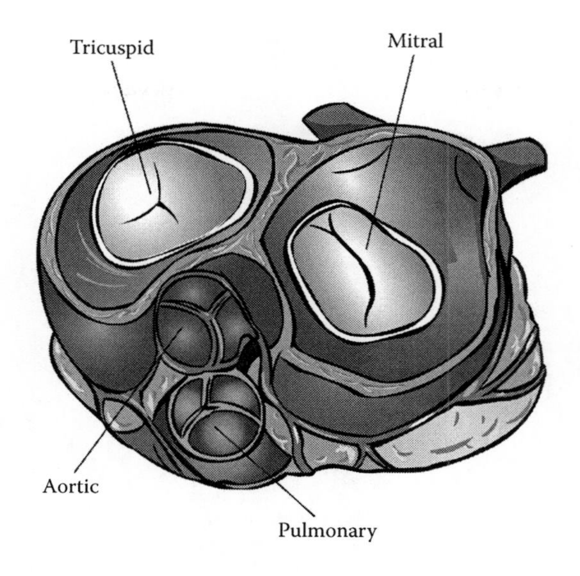

FIGURE 223
Heart valves.

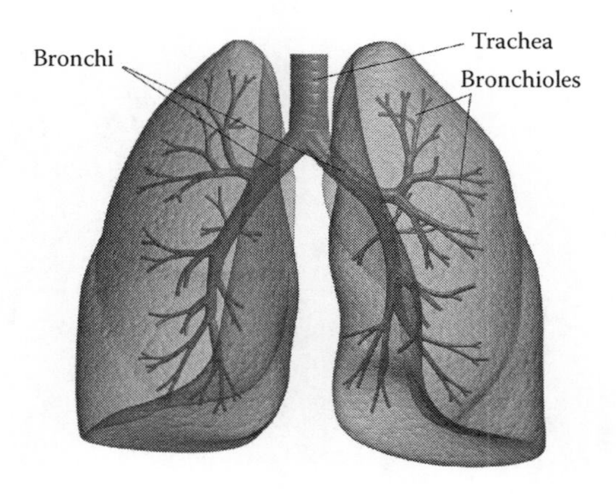

FIGURE 224
Respiratory system.

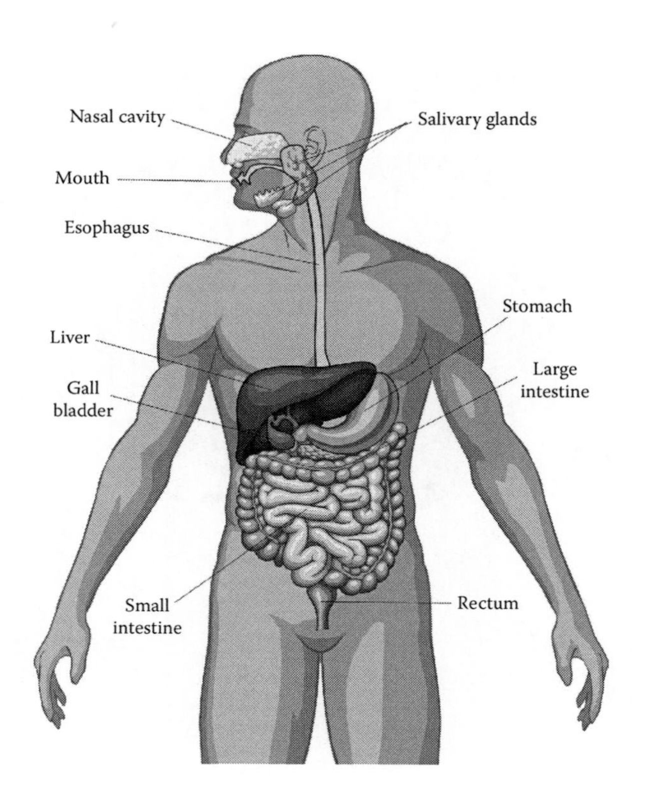

FIGURE 225
Digestive system.

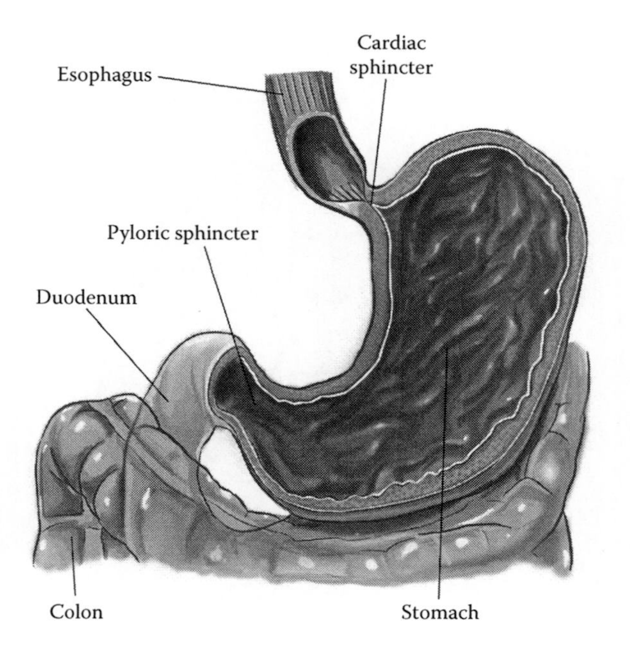

FIGURE 226
Stomach.

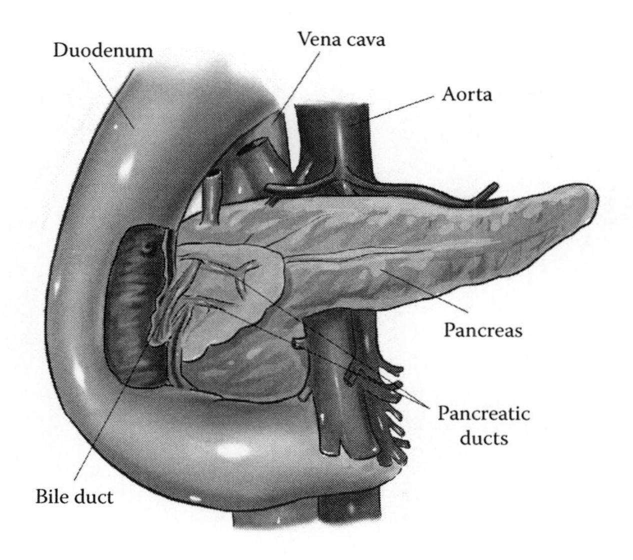

FIGURE 227
Pancreas.

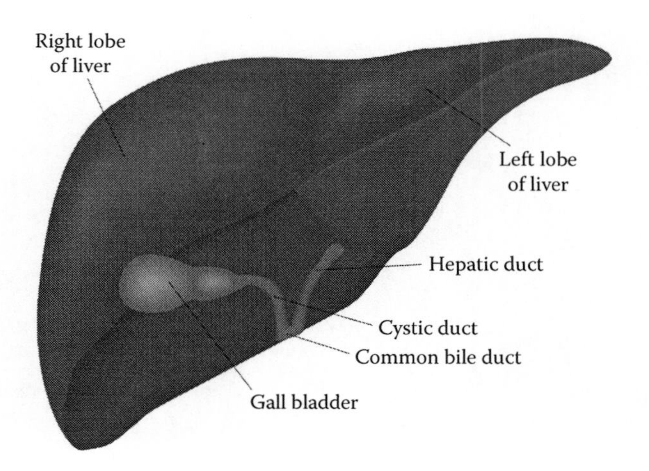

FIGURE 228
Liver and gall bladder.

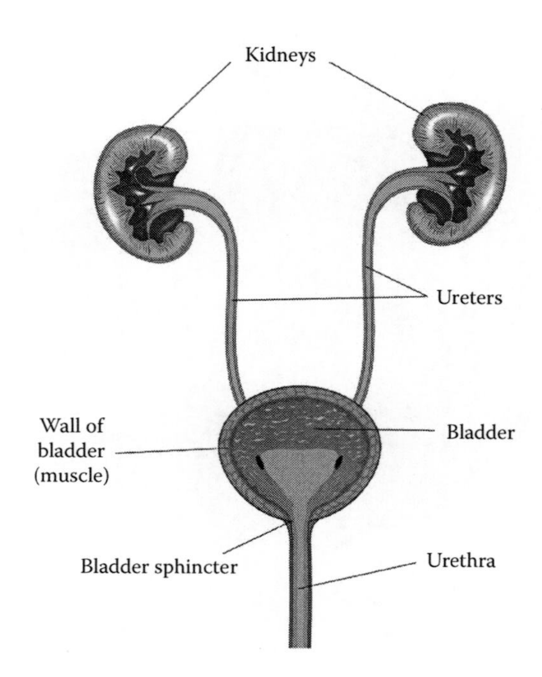

FIGURE 229
Urinary system.

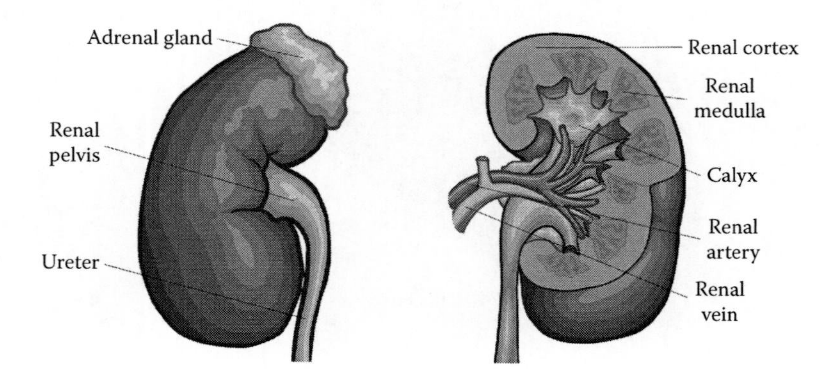

FIGURE 230
Kidneys.

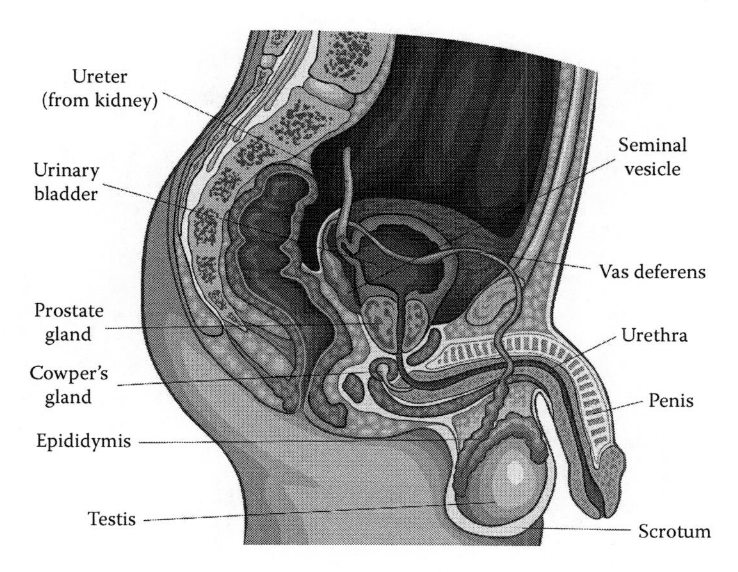

FIGURE 231
Male reproductive system.

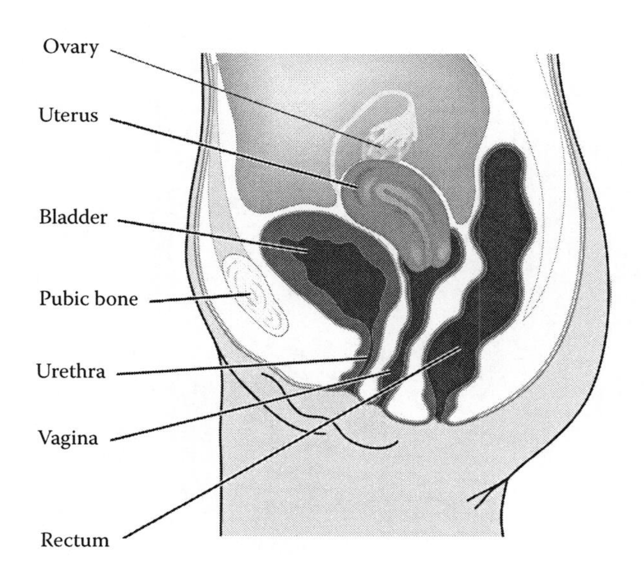

FIGURE 232
Female reproductive system.

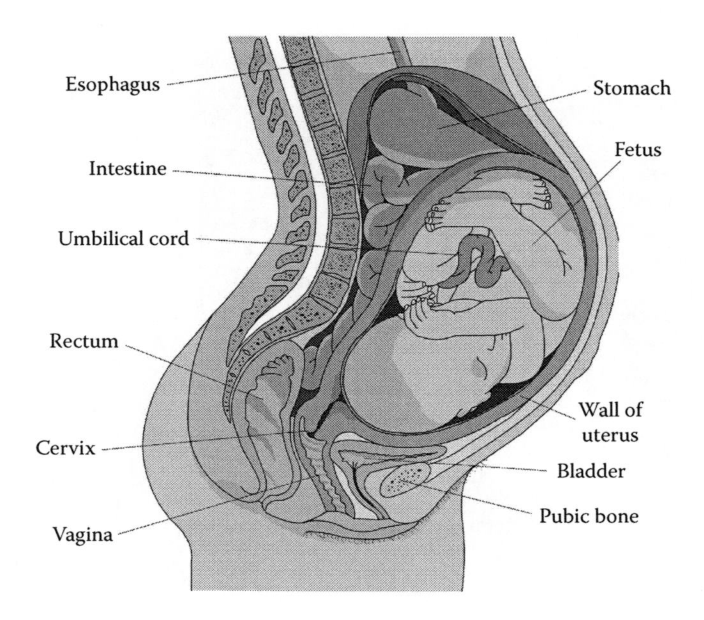

FIGURE 233
Pregnant anatomy.

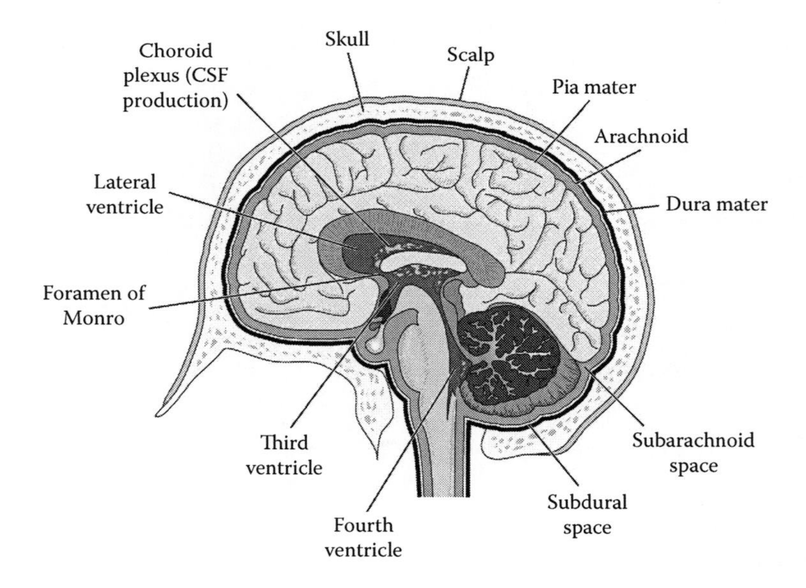

FIGURE 234
Brain cross section.

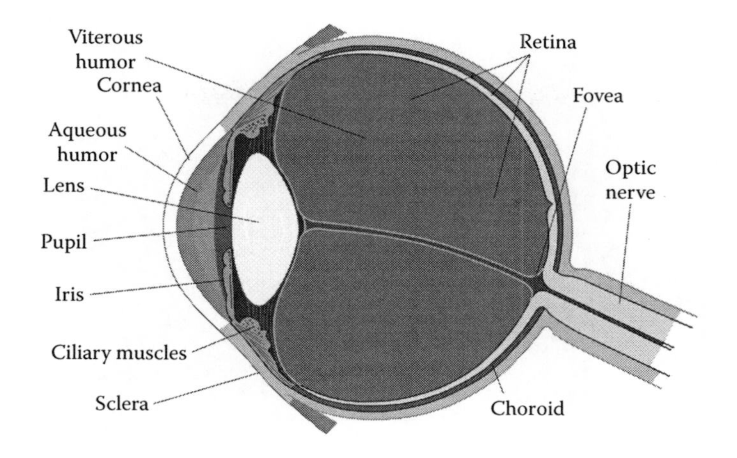

FIGURE 235
Eye cross section.

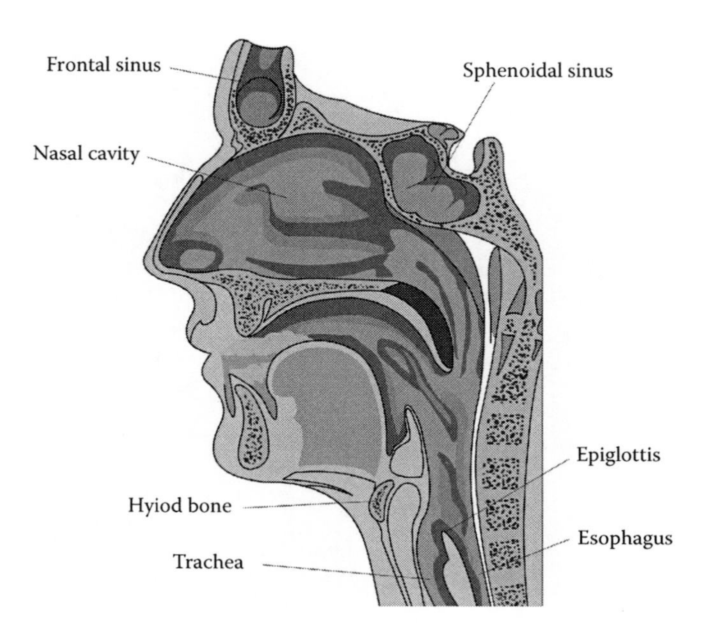

FIGURE 236
Nose, mouth, and throat cross section.

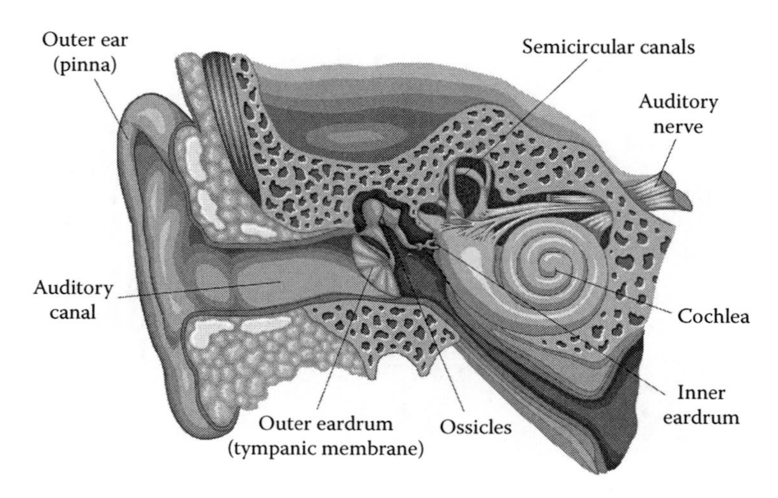

FIGURE 237
Ear cross section.

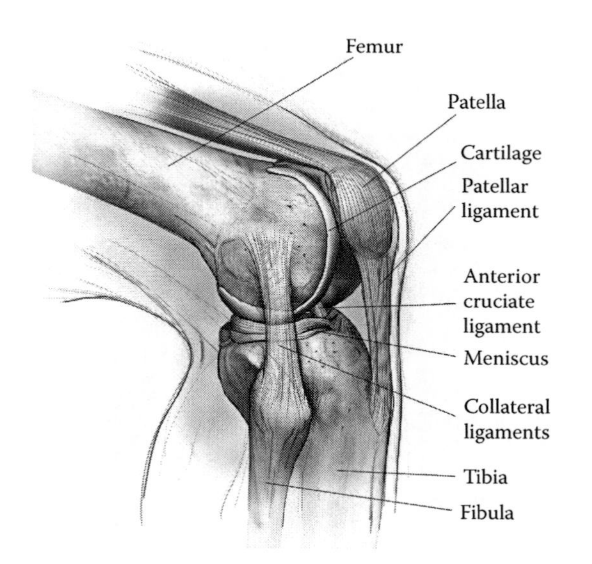

FIGURE 238
Knee joint.

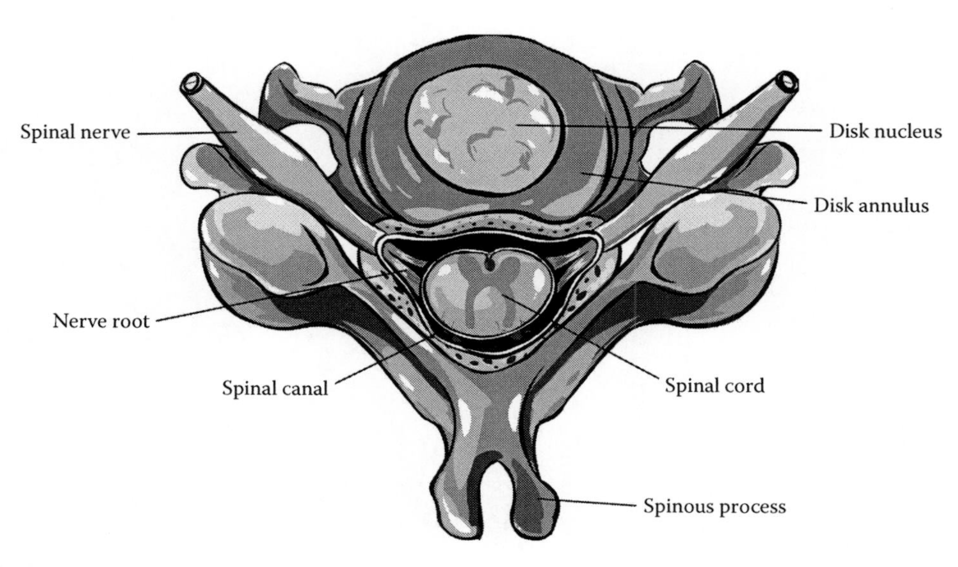

FIGURE 239
Spinal column lateral section.

Appendix D: Sterilization, Sterile Technique, and Isolation

Guiding principle—Any object, surface or person is assumed to be nonsterile until shown or made to be sterile. Anything is either sterile or nonsterile: there is no in-between. If in doubt, it is nonsterile.

Sources of infection in hospitals—Infectious microorganisms (IMs), which include bacteria, viruses, fungi, and parasites, usually originate from human sources. Patients, visitors, and staff may have these organisms present in their body, either as an active disease or as an otherwise healthy carrier. Infectious agents can pass from one person to another by a variety of means. Sneezing and coughing produce aerosol droplets that can contain IMs; these droplet can be inhaled by others, spreading the infection. IMs can also be found in body fluids such as mucus, and wiping one's nose or mouth and then touching a door handle or countertop can allow them to be passed on. Improper toilet practices can allow pathogens from feces to be introduced to a variety of locations, from which they can infect others. Blood droplets or spatters can be especially dangerous, if they are from an infected person. When electrosurgery machines or surgical lasers are used in surgery, smoke particles are produced which can contain active pathogens. Finally, harmful organisms may be present within an individual's body, but only cause problems when the person's immune system is weakened or when the normal balance of organisms is upset.

Nosocomial infections—Infections contracted by patients while in hospital are called nosocomial infections and are an increasingly serious problem. It is estimated that in the United States almost 2 million people annually contract nosocomial infections, and up to 100,000 of them die (The Centers for Disease Control, http://www.cdc.gov/ncidod/dhqp/hai.html). This makes them one of the leading causes of death in the United States, and the human and financial costs are enormous. Health care workers may also contract diseases from patients, visitors, or other staff members with communicable diseases. Given these factors, methods to reduce such infections are extremely important. New treatments are always being developed to fight infections, but prevention is by far the best tool.

Superbugs—Making the problem of infection more difficult is the emergence of antibiotic-resistant organisms, or superbugs. Bacteria can mutate rapidly and evolve ways of avoiding the effects of various antibiotics, changing with each new medication to seemingly always stay one step ahead. Some organisms are named for their resistance to specific antibiotics, such as *methicillin-resistant Staphylococcus aureus* (MRSA) and vancomycin-resistant enterococcus (VRE). In a kind of revere situation, the bacteria *Clostridium difficile* (usually referred to as *C. difficile*) causes diarrhea in patients who are being treated with antibiotics for other diseases. The antibiotics alter the balance of beneficial bacteria in the gut, allowing *C. difficile* to flourish and produce harmful symptoms, which can even lead to death.

Prevention—There are three means of reducing infections: having all people in the hospital always practice good sterile techniques; using effective isolation for infected patients;

and following proper cleaning procedures for rooms, fixtures, bedding and clothing, tools, and instruments.

Sterile Technique

- Basic sterile technique simply consists of keeping clean, washing hands thoroughly between patients, after using the restroom, before handling food or medications, and on a regular schedule otherwise. Hospitals have strict guidelines for staff to follow, and post signs in public areas, especially restrooms, reminding all people to wash properly and providing instructions and diagrams to follow. Cleaning materials are kept readily available, and hand sanitizer dispensers are placed almost everywhere with signs encouraging their use.
- Formal sterile technique procedures are intended for staff members and visitors who will be entering areas in which infection could easily occur, such as rooms of patients with comprised immune systems or open wounds or burns, or the operating room. Very specific rules involve handling supplies; contacting a patient's open tissue, mucous membranes or body fluids; disposing of infected or possibly infected materials; deciding what is and what is not sterile; using protective clothing such as gowns, masks, foot and hair coverings, and gloves; and hand washing. Staff members working in infection-prone areas will be thoroughly trained and periodically refreshed on these rules and techniques.

Isolation—Patients who have or may have contagious diseases, or who have compromised immune systems and are thus susceptible to infections, are placed in special isolation rooms. The stringency of isolation practices is determined by the virulence of the disease involved, or by the degree to which the patient's immune system is weakened.

All persons entering isolation rooms use formal sterile technique, and the rooms themselves are designed to minimize the possibility of disease transmission.

The fact that a room is in use for isolation is made clear with signs and labels, and staff members will be vigilant in ensuring that everyone follows proper protocols. Rooms may have automatic doors, possibly even two layers of doors in especially critical situations, which cannot be blocked or held open for more than a very short time without triggering an alarm.

Airflow in the room may be controlled to produce negative pressure within the room, so that whenever the door is opened, air flows into the room and not out. Alarms signal if this negative pressure is not maintained. Special filters clean the air of any infectious material before it is exhausted to the outside.

Extra care and special methods are used for cleaning the room and contents, and for caring for the patient.

Sterilization methods—Instruments and supplies used on patients must be sterile. If they are new, the manufacturer or distributor will be required to follow strict guidelines in preparing and packaging materials, and packaging will be marked with a date beyond which sterility cannot be assured. Any damage to sterile packaging materials causes them to be considered nonsterile. Packaging may include special indicators that change color if sterility is no longer assured.

Supplies or equipment to be reused on different patients must be sterilized between uses.

There are four basic methods of sterilization, which are required because different materials or devices may not be able to withstand some of the methods.

- Gas sterilization, in which articles are placed in a temperature- and humidity-controlled enclosure to which ethylene oxide (ETO) gas is added. The articles stay in the enclosure for several hours, after which they are packaged using sterile methods.
- Heat and steam sterilization, also called autoclaving, uses steam at high pressures, which allows for higher temperatures than the boiling point of water in open air. Most plastic or latex objects will not withstand autoclaving, so other methods must be used with them.
- Liquid sterilization, in which articles are immersed in a glutaraldehyde solution for a specified time. Glutaraldehyde is toxic, and so processes must be used following sterilization to remove all traces of it from the articles being treated, by rinsing with materials such as detergents, sterile water, and alcohol. Instruments may be cleaned by simple immersion in alcohol, but this does not constitute true sterilization.
- Radiation, in which objects are placed in a shielded container and exposed to ionizing radiation of a specified strength, for a specified time. This method requires the proper precautions for using radiation and is only used in some specific situations.

Careful and regular testing of the efficacy of any sterilization technique is critical to ensure that infectious organisms are reliably eliminated.

Appendix E: Bibliography

Selected Texts for Reference

Burkitt, H. George, Clive R.G. Quick, and Joanna B. Reed. *Essential Surgery—Problems, Diagnosis and Management,* 4th ed. Illustrated by Philip J. Deakin. Churchill Livingstone Elsevier, 2007. ISBN 9780443103469.

Frey, Kevin, and Teri L. Junge, Senior Editors. *Surgical Technology for the Surgical Technologist,* 2nd ed. Thomson Delmar Learning, 2004. ISBN 1-4018-3848-0.

Hansen, John T., and David R. Lambert. *Netter's Clinical Anatomy.* Illustrated by Frank H. Netter. Saunders, 2005. ISBN 1-929007-71-X.

Montgomery, Royce L. *Basic Anatomy for the Allied Health Professions.* Urban and Schwarzenberg. 1981. ISBN 0-8067-1231-7.

Scanlon, Valerie C. *Essentials of Anatomy and Physiology,* 5th ed. Illustrated by Tina Sanders. F.A. Davis Co., 2003. ISBN-13 978-0-8036-1546-5.

Snell, Richard S. *Clinical Anatomy by Systems.* Lippincott, Williams & Wilkins, 2007. ISBN-13 978-0-7817-9164-9.

Thibodeau, Gary A., and Kevin T. Patton. *Anatomy and Physiology,* 5th ed. Mosby, 2003. ISBN 0-323-01628-6.

Tortora, Gerard J. *Principles of Human Anatomy,* 5th ed. Illustrated by Leonard Dank. Biological Sciences Textbooks, Inc., 1989. ISBN 0-06-046685-5.

Waugh, Anne, and Allison Grant. *Anatomy & Physiology in Health and Illness,* 9th ed. Illustrated by Graeme Chambers. Churchill Livingstone, 2001. ISBN 0-443-06468-7.

Appendix F: Internet Resources

Selected Web sites for reference, with a brief description of the site from its introductory section, unless self-explanatory.

http://www.americanheart.org/

The mission of the American Heart Association is to build healthier lives, free of cardiovascular diseases and stroke.

http://www.ctc.nhs.uk/

Liverpool Heart and Chest Hospital.

http://www.prk.com/cataracts/history_of_lens_implants.html

Cataract Surgery and Lens Implantation: History of Intraocular Lens Implants.

http://www.medscape.com/

Medscape offers specialists, primary care physicians, and other health professionals the Web's most robust and integrated medical information and educational tools.

http://www.fmc-ag.com/internet/fmc/fmcag/neu/fmcpub.nsf/Content/Product_Portfolio

Fresenius Medical Care provides a complete line of dialysis services and products.

http://www.meditec.com/normal-lab-values.html

Meditec. Normal Lab Values.

http://www.lasersurgeryforeyes.com/cataracthistory.html

Laser Surgery for Eyes. The History of Cataract Surgery.

http://www.virtual-anaesthesia-textbook.com/index.shtml

The Virtual Anaesthesia Textbook is sponsored by GE Healthcare's Clinical Window information service. The goal of the Virtual Anaesthesia Textbook is to organize all Internet resources on anaesthesia into one concise, textbook-style Web site.

http://www.courseweb.uottawa.ca/medicine-histology/English/Renal/Default.htm

University of Ottawa. The Urinary System. Information on the renal system plus links to other body systems.

http://medlineplus.gov/

MedlinePlus will direct you to information to help answer health questions. MedlinePlus brings together authoritative information from the National Library of Medicine (NLM), the National Institutes of Health (NIH), and other government agencies and health-related organizations.

http://emedicine.medscape.com/

eMedicine is the most authoritative and accessible point of care medical reference available to physicians and other health care professionals on the Internet. The evidence-based content, updated regularly by nearly 10,000 attributed physician authors and editors, provides the latest practice guidelines in 59 medical specialties. The eMedicine Clinical Knowledge Base contains articles on nearly 7,000 diseases and disorders and is richly illustrated with some 30,000 multimedia files.

http://www.acc.org/

The mission of the American College of Cardiology is to advocate for quality cardiovascular care—through education, research promotion, development and application of standards and guidelines—and to influence health care policy.

http://health.allrefer.com/

AllRefer Health is a medical and health information resource containing outstanding database of health articles and reference materials. Consumers and health professionals alike can depend on it for information that is authoritative and up-to-date. AllRefer Health has extensive information from trusted sources on over 4,000 topics including diseases, tests, symptoms, injuries, surgeries, nutrition, poisons, and special topics.

http://www.adam.com/

A.D.A.M. creates online information and technology solutions for employers, benefits brokers, healthcare organizations, and Internet companies. Its Employer and Broker Solutions help employers and benefits brokers provide employees a better benefits experience while helping to manage workflow and cut costs. Its customizable Health Solutions help hospitals, managed care organizations, and consumer Web sites become an integral part of the online consumer healthcare experience. A.D.A.M. create the content and tools that help companies help people. Its award-winning health content and technology solutions are an integral part of helping employees and healthcare consumers manage important aspects of their health and make important benefits and health decisions. Whether clients are looking for information about health coverage, a healthcare provider, or searching for credible health information, A.D.A.M. helps connect them to the right information.

Appendix G: Image Credits

The copyrights for the following images are held by the artists listed (as they are named on the Shutterstock web site), 2010. Used under license from Shutterstock.com. Some images have been modified by the author.

Page Number	
158	Aaliya Landholt
52	Alexander Kalina
31, 77	Alexandru Cristian Ciobanu
189	Alexey Averiyanov
132, 148	Alexonline
2, 22, 113, 114, 115, 117, 188	beerkoff
32, 50, 79, 87, 89, 101, 116, 118, 129, 150, 157, 159, 166, 229, 232, 234	Blamb
67, 162	Bork
1	Brasiliao
122	Carolina K. Smith, M.D.
3, 109, 175	Condor 36
41	Curt Ziegler
58, 59	Derek L Miller
60, 136	emin kuliyev
70, 187	Farferros
33	Floris Slooff
24, 30, 34, 36, 38, 39, 40, 43, 51, 53, 61, 84, 92, 95, 96, 103, 107, 112, 119, 135, 138, 139, 144, 221, 223, 226, 227, 239	hkannn
74, 124, 146, 237	Oguz aral
80, 81, 82, 83, 85, 111	kalewa
42	Kanwarjit Singh Boparai
55, 56, 145	LesPalenik
44, 69	Lorelyn Medina
23	Milos Luzanin

26, 28, 57, 62, 76, 94, 104, 127, 133, 137, 140, 143, 156, 164, 233, 235, 236	mmutlu
66, 161, 173, 198, 202	Monkey Business Images
78, 154, 220	n.n
4	Norman Chan
71, 93, 121, 125, 126, 130, 131, 142, 152, 163, 165, 167, 225, 230, 231	Oguz Aral
8, 12	olly
27, 196	Rd
120	Regien Paassen
68	saginbay
65	SandiMako
35, 45, 46, 47, 48, 49, 72, 73, 75, 88, 153, 224	Sebastian Kaulitzki
21	Tiplyashin Anatoly
110	tlorna
54	Tomas Hlavacek
134	Tonylady
86	Travis Hilliard
5, 29	vadim kozlovsky
17	vhpfoto
99	YAKOBCHUK VASYL

The following images are modified from Inmagine Corp., www.123rf.com, with permission.

195, 197, 200, 201, 212

All remaining images are by the author.

Index

B

K

L

O

P

Q

R

S

T

U

V

W

X